Springer Series in Materials Science

Volume 348

The Springer Series in Materials Science covers the complete spectrum of materials research and technology, including fundamental principles, physical properties, materials theory and design. Recognizing the increasing importance of materials science in future device technologies, the book titles in this series reflect the state-of-the-art in understanding and controlling the structure and properties of all important classes of materials.

Peter Ozaveshe Oviroh
Sunday Temitope Oyinbo • Sina Karimzadeh
Patrick Ehi Imoisili • Tien-Chien Jen

Emerging Atomic Layer Deposition for Hydrogen Energy

Springer

Peter Ozaveshe Oviroh
Mechanical Engineering Science
University of Johannesburg
Johannesburg, Gauteng, South Africa

Sunday Temitope Oyinbo
Mechanical Engineering Science
University of Johannesburg
Johannesburg, Gauteng, South Africa

Sina Karimzadeh
Department of Mechanical Engineering
University of Johannesburg
Johannesburg, Gauteng, South Africa

Patrick Ehi Imoisili
Mechanical Engineering Science
University of Johannesburg
Johannesburg, Gauteng, South Africa

Tien-Chien Jen
Mechanical Engineering Science
University of Johannesburg
Johannesburg, South Africa

ISSN 0933-033X ISSN 2196-2812 (electronic)
Springer Series in Materials Science
ISBN 978-3-031-67773-1 ISBN 978-3-031-67774-8 (eBook)
https://doi.org/10.1007/978-3-031-67774-8

This Springer imprint is published by the registered company Springer Nature Switzerland AG
The registered company address is: Gewerbestrasse 11, 6330 Cham, Switzerland

Preface

Hydrogen is a promising and environmentally friendly energy source, with a demand of 94 million tonnes in 2021, equivalent to 2.5% of global energy consumption; however, this is mostly fossil fuel based which is not sustainable due to the environmental impact. Hydrogen has high yield of energy than gasoline and other fuel types, which makes it a significant fuel. Hydrogen is not an alternative fuel but a carrier of energy that needs to be produced using other forms of energy. Key new applications include steel projects, hydrogen fuel cell trains, shipping, and power sector use.

Wind, photovoltaic, and concentrated solar power can be used to generate large amounts of electric power but are not well positioned for that at the moment to be competitive with fossil fuel–based energy sources. Excess electricity from renewable energy sources can be used to produce hydrogen through electrolysis, crucial for climate neutrality. This hydrogen can store surplus renewable power, decarbonize sectors, and replace fossil fuels as a zero-carbon feedstock. The decreasing cost of renewable electricity makes green hydrogen cost-effective in the near future. The hydrogen-based economy will depend on water electrolysis and various methods, with the potential for green hydrogen production with zero carbon emissions.

To create a hydrogen-based economy, improvements in hydrogen generation, purification, storage, transportation, and distribution are needed. Scaling renewable energy generation, enhancing electrolysis technology, cost reduction, and innovations in materials and manufacturing processes can help. Advanced hydrogen storage technologies, transportation infrastructure, and atomic layer deposition (ALD) techniques can improve material performance in various energy-related devices. ALD techniques offer precise thickness control, excellent uniformity, tunable composition, and low deposition temperature, making them powerful for surface and interface engineering in energy-related devices. This book provides insight on how the ALD process can enhance the integration of the ALD technology into the green hydrogen process.

Chapter 1 provides a review on the advancements ALD approaches, factors affecting deposition, and methods for optimizing the ALD processes that have applications for surface and interface engineering in energy-related devices, with particular potential for hydrogen energy. Chapter 2 focuses mainly on environmentally friendly methods of producing hydrogen. Chapter 3 elaborates on the recent developments of ALD in hydrogen production technologies, with the recent advance development of ALD in designing catalytic materials for hydrogen production processes. Chapter 4 provides an overview of the current state of various materials and techniques used for hydrogen storage. In addition, the mechanism of hydrogen adsorption phenomena highlights different methods for storing hydrogen, such as high-pressure and cryogenic-liquid storage, adsorptive storage by high-surface-area adsorbents, chemical storage in metal hydrides and complex hydrides, etc. In addition, this chapter outlines a range of hydrogen storage techniques, together with their advantages and disadvantages, as well as the latest advancements in hydrogen technology. Furthermore, this chapter will demonstrate instances where the use of ALD, specifically in vapor-phase functionalization, has enhanced the material characteristics of solid materials such as 2D-materials, metal hydrides, and organic frameworks for the purpose of H_2 storage. Moreover, different hydrogen storage measurement techniques were highlighted. Finally, the future prospects and economy aspect of hydrogen storage were discussed. Chapter 5 introduces the comprehensive exploration of ALD analysis/characterization techniques and their application in the realm of hydrogen energy. We delve into the intricate interplay between ALD processes, material properties, and device performance, shedding light on the underlying mechanisms and opportunities for optimization. Through rigorous analysis and characterization, we aim to elucidate the fundamental principles governing ALD for hydrogen-related applications, paving the way for enhanced efficiency and efficacy in energy conversion and storage systems. This chapter explores the intricate ALD processes for hydrogen-related applications, utilizing interdisciplinary insights from materials science, chemistry, and engineering. It delves into surface reactions, precursor chemistries, and film growth kinetics. Chapter 6 provides some insight on what some countries are doing, the place of materials innovations, process innovations, industry 4.0, and future perspectives.

Gauteng, South Africa

Peter Ozaveshe Oviroh
Sunday Temitope Oyinbo
Sina Karimzadeh
Patrick Ehi Imoisili

Johannesburg, South Africa

Tien-Chien Jen

Acknowledgments

Dr. Oviroh Peter Ozaveshe would like to acknowledge all who have supported me in my academic journey, including my parents, Late Pastor D. A., and Deaconess Oviroh. Thanks to my wife Omolola Idowu Oviroh and daughter Hadassah Oviroh for their sacrifice. Your continued encouragement and support have spurred me to continue. I am very grateful. God bless you all.

He also thanks all his colleagues in the JENANO Research group, University Research Committee (URC), Global Excellence and Stature (GES) at the University.

Dr. Oyinbo ST would like to acknowledge the valuable encouragement, inspiring discussions, and steady monitoring from Professor Tien-Chien Jen, his PDRF host at the University of Johannesburg. The work found in this book was provided through his prompt assistance, constructive criticism, and conscientious financial and moral support. Dr. Oyinbo ST acknowledges his wife, Beauty, for her love and support during this period, when she had to take the place of being both the mother and father to his kids, Greg, Chloe, and Kim. Her prayers and encouragement gave him a reason to type every letter in this book. He also thanks all his colleagues in the JENANO Research group.

Dr. Patrick Ehi Imoisili thanks his wife and kids for their love, patience, understanding, encouragement, and prayers through this journey.

Professor Tien Chien Jen would like to thank all members of the JENANO Research group, University Research Committee (URC), Global Excellence and Stature (GES), The South Africa's National Research Foundation (NRF), and also SASOL for the Green Hydrogen SARChI Chair and other projects.

Introduction

Hydrogen as an energy source is viewed as environmentally friendly and potentially promising energy for the foreseeable future [1,2]. In 2021, the demand for hydrogen reached 94 million tonnes (Mt), which corresponds to approximately 2.5% of the total global final energy consumption [3]. The high yield of energy that is produced upon combustion of hydrogen (122 kJ/mol), which is far higher than that of gasoline or any other form of fossil fuel, is the primary reason for the significance of hydrogen as a fuel [4,5]. Hydrogen is not an alternative fuel; rather, it is a carrier of energy that needs to be produced by utilizing some other form of energy, such as fossil fuel, renewable energy, nuclear energy, etc. Several emerging hydrogen uses are making notable advancements. The steel industry is experiencing rapid growth, with new projects emerging. In Germany, the first fleet of trains powered by hydrogen fuel cells has begun operating. Hydrogen and its derivatives are also being utilized in the shipping industry, and major companies are forming strategic partnerships to ensure a steady supply of these fuels. Additionally, there is increasing interest in the use of hydrogen and ammonia in the power sector. The total capacity of announced projects in this field is almost 3.5 GW of potential capacity by 2030 [3].

The creation of a hydrogen-based economy in the coming decades will almost certainly depend on water electrolysis and several methods. This process will require massive amounts of electric power, which will need to be created first. Although most hydrogen is currently produced through gasification processes, which are combined with carbon capture and storage methods to reduce emissions, there is a growing recognition that solar and wind energy have significant potential to lead the way in the development of green hydrogen production with no carbon emissions. Due to the declining costs associated with renewable energy, it is anticipated that green hydrogen will soon become economically viable. There is a requirement for electrolyzers that utilize power with minimal emissions in order to generate hydrogen with low emissions. Currently, the manufacturing capacity for electrolyzer stands at around 8 GW/year. However, according to industry reports, this capacity might surpass 60 GW/year by 2030. The number of initiatives aimed at producing low emission hydrogen is expanding rapidly. If all projects presently in progress are completed, the annual production of low-emission hydrogen might reach 16–24 million metric tonnes by 2030. Approximately 9–14 million metric tonnes would be produced by electrolysis, while 7–10 million metric tonnes would be produced using fossil fuels with carbon capture, utilization, and storage (CCUS). If all the projects in the pipeline for electrolysis are successfully implemented, the total installed capacity of electrolyzers might reach 134–240 GW by 2030.

Creating power by burning fossil fuels is obviously not a viable option and cannot be considered reasonable looking at the environmental impact. Wind, photovoltaic, and concentrated solar power has the capacity to produce substantial quantities of electrical power. Still, except for wind, these technologies are not sufficiently developed and are too expensive to take on such a large challenge in such a short time [2]. In times of excess electricity production from wind farms, instead of curtailing the electricity as is commonly done, it is possible to use this excess electricity to produce hydrogen through electrolysis. Climate experts consider green or renewable hydrogen, produced by the electrolysis of water using solar or wind power, to be vital to achieving climate neutrality. Conceptually, it has the potential to accomplish three objectives: firstly, to store excess renewable energy when the power grid is unable to accommodate it; secondly, to facilitate the reduction of carbon emissions in sectors that are hard-to-electrify or challenging to electrify, such as long-distance transportation and heavy industries; and thirdly, to substitute fossil fuels as a zero-carbon feedstock in the production of chemicals and fuels. In order to fully create an economy based on hydrogen, there must also be improvements made to the ways in which hydrogen is generated, purified, stored, transported, and distributed. Scaling renewable energy generation, enhancing electrolysis technology, cost reduction, innovations in materials, manufacturing processes, and system designs can help drive down the cost of electrolyzers, balance of plant components, and overall hydrogen production systems. We balance the business and service delivery. Innovations in hydrogen storage technologies, such as advanced compressed hydrogen or solid-state storage systems, can enable efficient and safe storage of hydrogen for use in various applications, advancements in transportation and distribution infrastructure, including pipelines and hydrogen refueling stations, repurposing and retrofitting of current fossil-based system. ALD, as a high technology, has promise for material improvements. This finds applications in hydrogen production (electrolyzers), hydrogen purification, and hydrogen storage; batteries, supercapacitors, fuel cells, solar cells, and photoelectrochemical water splitting are some examples of energy storage and conversion technologies that have played critical roles in the decrease of the use of fossil fuels, addressing environmental issues, and the creation of electric cars [6]. To achieve optimal performance for various energy-related devices, it is necessary to fabricate electrode materials with refined structures and engineer their surfaces and interfaces of different materials that find applications. Atomic layer deposition (ALD) techniques have become highly valuable for surface and interface engineering in energy-related devices and hydrogen energy applications. This is due to their exceptional ability to precisely control thickness, achieve excellent uniformity and conformity, tune composition, and operate at relatively low deposition temperatures.

References

1. P. Ganguly et al., 2D nanomaterials for photocatalytic hydrogen production. ACS Energy Lett. **4**, 1687–1709 (2019)
2. N. Armaroli, V. Balzani, The hydrogen issue. ChemSusChem **4**. Preprint at https://doi.org/10.1002/cssc.201000182 (2011)
3. IEA, *Global Hydrogen Review 2022* (International Energy Agency, Paris, 2022). https://doi.org/10.1787/a15b8442-en
4. W. Peschka, *Liquid Hydrogen: Fuel of the Future* (Springer Science & Business Media, 2012).
5. Y. Li, G.A. Somorjai, Nanoscale advances in catalysis and energy applications. Nano Lett. **10**, 2289–2295 (2010)
6. Y. Zhao et al., Atomic/molecular layer deposition for energy storage and conversion. Chem. Soc. Rev. **50**, 3889–3956. Preprint at https://doi.org/10.1039/d0cs00156b (2021)

Contents

About the Authors

Patrick Ehi Imoisili is a senior research fellow at The Mechanical Engineering Science Department, Faculty of Engineering and the Built Environment. University of Johannesburg, South Africa. He is a 4IR (Fourth Industrial Revolution) Nanotechnology expert with the JENANO Research Group and head of the smart material sub-research group. He received his PhD in Industrial Chemistry from the Federal University of Technology, Akure, Nigeria. He is a charted Chemist, with the Institute of Charted Chemists of Nigeria (ICCON) and a registered member of the Polymer Institute of Nigeria (PIN), Chemical Society of Nigeria (CSN), South Africa Chemical Institute (SACI), and South African Council for Natural Scientific Professions (SACNASP). He is the recipient of "Outstanding Achievement in Research and Academic Citizenship" by the University of Johannesburg, in 2020 as the best postdoctoral research fellow from Mechanical Engineering Science/ Faculty of Engineering, First prize winner in PDRF category UJ PG School Conference 2022, Innovative Publication award winner for the Post-Doctoral Category by the Congress of Nigerian Student in University of Johannesburg in 2022, Volunteer Champion 100+ Award, and for Community Engagement by the University of Johannesburg in 2021, among other awards. He has published over 40 peer-reviewed articles since joining the University of Johannesburg in 2020. His research interest is in but not limited to chemistry, polymers, natural fibers polymer composite, hybrid composite, nanocomposite, nanoparticle synthesis for drug delivery, and wastewater treatment, renewable energy, solar cells, thin films, green hydrogen, and atomic layer deposition (ALD) research and innovations.

Tien-Chien Jen is a full professor at the University of Johannesburg since 2015. Prof. Jen is the Director of the Joint Research Centre with the Nanjing Tech University of China on the "Sustainable Materials and Manufacturing." Prof. Jen is also the Director of the Atomic Layer Deposition Research (ALD) Center, which is the first ALD center in South Africa.

In 2011, Prof. Jen was elected as a Fellow of the American Society of Mechanical Engineers (ASME), which recognized his contributions to the field of thermal science and manufacturing.

In 2020, Prof. Jen was inducted as the Academician of the Academy of Science of South Africa, which recognizes him as one of the top scholars in South Africa. Very recently, Prof. Jen has been appointed as the South Africa Research Chair (SARChI) on Green Hydrogen in 2023. Prof. Jen has written a total of over 620 peer-reviewed articles, including 300 peer-reviewed journal papers, published in many prestigious journals published in Elsevier, ACS, Taylor and Francis, Springer Nature, Wiley, etc. Some of the journals include: *Electrochemical Energy Reviews*, *npj Clean Water*, *Journal of Energy Storage*, *International Journal of Hydrogen Energy*, *Nano-Micro Letters*, *Coordination Chemistry Reviews*, *Nanotechnology Review*, *Journal of Molecular Liquids*, etc.

He has written 16 book chapters and has 7 books published with the latest titled "Machine Learning-Based Modelling in Atomic Layer Deposition Processes," published by CRC Press-Taylor Francis, December 2023.

Sina Karimzadeh is a postdoctoral research fellow at the University of Johannesburg. He holds a PhD degree in Mechanical engineering from the University of Johannesburg in 2024. His current research interest focuses on the development of energy storage systems including Li-ion battery and Hydrogen Storage by using atomic layer deposition (ALD) technique. He has also been involved in a number of projects including Hydrogen Generation, Thin Films and Nanotechnology, Drug Delivery, Heat Transfer, Water Purification Membrane, and Computational modeling and simulation. He is currently the Head of the ALD and innovation sub-research JENANO group and the Lead experimentalist at the ALD facility in the University of Johannesburg.

Peter Ozaveshe Oviroh is a research fellow and also lectures on sustainable energy technologies at the Mechanical Engineering Science Department, University of Johannesburg, South Africa. He obtained his PhD from the same department. He is a member of the American Society of Mechanical Engineers (ASME) and also a member of the Council of Engineers for the Energy Transition (CEET)—an independent advisory council to the United Nations Secretary-General. He has research interests in nanotechnology, particularly applying atomic layer deposition (ALD) to develop materials for water desalination applications. In addition, he is working on the applications of green hydrogen. He has been involved in several other projects that involve energy modeling, system optimization, and, in general, sustainable energy technologies. Dr. Oviroh has published in several journals and conferences.

Sunday Temitope Oyinbo received his PhD in Mechanical Engineering Science from the University of Johannesburg (UJ). He is currently a postdoctoral research fellow at the University of Johannesburg. As a member of the JENANO Research Group at the University of Johannesburg, his research interest focuses on nanotechnologies and associated processes (e.g., atomic layer deposition (ALD) and cold gas dynamic spray (CGDS)). He has accumulated rich research experience in the microstructure reconstruction of metal matrix composite and hydrogen purification technology. He has research expertise in computational modelling and simulation, materials and microstructural characterization, finite element analysis (FEA), molecular dynamics (MD), density functional theory (DFT), and proficient knowledge in programming (LAMMPS code, Python script, VASP code). At present, the ultra-thin film composite metal membrane experiment platform has been built through computational modeling and surface functionality in the absorption and separation of gas through the nanoporous membrane has been achieved. He has published various research papers in international journals and conferences in the areas of nanotechnology and hydrogen economy.

List of Figures

List of Tables

Chapter 1
Introduction to ALD Technologies

Concerns about depletion of resources, climate change, and pollution drive the increasing importance of sustainable processes. Atomic layer deposition (ALD) is a highly versatile process that enables precise application of thin films onto difficult substrates, with the ability to modify thickness at the nanoscale level. The reason ALD is extensively utilized in many applications is due to its exceptional capacity to nanoengineer interfaces and surfaces. Although the ALD offers promise in addressing environmental issues, it is essential to prioritize efficiency and minimize the carbon footprint of renewable energy devices to guarantee their long-term sustainability [1]. Atomic layer deposition (ALD) is a thin film deposition process that sequentially introduces chemical precursors to the surface of a substrate, where they chemically react directly with the surface to generate submonolayers of film. As the name implies, it is atomic in nature, resulting in the precise deposition of films on the surface of a specified substrate at the atomic scale [2–7].

1.1 Background Development of ALD

Finland and the Soviet Union independently invented atomic layer deposition (ALD), a gas-phase thin film deposition technology, in the 1960s and 1974, respectively [8–10].

Prof. V.B. Aleskovskii first proposed the concept of atomic layer deposition (ALD) in his doctoral thesis in 1952. In 1960, Prof. Kolt'sov and his team released a paper titled "Molecular Layering," which further explored the concepts of ALD [9].

Dr. Tuomo Suntola and other researchers invented an industrial thin film deposition process known as "atomic layer epitaxy" (ALE) in 1970 [11]. This technique was initially used in thin-film electroluminescent (TFEL) flat panel displays [11], the desire for uniform films in large-area displays and precise coatings in the

P. O. Oviroh et al., *Emerging Atomic Layer Deposition for Hydrogen Energy*,
Springer Series in Materials Science 348,
https://doi.org/10.1007/978-3-031-67774-8_1

microelectronics sector led to the creation and advancement of ALD [12] and the technology and materials used have evolved to find applications in catalysis, photovoltaics, semiconductor devices, and many other fields [13].

1.2 Principles of ALD

ALD is a method for depositing thin films by controlling their growth at the surface. It is a layer-by-layer, surface-controlled method that deposits thin films one atomic layer at a time. Layers are formed during reaction cycles by pulsing precursors and reactants alternately and purging with inert gas between each pulse [2, 3, 7, 14, 15]. Using successive, self-limiting surface reactions, ALD can meet the requirements for conformal deposition and atomic layer control [6].

In ALD, the reactants are presented to the heated surface at separate times. Step one introduces the first reactant to the substrate before it is pumped out. During this exposure, a single layer of the initial reactant adheres to the substrate and remains there when the chamber is evacuated. Subsequently, an additional reactant is introduced into the chamber and subsequently undergoes a chemical reaction with the monolayer of the initial reactant. This subsequently creates the first layer of the desired solid film, which is often less than a full monolayer. The final reactant and any gaseous by-products of the reaction are then removed from the reaction chamber using a pump. In order to achieve the desired thickness, the operation is repeated as necessary in order to create a film of the appropriate thickness [2, 7].

The following are the steps in a single-cycle deposition process, as shown in Fig. 1.1:

1. Step 1, in the first step of the process, the initial reactant A, which is TMA, interacts in a self-terminating manner with the functional groups that are accessible on the OH-terminated surface.
2. Step 2 involves purging or pumping away any surplus reactant A (TMA) and any gaseous by-products (CH_4) that have been produced.
3. The third step involves the reaction of the second reactant, B (H_2O), with the adsorbed species (A) on the surface. This reaction is self-terminating.
4. Step 4 involves purging or pumping away any surplus reactant B (H_2O) and any gaseous by-products (CH_4) that have been produced.

Among the several characteristics of ALD are conformity, uniformity, and crystallinity.

1.2.1 Conformality of ALD

Despite being frequently overlooked in ALD process development, it is crucial to consider the conformality of the ALD film. Conformality refers to the capacity to

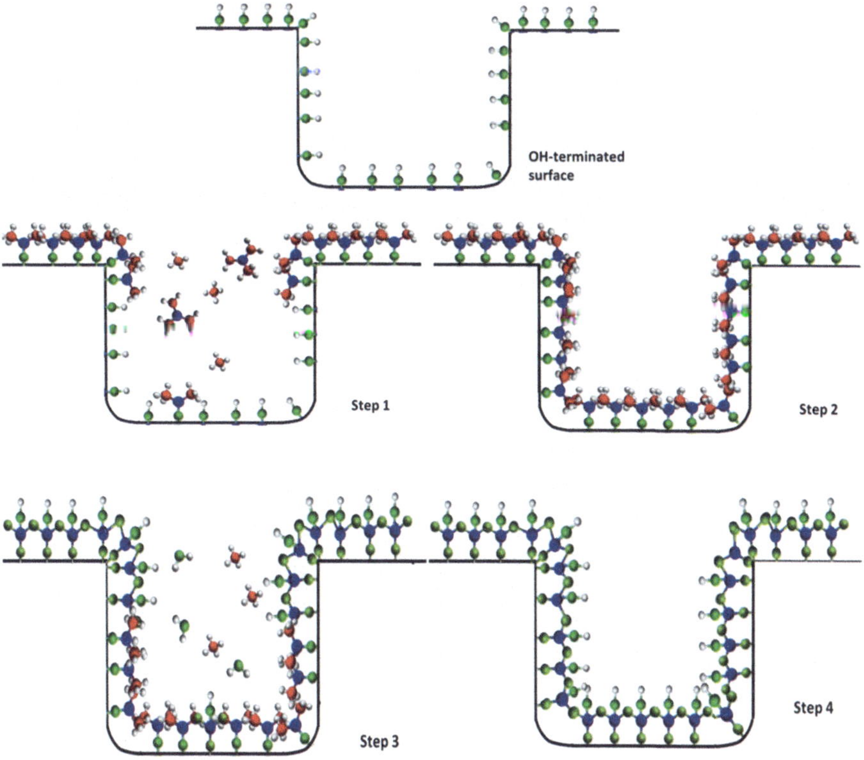

Fig. 1.1 An exemplary model for ALD which involves depositing TiO_2 onto a substrate that has been functionalized with hydroxyl groups. This technique utilizes $TiCl_4$ and H_2O as precursors [16] as shown in Fig. 1.2

uniformly deposit a film on three-dimensional (3-D) structures, ensuring no variation or minimal variation in thickness across the structure [17, 18]. An excellent method for coating high aspect-ratio materials is ALD [19]. In ALD, gas-phase precursor molecules react with a solid surface in a binary sequence. Due to the digital nature of the ALD growing process, the film thickness may be precisely controlled, thus leading to excellent uniformity and conformity in the process. ALD is unparalleled in its ability to achieve exceptional conformality on high-aspect structures, surpassing any other thin film method [6]. ALD can meet the requirements for precise control over atomic layers and uniform deposition by employing sequential, self-limiting reactions on the surface. Powders, tubular gas separation membranes, porous silicon, anodized alumina, and anodic alumina are just a few of the porous media that have benefited from ALD coating [19].

Because of the need of a conformal coating on 3D nanostructured surfaces, ALD methods are frequently selected for their conformality, which plays a pivotal role in their practical applications. ALD allows for the control of coating thickness down to the submonolayer level by alternating gaseous reactants with an exposed solid

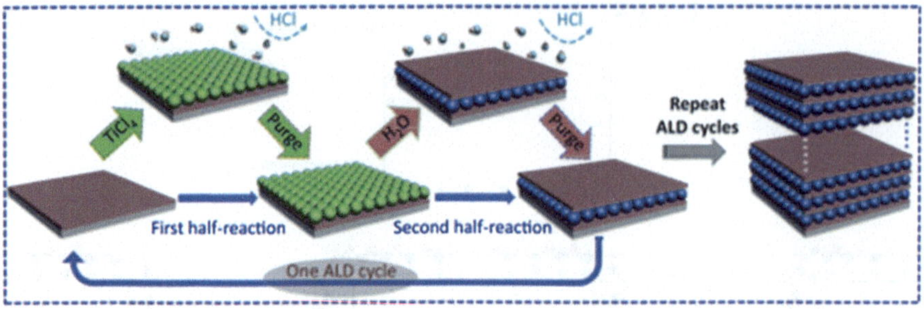

Fig. 1.2 A model ALD method utilizing $TiCl_4$ and H_2O as precursors is described for the deposition of TiO_2 on a substrate functionalized with hydroxyl groups

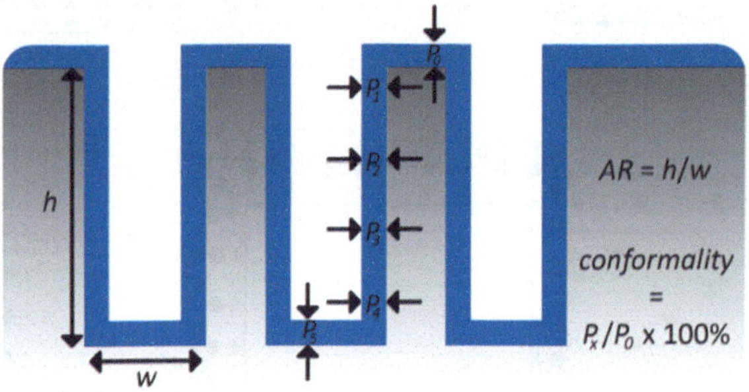

Fig. 1.3 The schematic diagram illustrates the concept of conformality, which refers to the variation in film thickness at different locations inside a trench relative to the thickness at the top of the trench. The aspect ratio (AR) of a trench is determined by dividing its height by its width

surface, resulting in extremely conformal coatings [20]. ALD is able to produce films that are homogeneous and conformal in 3D structures that have complicated forms and have a large depth-to-width ratio, or, in more general words, a large aspect ratio. *The thickness (and properties) of conformal films are constant even inside a 3D structure.*

To measure conformality, unique samples with trenches or vias are required rather than ordinary planar coupons (Fig. 1.3). In order to evaluate the degree of conformity, it is necessary to deposit material in a vertical trench or via with a defined aspect ratio (AR). After creating a cross-section, the thickness ratio at various locations may be calculated [18]. Markus et al. [18] illustrated this method, as shown in Fig. 1.4 is the sectional scanning electron micrograph (SEM) illustrating conformity.

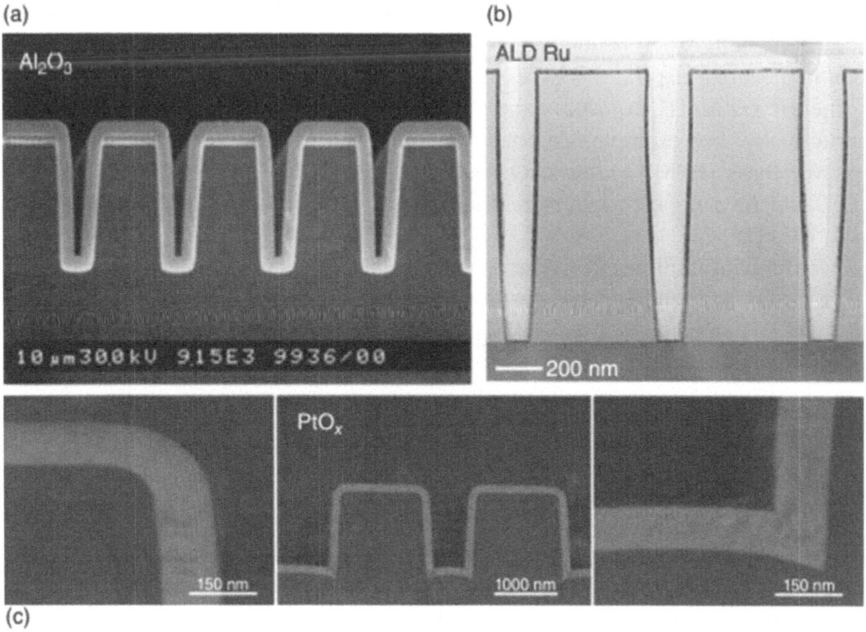

Fig. 1.4 (**a**) Sectional scanning electron micrograph of a 300 nm thick Al_2O_3 film grown on a Si wafer with trench structures. (**b**) TEM slicing through a trench-patterned substrate reveals a conformal RuALD coating. (**c**) Images captured using a field-emission scanning electron microscope (FESEM) of a conformal PtOx film with trench features formed on a Si substrate

1.2.1.1 Conformality Analysis of Thermal, Plasma-Enhanced (PEALD), and Ozone-Based Atomic Layer Deposition (ALD) Processes

Conformal coatings are becoming more and more important in a broad range of technologies. For instance, films of insulators, conductors, diffusion barriers, adhesion layers, etch stops, and protective layers need to be positioned inside trenches and holes that are getting deeper and more precise every year in the field of microelectronics [21]. ALD has emerged as the preferred technique for the conformal deposition of thin films within structures characterized by high aspect ratios (AR). ALD is an excellent method for coating materials with a large aspect ratio.

Because the nanopores have an exceptionally high aspect ratio of $L/d \approx 10^3$, depositing conformal coatings inside them is exceedingly challenging. The optimization of ALD procedures often involves prioritizing planar geometries prior to attempting deposition on 3D objects. In order to attain conformality in moderate aspect ratios (ARs), it is sometimes unnecessary or requires minimal adjustments to precursor exposure as compared to the exposure times typically optimized for deposition on flat surfaces. Atomic layer deposition (ALD) is a thin film generation process that can generate extremely homogeneous films with monolayer thickness control on high aspect-ratio substrates [19].

When transitioning to larger aspect ratio (AR) structures, it has been demonstrated via both experimental and modeling endeavors that a substantial augmentation in precursor exposure is necessary in order to achieve conformality [22]. With sufficient reactants' exposure time, nanopores with small length can be coated conformally. For example, Zn concentration profiles evaluated by electron probe microanalysis (EPMA) after ZnO ALD revealed gradual penetration of the ZnO ALD into the nanopores with increasing exposure durations for aspect ratios up to $L/d \sim 5000$ [19].

Conformality of Plasma-Assisted ALD: Achieving conformal deposition in high aspect-ratio structures is more challenging in plasma-assisted atomic layer deposition (ALD) compared to thermal ALD, mostly due to the surface recombination loss of plasma radicals [22].

1.2.2 Uniformity of ALD

ALD is a very promising approach for growing superconducting thin films with excellent uniformity, thickness control, and conformality. It is advantageous for superconducting quantum photonic applications, surpassing other methods like reactive sputtering.

Some of the factors contributing to nonuniformity in ALD processes are the overlapping of pulses, the thermal self-decomposition of precursors, and the uneven distribution of gas [23]. In addition, there is also the hardware as a source of nonuniformity; there are also the downstream surface reactions of gaseous by-products as highlighted in the work of Elers et al. [23] on film uniformity in atomic layer deposition. Even on trench structure, as evidenced in the work of Choi et al. [24], ALD has been shown to exhibit uniformity. Using the Plasma-Enhanced Atomic Layer Deposition (PEALD), they were able to show the uniformity of HfO_2 thin films prepared on trench structures. For more recent applications, ALD is used for high uniformity in superconducting niobium nitride thin films for quantum photonic integration [25, 26].

Several studies have been done that need uniformly thin films; for example, when upgrading from single-pixel detectors to arrays with a big active area, Knehr et al [25], for instance, used wafer-level uniformity of atomic-layer-deposited niobium nitride (NbN) thin films for quantum devices. The findings indicate that ALD is well-suited for producing NbN thin films, which may be used as a material for large-scale detector arrays. Additionally, it is suitable for creating new detector designs and technologies that require uniform superconducting thin films with excellent thickness control. Every point along an even substrate in a uniform film will have the same thickness, composition, and other properties.

As highlighted by Martijn et al. [18], ALD also has the advantage of producing films that have consistent thickness (uniformity) and composition across wide surface areas. Frequently, saturation is confirmed by employing coupon samples positioned at the center of the substrate table. However, this does not imply that the dosage of precursor or co-reactant is enough for the whole area. Furthermore, nonuniformity is typically a more distinct indication of a CVD component compared to the information received from saturation curves. Therefore, a high level of uniformity serves as proof that the process is a well-behaved ALD process. Thus, depositing a film on the biggest substrate that can be accommodated in the ALD reactor is advisable. Crucially, the variance in thickness may be readily identified, either manually or by employing an automated mapping stage. Nevertheless, certain material characteristics, such as composition and resistivity, can also play a significant role in ensuring consistency. For instance, the deposited film may exhibit high thickness uniformity, although the resistivity of the film shows substantial variation across the substrate.

1.2.3 *Crystallinity of ALD*

This section analyzes the experimental data to ascertain the crystallinity and phase of inorganic materials that are deposited by ALD using different reactants and temperatures as detailed by Miikkulainen et al. [27]in their work on "Crystallinity of inorganic films grown by atomic layer deposition: Overview and general trends."

In the study of Cai et al. [5] on "Atomic Layer Deposition of Two-Dimensional Layered Materials: Processes, Growth Mechanisms, and Characteristics," they highlighted how ALD enhances the crystallinity of 2D materials. The as-deposited films were amorphous, a posttreatment for better crystallinity. Annealing significantly improved their crystallinity [5]. By modifying the substrates and changing the deposition temperature, ALD has the ability to tune the composition and crystallinity of compounds [28].

1.3 Types of ALD

ALD methods may be classified into many categories, mostly depending on the chemical reactions used for deposition. Thermal ALD and plasma-enhanced ALD are the two most common types of ALD [2]. Radical-enhanced or plasma-assisted atomic layer deposition is often referred to as PEALD. These different ALD processes are discussed below.

1.3.1 *Thermal ALD*

Thermal ALD is primarily a process driven by surface reactions. This characteristic allows for precise control over the thickness and uniformity of the deposited material, regardless of the shape of the substrate or the design of the reactor. Thermal atomic layer deposition (ALD) techniques require elevated temperatures, generally ranging from 150 °C to 350 °C. This temperature requirement imposes a constraint on their potential applications. However, this limitation may be overcome by employing PEALD [29].

1.3.1.1 Initial Surface Reactions

During each ALD cycle, a limited number of active sites exist on the surface of the substrate. During the ALD process, the initial surface reactions result in the coverage of active sites by the reactants. These active sites are depleted at the end of each half-cycle, and new active sites are subsequently generated in the following half-cycle. The rate of initial development in ALD is determined by the quantity of nucleation sites and may be classified into three categories: linear, surface-enhanced, and surface-inhibited [30, 31] (Fig. 1.5).

1.3.1.2 Reaction Pathways

Quartz crystal microbalance (QCM), Quadrupole mass spectrometry (QMS), and infrared (IR) spectroscopy are the most often used methods for studying ALD chemistry in real time and under normal process circumstances. Additional optical

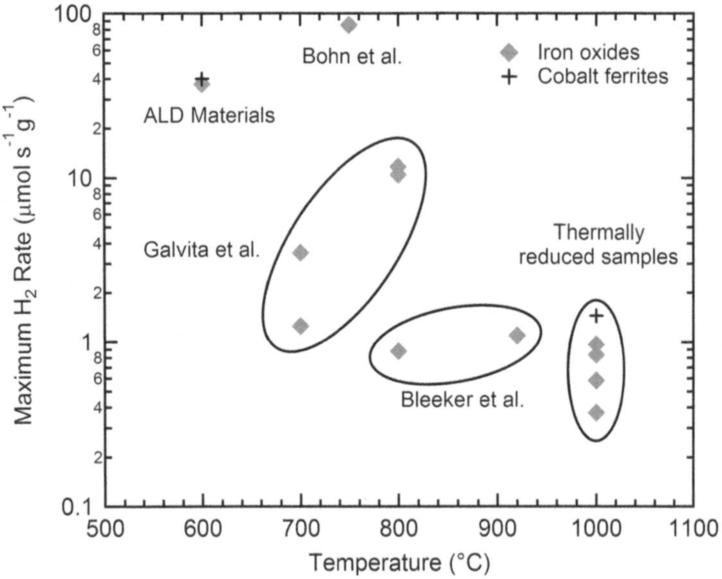

Fig. 1.5 The maximum hydrogen (H_2) production rates of atomic layer deposition (ALD) materials are analyzed in relation to their temperature dependence and compared to rates reported in existing literature for other materials [32]

methods, including ellipsometry, have also been employed to investigate ALD processes. However, these approaches offer limited insights into the underlying reaction mechanisms [33]. The ALD reaction process is represented by the balanced equations for the ALD half-reactions, which are elucidated by expert in situ research. ALD processes involve ligand exchange reactions, in which the ligands of the metal precursor are eliminated by bonding to the Lewis acids of the nonmetal precursors. However, there are exceptions to this rule that use different chemical approaches, such as using oxygen, oxygen plasma, or ozone as oxygen or co-reactant or precursors. Hydrogen is commonly utilized as the nonmetal precursor in these processes. To understand the chemical process, it is necessary to know the precise ratios of these volatile by-products, which are often created during both precursor pulses [33]. In Atomic Layer Deposition (ALD), the half-reactions take place consecutively on the same substrate at various time intervals. However, in combustion chemistry, the half-reactions happen simultaneously and continuously in different vessels. The temporal isolation is accomplished by either pulsating the precursors in an alternating fashion on a stationary substrate or by shifting the substrates from one precursor stream to the other. The latter technique is being progressively employed in ALD reactors that are physically isolated [15].

Some ALD processes or precursors may need more than just thermal activation, hence limiting the range of materials that may be synthesized. Thermal ALD is primarily affected by the temperature of the substrate and, to a lesser degree, the pressure of the precursor vapor. These factors have an impact on the ALD process for a certain precursor chemistry. Certain widely used oxidizing agents, such as water, tend to adhere to the walls of the reactor, necessitating higher temperatures to decrease the duration of the reaction cycle. Thermal atomic layer deposition (ALD) techniques typically have a very slow deposition rate, achieving around one atomic layer per second.

1.3.2 PEALD

Thermal ALD procedures require high temperatures, generally between 150 °C and 350 °C. However, this temperature requirement creates restrictions on their potential applications. One possible solution to overcome this constraint is the implementation of plasma-enhanced ALD (PEALD) [29]. A plasma is employed in PEALD to increase the reactivity of precursors, allowing for lower deposition temperatures and better film characteristics. Plasma may be used with several ALD techniques.

PEALD, or plasma-assisted ALD (PA-ALD), utilizes highly reactive plasma species to lower deposition temperatures while maintaining film quality. Particularly desired is the deposition of thin films on materials that are sensitive to temperature. PEALD may potentially lead to improved material properties. Furthermore, the high reactivity of plasma enables the use of a wider variety of precursors and the deposition of materials that are challenging or impossible to achieve using thermal techniques [34].

Reactants can be activated not only by the thermal ALD process but also by using a plasma source, referred to as plasma-enhanced ALD or PEALD. A plasma is a gaseous state of matter consisting of a combination of neutral and charged particles, which overall have a neutral charge at a macroscopic level. PEALD employs the species produced in plasma as reactants during the ALD process. Radicals and ions that reach the surface of the sample can impart extra energy to the surface, resulting in physical changes in the sample. This creates a materials processing environment that cannot be easily achieved via chemical means alone. Primarily, plasmas consisting of O_2, N_2, NH_3, and H_2 (or combinations thereof) have been employed for the synthesis of oxides, nitrides, and metal films [35]. PEALD enables deposition at comparatively low temperatures, resulting in film characteristics that surpass those achieved with thermal ALD [36].

Some of the advantages of PEALD, as highlighted are [2, 35, 36]

1. The plasma's strong reactivity lowers the required thermal energy for surface activation. This enables the deposition process to occur at lower temperatures than what is achievable with thermal processing; this also allows for the deposition on surfaces sensitive to heat.
2. PEALD allows for deposition at comparatively low temperatures, and the properties of the film are superior than those achieved with thermal ALD.
3. PEALD also provides highly reactive species or catalysts that improve the process, enabling lower temperatures and a broader range of substrates and precursors for deposition [37].
4. Plasmas offer greater flexibility in ALD processing. The introduction of many additional factors, such as plasma power and composition, allows for modification to affect the process chemistry and resultant characteristics of the film.
5. PEALD enables the use of many precursors and the deposition of novel materials. PEALD enables the use of thermally stable precursors with higher reactivity, making it feasible to employ PEALD for systems that lack a thermal ALD process, such as some elemental metals. PEALD offers several additional advantages, such as a wider range of precursors, a higher growth rate, and improved process flexibility.
6. Certain plasma-enhanced atomic layer deposition (ALD) techniques have a higher rate of development when compared to their thermal equivalents.

Although the process of PEALD offers benefits, it also triggers undesirable side reactions. To obtain the most benefits, it is crucial for the ALD process to operate within the specified temperature range (ALD temperature window). The "ALD temperature window" refers to the temperature range in which the growth is at its maximum level, and this range is determined by the specific ALD technique being used. On the other hand, unfavorable consequences such as reduced growth rates, slow reaction kinetics, or the occurrence of precursor condensation and heat degradation may occur [38]. There are several advantages, but it is not also without its disadvantages. These include limited conformality, more complex reaction chemistry, reactor design costs, and complications, damage to films in the ALD process, lower throughput, potential poor film conformality, and additional requirements for

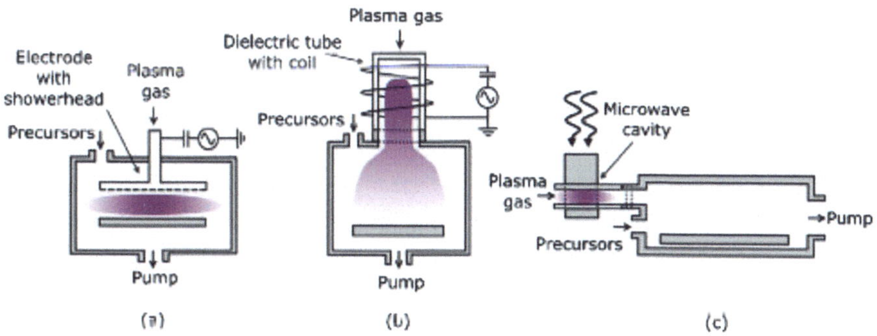

Fig. 1.6 The diagram illustrates the three different types of plasma-assisted atomic layer deposition: (**a**) direct plasma, (**b**) remote plasma, and (**c**) radical enhanced for each different hardware configuration and plasma source type

growth parameters [2]. Hence, PEALD proves advantageous for depositing thin films on substrates susceptible to heat damage. PEALD has the capability to generate films of higher purity compared to those created by thermal ALD [36].

Figure 1.6 illustrates three different plasma sources utilized in low-pressure plasma-enhanced atomic layer deposition (PEALD). The first setup, shown as configuration (a), is referred to as the "direct plasma" due to the exposure of the substrate to high-energy ions or the direct contact between the deposition surface and the active plasma. (b) is the second configuration that is referred to as a distant plasma system. The remote plasma ALD method is characterized by the absence of direct interaction between the plasma and the substrate, as well as the utilization of low-energy ions. This distinguishes it from radical-enhanced and radical-assisted ALD methods. This is illustrated in Fig. 1.6c. Radical-enhanced ALD has the advantage of minimizing or not causing substrate or film damage caused by energetic plasma species [36], as only neutral plasma species are involved in the deposition process.

Figure 1.7 compares the advantages of remote plasma-enhanced ALD using some materials.

Donnell et al. [40] have emphasized that silicon has been studied as a photoanode material because of its low cost as well as the suitable band gap. However, its lack of stability during the oxygen evolution reaction (OER) renders it unsuitable for extended use. Titanium dioxide (TiO_2) films with an ultra-thin film have been employed as protective coatings that enhance the lifetime of silicon photoanodes and boost their ability to convert solar energy into hydrogen. This study examines the nucleation, growth chemistry, and performance of TiO_2 thin films prepared via thermal and plasma-enhanced atomic layer deposition (ALD) using titanium isopropoxide and tetrakis(dimethylamido)titanium as precursor materials. The effects of plasma and air annealing treatments were also studied. TiO_2 grown by plasma-enhanced ALD results in cleaner films with reduced carbon incorporation. The findings of our study reveal significant differences in the chemical composition

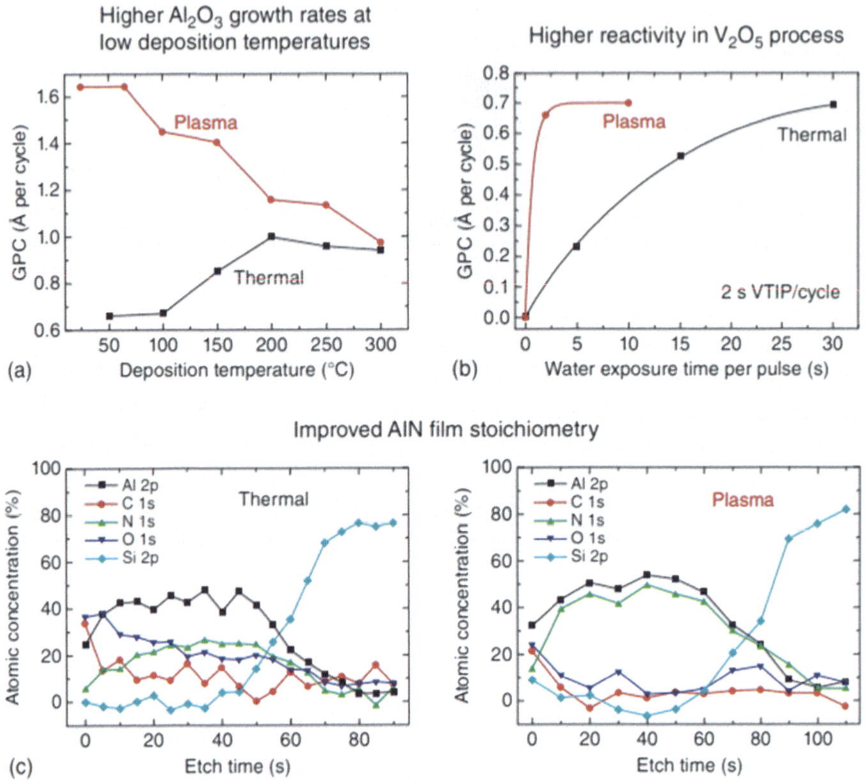

Fig. 1.7 Comparison of the advantages of remote plasma-enhanced atomic layer deposition (PEALD) with thermal ALD [39]

of films produced using PEALD and T-ALD methods [40]. Thermal ALD has limitations, and that gives rise to PEALD.

1.3.3 Spatial ALD (SALD)

In traditional ALD, precursors are introduced to a substrate one after another in a sequential manner, with short pulses, while being physically separated by an intermediary purging process. Spatial Atomic Layer Deposition (SALD) is a modified form of ALD where precursor materials are consistently introduced in separate locations and separated by an inert gas area. Spatial atomic layer deposition (SALD) has been created that divides the two precursors spatially (space) as opposed to temporally (time). Film growth is grown by exposing the substrate to regions containing distinct precursors. The process becomes faster by eliminating the purging phase, making it compatible with fast-throughput processes like roll-to-roll

(R2R). Spatial atomic layer deposition (ALD) has the capability to significantly reduce ALD cycle times and may be used for roll-to-roll production [15].

Additionally, it becomes more adaptable, simpler to scale up, and more cost-effective. Furthermore, a notable advantage of SALD is its capability to be conducted at ambient pressure conditions, even in the open air, without the need for a deposition chamber. This is achieved without any negative impact on the rate at which the deposition occurs [41]. ALD, commonly used in research laboratories, has the potential to be applied in various industrial applications. However, for this to occur, it is necessary to achieve high deposition rates, also known as high-throughput (HT), and to avoid the use of vacuum processing [12]. This is because most applications do not require the same level of precision as microelectronic products. Spatial atomic layer deposition (ALD) entails the separation of precursors into different zones, with the substrate moving between these zones. This enables simultaneous exposure to several precursors, which has the potential to enhance deposition rates. SALD is a high-throughput ALD process. They provide the opportunity to reduce the time needed for layer preparation, along with other advantages [1]. When an ALD reactor's surface area and volume are increased, the pulse and purge time are extended, which in turn results in a reduction in the throughput of the reactor. Spatial ALD (SALD) improves the technique by removing pulse purge chambers and use a spatially resolved head to selectively expose the substrate to a specific precursor based on its position. However, this may result in a deterioration in the quality of the thin film [42]. The SALD technology may produce deposition rates of around 3600 nm/h [38]. Depending on the amount of precursor intermixing, SALD may be used to build films in either the atomic layer deposition or the chemical vapor deposition regime. Precursor intermixing is heavily impacted by SALD head design and operating conditions, both of which have complex effects on film growth, making it difficult to accurately project the growth regime before depositions [43].

Some of the advantages of SALD, as highlighted [44], are (1) by using ambient pressure for the deposition process, the high capital and operating costs of vacuum-based techniques in current ALD, CVD, and PVD technologies are eliminated (2). Capable of depositing controlled and conformal nanoscale films, from sub-10 nm to >20 nm thick, with resolution of ~1 nm (3). The design is modular, enabling scalability for achieving high throughput, with the potential to process more than 1000 wafers per hour. Additionally, it allows for deposition across vast regions, specifically exceeding dimensions of 15 cm × 30 cm (4). Ability to deposit various materials on a variety of substrate types, thicknesses, and geometries (e.g., SnO_2, WO_3, Al_2O_3, AZO, etc.) (5). Flexible substrates can be used with low-temperature deposition.

1.3.4 Sequential ALD

In this type of ALD, the precursors are introduced in a sequential manner, with each one at a time. Before the subsequent precursor is introduced, every precursor undergoes a self-limiting reaction with the surface—the precise control results in very consistent and uniform film deposition.

1.3.5 Photo-Assisted ALD

Photo-assisted atomic layer deposition (photo-ALD) is a type of version of the ALD process that employs ultraviolet or visible photons to provide energy and alter the surface reactions of ALD [45]. Photo-assisted ALD has been used for hafnium and zirconium-based precursors [46].

A study investigates the process of low-temperature, plasma-free atomic layer etching (ALE) of PdO. A vacuum ultraviolet (VUV) light source with a wavelength range of 115–400 nm is combined with regulated exposure to O_2 gas to create PdOx at a temperature of 100 °C. The quantity of PdOx that is produced is contingent upon the length of time that O_2 is exposed to VUV irradiation at a pressure of 1 Torr. In order to partly oxidize 2 nm Pd, a minimum coexposure time of 1 minute is necessary. However, for 20 nm Pd films, a longer coexposure time of 3 minutes is required. This information has been confirmed in real-time using X-ray photoelectron spectroscopy (XPS) [47].

In a separate investigation conducted by Ville et al. [45], the authors provided a comprehensive account of employing photo-assisted atomic layer deposition (ALD) techniques for the deposition of titanium, zirconium, hafnium, niobium, and tantalum oxides. They achieve this by using liquid, volatile metal alkoxides as precursors in a single-source method without the need for any additional reactants. The ALD reactor was provided with a light source that emitted photons throughout a continuous spectrum ranging from 190 to 800 nm in wavelength. The deposition process comprised a precursor pulse, followed by a purge, a photon exposure, and another purge. The study examined the attributes of the procedure and the qualities of the film. The Nb_2O_5 and Ta_2O_5 films were determined to have an amorphous structure, whereas the TiO_2, ZrO_2, and HfO_2 films displayed either an amorphous or polycrystalline structure, depending on the deposition conditions. Shadow masking allows for the precise deposition of materials in specific areas using photo-ALD. The deposition experiments employ patterned substrates and optical filtering to investigate the nature of the growth process, focusing on whether it is driven by photolytic or photothermal processes. The electrical study of photo-ALDHfO_2 reveals advantageous dielectric properties.

1.4 Deposition Characteristics

1.4.1 Growth

ALD is a precisely controlled process where precursors are added one at a time to the growing surface. Each precursor undergoes a reaction that stops after the surface is fully covered, and this cycle is repeated [48]. The initial growth ALD is particularly important as it establishes the nucleation behavior and the number of cycles required to reach a closed layer. Growth per cycle (GPC) is a measure of growth rate. Since the GPC is often reported as an average value of at least a few hundred cycles, nucleation inhibition or enhancement may lead the initial GPC for ALD procedures of a particular material system onto a particular substrate to deviate from its (steady-state) literature value [49]. By repeatedly saturating at least two gaseous reactants separately with a solid substrate, ALD is able to grow materials. The growth obtained per ALD reaction cycle when surface saturation is systematically exploited is a well-defined quantity that relies on (i) the reactants used, (ii) the temperature during ALD processing, and (iii) occasionally the substrate material [50]. A key attribute of ALD is its ability to deposit a consistent amount of material in each cycle, which enables precise control over the final thickness. To verify this, an increase in thickness or material is measured for each cycle, referred to as the growth per cycle (GPC). Two approaches may be used to determine the GPC: ex situ, which requires depositing many samples with different numbers of cycles, and in situ, which includes monitoring the material increase during deposition [51].

The growth per cycle number does not accurately reflect the chemical reaction kinetics that occur throughout the deposition process. Instead, the outcome is determined by the quantity of chemisorption sites present on the growth surface, which is affected by the reactivity and number of accessible surface sites, as well as the form of the surface.

The GPC can be affected by the substrate at the initial deposition process, and it requires a specific number. This phenomenon occurs due to the disparity in reactivity between the chemical sites on the original substrate and those on the surface of the as-grown substance. Substrate inhibition and the ensuing delay in film nucleation/growth may appear to be disadvantageous at first, but in the case of area-selective ALD, they may really be highly helpful [17].

A theoretical model derived from the chemistry of the growth when compounds are utilized as reactants was used to describe the GPC in ALD [50, 51] (Fig. 1.8).

The GPC, or Growth per Cycle, represents the rate at which the thickness increases with each cycle. The GPC is determined by measuring the atoms or mass deposited as a function of cycles and is represented as the amount of atoms or grammes per unit area deposited each cycle.

ALD development is typically shown as the formation of an ideal monolayer of deposited material after each half-cycle, which then grows into a perfect layer-by-layer (LBL) film. However, in practical terms, there are occasional deviations from this, which might result in the formation of various submonolayers. This is could

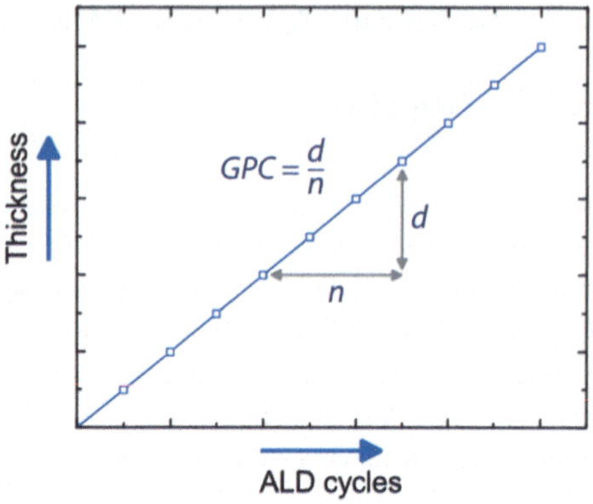

Fig. 1.8 An example which demonstrates the relationship between the thickness of a film and the quantity of atomic layer deposition (ALD) cycles, displaying the typical linear growth trend that is frequently observed in ALD

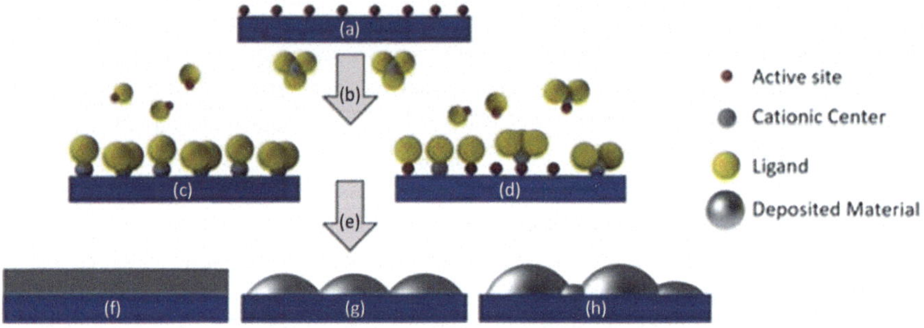

Fig. 1.9 Illustration of chemisorption and growth at several stages of an ALD process. (**a**) Substrate without any active sites, (**b**) initial introduction of precursor material, (**c**) complete reaction occurring between the precursor and the active sites, (**d**) partial reaction occurring between the precursor and the active sites, (**e**) multiple cycles of ALD, (**f**) growth of a film that conforms to the shape of the substrate, (**g**) uniform growth of nanoparticles, and (**h**) nonuniform growth of nanoparticles [48]

have been caused by the steric effects which are from the ligands blocking of the actives sites, chemisorption, and others. Moreover, several factors such as limited nucleation sites or diffusion might lead to the creation of islands or nanoparticles (NPs) instead of a conformal layer. This is illustrated in Fig. 1.9 [48].

In an actual ALD process, the adsorption of precursor is self-limited to a maximum of one monolayer after each exposure. Adsorption can take place through one of two broadly characterized mechanisms: Physisorption and chemisorption are two types of adsorption processes [48]. Chemisorption is the process in which a chemical bond is formed between the precursor and the growth substrate while Physisorption is adsorption by van der Waals force, which is a weak intermolecular attraction that takes place below the critical temperature of the adsorbate and can result in the development of a monolayer or multilayer.

Chemisorption is a chemical reaction between a precursor and substrate's surface, forming a new bond at the expense of another metal–ligand bond. ALD processes have a self-limiting characteristic, as precursors react with the growth surface but not themselves. When volatile by-products are produced, they desorb and be purged away, making chemisorption irreversible [50]. Nevertheless, in cases when by-products are not produced or fail to desorb, chemisorption may exhibit reversibility. Adsorption should be exhibit irreversibility during the whole ALD cycle.

Although several atomic layer deposition (ALD) processes have been documented in scientific literature, it is widely acknowledged that a substantial proportion of these processes adhere to one or more chemisorption mechanisms. These pathways are typically known as ligand exchange, dissociation, association, and oxidation [48].

The ligand exchange mechanism is the main mechanism for atomic layer deposition (ALD), and it may be characterized by the following general reaction sequence as highlighted by Li et al. [52]

$$\| -a + ML_n(g) \rightarrow \| -ML_{n-1} + aL(g)$$

|| represent the surface, a represent a surface group, M represent a metal atom, and L represent a ligand of the precursor. During the ligand-exchange process, a ligand (L) from the gaseous precursor molecule (ML_n) is replaced by the surface species ($\| - a$), resulting in the formation of a volatile by-product of L. At the same time, the remaining part of the precursor molecule binds to the surface, generating a chemisorbed new surface species of $\| - ML_{n-1}$.

The ligand-exchange process can persist between the chemisorbed $\| - ML_{n-1}$ and a neighboring $\| - a$ surface group, resulting in the formation of numerous chemical bonds between the precursor and surface, as follows:

$$\| -ML_{n-1} + \| -a \rightarrow \| -ML_{n-2} - \| + aL(g)$$

In the true ALD process, the adsorption of the precursor is limited to a maximum of one monolayer of adsorbed precursor after each exposure. Adsorption can occur through one of two generally categorized mechanisms: Physisorption and chemisorption are two types of adsorption processes.

The slow adaptation can be attributed to the very low deposition rate of ALD, particularly at low temperatures, typically about 0.15 nm/s. This rate is significantly slower compared to physical vapor deposition (PVD) or chemical vapor deposition (CVD) processes [12].

The growth rate of ALD is directly affected by the flow of the precursor on the substrate. Determining the growth rate may be a complex task since it is influenced by several factors such as the temperature window, which is the range of temperatures where the precursor concentration per pulse and purge durations are in balance, resulting in a stable growth per cycle (GPC) [2].

Outside the ALD window, the GPC can be reduced owing to slow reaction kinetics at low temperature or due to desorption during the purge stage at high temperature. Conversely, the GPC can be increased due to the condensation of

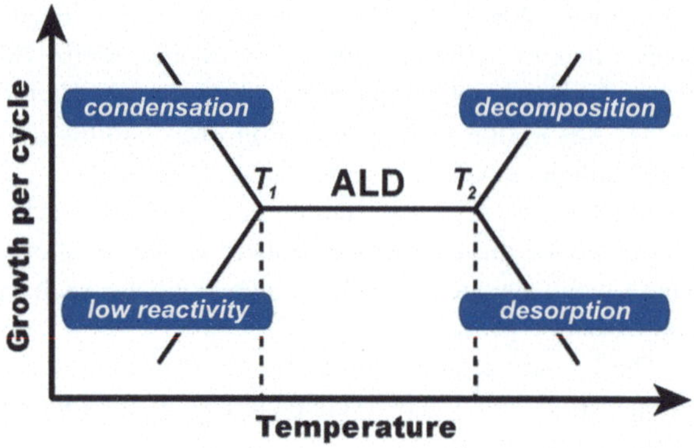

Fig. 1.10 The growth per cycle (GPC) varies depending on the deposition temperature

precursor molecules at low temperature or due to the thermal breakdown of precursors, at high temperature.

As highlighted by Sanders et al. [53], a general temperature window of an ALD process is shown in Fig. 1.10. To avoid the condensation of any reactants, the substrate temperature should exceed T1, as shown in the Fig. 1.10. Undesirable or uncontrolled reactions may occur if condensation takes place during an ALD cycle, leading to the creation of porous and impure films. In addition, for surface reactions to occur, the activation energy need to be exceeded. Hence, the ALD process requires a minimum substrate temperature (T1) for its process. However, an undesirable degradation of a reactant can occur if the temperature exceeds T2 as illustrated in Fig. 1.10.

The poor deposition rate is attributed to the cyclic nature of ALD, characterized by extended injection/saturation pulses and even lengthier purging rates. The situation becomes more unfavorable for larger substrates, as the size of the ALD reactor and the duration required for pulsing and purging both rise. Furthermore, the requirement for most ALD procedures to be conducted under vacuum, both to expedite the purging process and due to the conventional usage of ALD in microelectronics, presents an additional disadvantage for industrial use. This is due mainly to the substantial expenses incurred when scaling up vacuum equipment [12].

When it comes to constructing reactors and process engineering, fluid dynamics are now essential.

After each ALD cycle is finished, a specific amount of material (coating) is added to the surface. This amount is called the growth per cycle (GPC). The GPC can be expressed using several measures, such as thickness (e.g., nm), increase in areal density (e.g., atoms/nm^2), and gain in mass (e.g., g) [20].

Research is now being conducted to study the modeling and simulations of ALD at various lengths and timescales [2, 54–57]. At the atomic level, it is frequently used to employ ab initio methods, such as density functional theory (DFT), to investigate the binding of precursor molecules to the surface and analyze the structural and electronic characteristics of these precursors at the smallest scales [58]. Computational fluid dynamics (CFD) is commonly used to study the behavior of precursors at a large scale and how the equipment settings affect the distribution of gas inside the ALD reactor [59].

1.5 Different Types of Reactors

An optimized ALD chamber design incorporates many geometric enhancements, such as enhanced intake tubes, feed nozzles, showerhead, and the primary chamber. The dimensions and shape of the reactor will have a substantial impact on the movement of gas and the speed of surface reactions, which are essential for optimizing the efficiency of the ALD process [12]. Moreover, as the energy consumption of an ALD process is mostly determined by the operating temperature, it is essential to have efficient thermal management of the heated chamber and heating components in order to reduce power usage and lower carbon emissions. By improving the design of the ALD reactor and fine-tuning the processing parameters, it is feasible to reduce the usage of precursors and the release of by-products. Moreover, this can substantially decrease the duration of the ALD cycle and the thermal budget.

Depending on a number of criteria, including the substrate to be coated, materials to be deposited, throughput, and in situ characterization capabilities, ALD reactors can be configured in a wide variety of ways. Reactors may generally be classified according to the kind of ALD reaction they conduct—thermal or plasma—and how precursors are added to the substrate—spatially or temporally.

Design of the ALD reactor differs in terms of the application and type of precursor to be used.

There are several types of reactors, as highlighted in Fig. 1.11; they include:

1. Flow-type ALD reactor is temporal ALD, Pulse-train of precursors, Reactor pressure 1–10 Torr, and finds application in semiconductor.
2. Showerhead ALD reactor.
3. Batch ALD reactor: Batch atomic layer deposition (ALD) enables the simultaneous coating of many samples, as opposed to the deposition of a single substrate.
4. Energy-Enhanced ALD reactor: Energy-enhanced atomic layer deposition (ALD) is employed for ALD processes that require additional energy beyond thermal energy to achieve a high-quality film or a viable process. In this case, the co-reactants refer to species such as plasmas or ozone that have a limited lifespan. To do this, it is necessary to incorporate machinery that generates the species in close proximity to the substrate [60]. Energy-enhanced ALD offers several benefits over thermal ALD, including improved material properties, better stoichiometry control, increased growth rate due to shorter cycle times, and the possibility of in situ treatments with plasma sources or ozone generators. Additionally, energy-enhanced ALD provides more precursor freedom and allows deposition at reduced substrate temperatures, making it a more efficient and cost-effective alternative to thermal ALD [61].
5. Spatial ALD reactor.

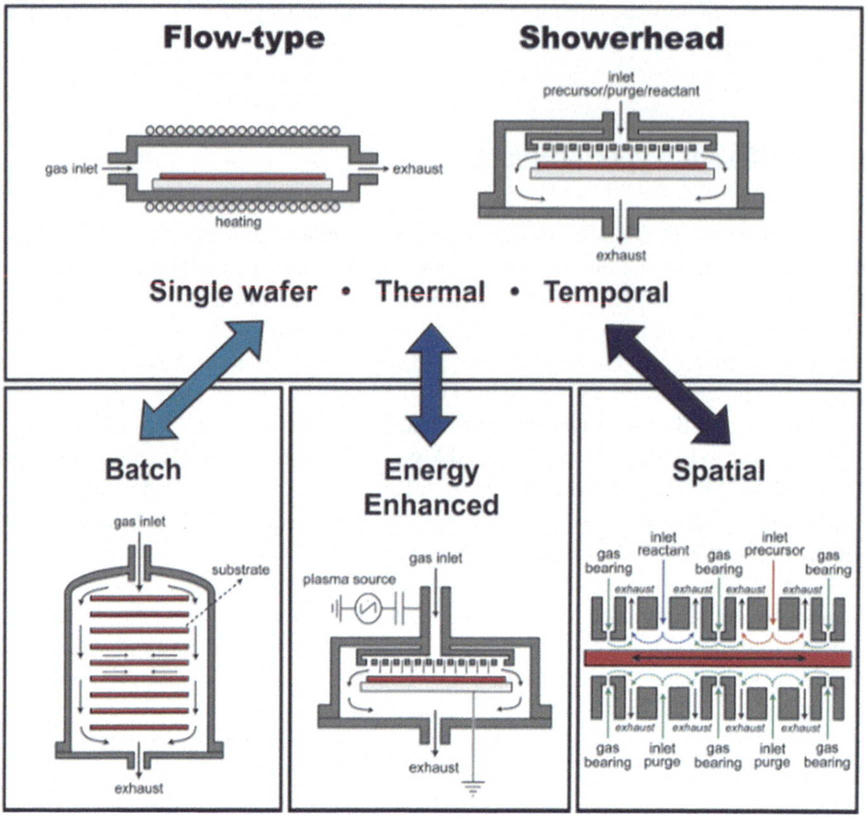

Fig. 1.11 An illustration of the several types of ALD reactors [60]

1.6 Overview of ALD for Hydrogen Energy

Hydrogen has been regarded as a promising fuel due to its high-energy density and environmental sustainability. Hydrogen generated by the process of water splitting is free from the disadvantage of undesirable emissions [62]. However, it does need the supply of energy from an external source. Higher efficiency in energy capture, storage, and consumption applications is fueling a comparable demand for atomically precise materials [63]. Nanoscale materials have greatly revolutionized the domain of renewable energy storage and conversion devices, including solar cells, water-splitting systems, fuel cells, batteries, and supercapacitors. Optimizing the structure and interface qualities is essential for achieving exceptional performance and long-lasting durability [64]. ALD is a crucial technique in nanofabrication, enabling the creation of films with no defects and precise control over their thickness and composition at the atomic level [2]. It is particularly effective on surfaces with a high aspect ratio [19].

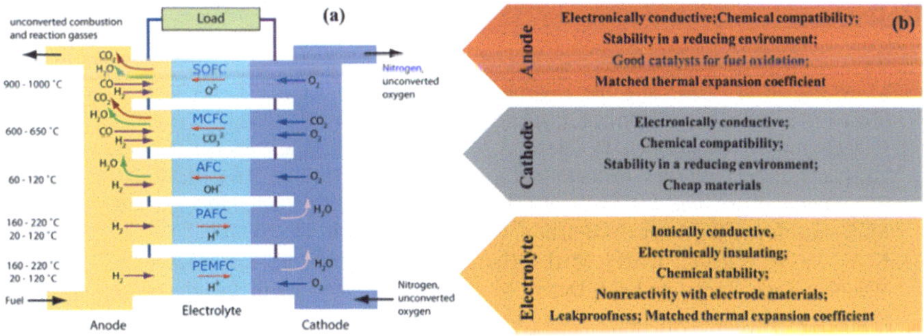

Fig. 1.12 is a description of the different fuel types and some of the working conditions. (**a**) Alkaline fuel cell (AFC), Molten carbonate fuel cell (MCFC), Phosphoric acid fuel cell (PAFC), Proton exchange membrane or polymer electrolyte membrane fuel cell (PEMFC); Solid oxide fuel cell (SOFC). (**b**) Material requirement for SOFC

As highlighted by Arun et al. [63], there are roles ALD can play in the study and synthesis of energy storage materials:

1. Energy materials fabrication using ALD.
2. Modification of surface properties.
3. Physical/Chemical Barriers and Encapsulation.
4. Fundamental and Mechanistic Insights.

Sustainable energy technologies are being advanced through three primary classifications: energy capture, energy storage, and efficient energy use [63].

There are several considerations when it comes to green hydrogen, some of which are the renewable energy source, the production of green hydrogen, and the end-user applications. ALD finds applications, particularly in electrolyzers and fuel cells.

Electrodes, electrolytes, and the electrode/electrolyte interface can be fine-tuned on an atomic scale using ALD, which has helped to address underlying problems in fuel cell technology like poor redox kinetics, electrode durability during long-term cycling, and low electrolyte ionic conductivity [64].

With their potential for direct conversion of chemical energy into electrical energy, low operating energy losses, and environmentally friendly emissions, fuel cells have recently emerged as a dependable and clean energy technology [64, 65] (Fig. 1.12).

The anode [66–68], cathode [69–71], and electrolyte, which are crucial components affecting fuel cell performance, are enhanced by the use of ALD.

The Levelized Cost of Hydrogen (LCOH) serves as a valuable tool for policymakers and industry stakeholders, offering a techno-economic foundation upon which to base their decision-making processes for hydrogen projects. This information can facilitate the early adoption and advancement of the hydrogen market. The subsequent section will focus on addressing some hurdles, particularly those associated with technological concerns.

References

1. M. Weber et al., Assessing the environmental impact of Atomic Layer Deposition (ALD) processes and pathways to lower it. ACS Mater. Au **3**, 274–298 (2023)
2. P.O. Oviroh, R. Akbarzadeh, D. Pan, R.A.M. Coetzee, T.-C. Jen, New development of atomic layer deposition: processes, methods and applications. Sci. Technol. Adv. Mater. **20**, 465 (2019)
3. P.O. Oviroh, T.-C. Jen, J. Ren, A. van Duin, Towards the realisation of high permi-selective MoS_2 membrane for water desalination. NPJ Clean Water **6**, 14 (2023)
4. Keren Jacobs Kanarik, Jeffrey Marks, Harmeet Singh, Samantha Tan, Alexander Kabansky, Wenbing Yang, Taeseung KIM, Dennis M. Hausmann, Thorsten Lill. Integrating Atomic Scale Processes: ALD (Atomic Layer Deposition) and ALE (Atomic Layer Etch). (Springer Nature, 2017)
5. J. Cai, X. Han, X. Wang, X. Meng, Atomic layer deposition of two-dimensional layered materials: processes, growth mechanisms, and characteristics. Matter **2**, 587–630 (2020), Preprint at https://doi.org/10.1016/j.matt.2019.12.026
6. S.M. George, Atomic layer deposition: an overview. Chem. Rev. **110**, 111 (2010). https://doi.org/10.1021/cr900056b
7. T. Kääriäinen, D. Cameron, M.L. Kääriäinen, A. Sherman, *Atomic Layer Deposition: Principles, Characteristics, and Nanotechnology Applications* (Wiley, 2013). https://doi.org/10.1002/9781118747407
8. E. Ahvenniemi et al., Review article: recommended reading list of early publications on atomic layer deposition – outcome of the "Virtual Project on the History of ALD". J. Vac. Sci. Technol. A **35**, 010801 (2017)
9. R.L. Puurunen, A short history of atomic layer deposition: Tuomo Suntola's atomic layer epitaxy. Chem. Vap. Depos. **20**, 332–344 (2014)
10. A.A. Malygin, V.E. Drozd, A.A. Malkov, V.M. Smirnov, V.B. From, Aleskovskii's "Framework" hypothesis to the method of molecular layering/atomic layer deposition. Chem. Vap. Depos. **21**, 216–240 (2015)
11. T. Suntola, Atomic layer epitaxy. Thin Solid Films **216**, 84 (1992). https://doi.org/10.1016/0040-6090(92)90874-B
12. D. Muñoz-Rojas et al., Speeding up the unique assets of atomic layer deposition. Mater. Today Chem. **12**, 96–120 (2019)
13. P.H. Li, P.K. Chu, Thin film deposition technologies and processing of biomaterials, in *Thin Film Coatings for Biomaterials and Biomedical Applications*, (2016). https://doi.org/10.1016/b978-1-78242-453-6.00001-8
14. R.J. Narayan et al., Atomic layer deposition-based functionalization of materials for medical and environmental health applications. Philos. Trans. R. Soc. A Math. Phys. Eng. Sci. **368**, 2033 (2010). https://doi.org/10.1098/rsta.2010.0011
15. P. Poodt et al., Spatial atomic layer deposition: a route towards further industrialization of atomic layer deposition. J. Vac. Sci. Technol. A **30** (2012). https://doi.org/10.1116/1.3670745
16. B. Yan et al., A review of atomic layer deposition providing high performance lithium sulfur batteries. J. Power Sources **338**, 34 (2017), Preprint at https://doi.org/10.1016/j.jpowsour.2016.10.097
17. J. Dendooven, C. Detavernier, Basics of atomic layer deposition: growth characteristics and conformality, in *Atomic Layer Deposition in Energy Conversion Applications*, (Wiley, 2017), pp. 1–40. https://doi.org/10.1002/9783527694822.ch1
18. A.M.E.K. Martijn Vos, Atomic layer deposition process development – 10 steps to successfully develop, optimize and characterize ALD recipes. AtomicLimits (2019), Preprint at https://www.epfl.ch/research/facilities/cmi/wp-content/uploads/2020/07/ALD-Process-Development.pdf (2019)
19. J.W. Elam, D. Routkevitch, P.P. Mardilovich, S.M. George, Conformal coating on ultrahigh-aspect-ratio nanopores of anodic alumina by atomic layer deposition. Chem. Mater. **15**, 3507 (2003)

20. V. Cremers, R.L. Puurunen, J. Dendooven, Conformality in atomic layer deposition: current status overview of analysis and modelling. Appl. Phys. Rev. **6** (2019), Preprint at https://doi.org/10.1063/1.5060967
21. R.G. Gordon, D. Hausmann, E. Kim, J. Shepard, A kinetic model for step coverage by atomic layer deposition in narrow holes or trenches. Chem. Vap. Depos. **9**, 73 (2003)
22. H.C.M. Knoops, E. Langereis, M.C.M. van de Sanden, W.M.M. Kessels, Conformality of plasma-assisted ALD: physical processes and modeling. J. Electrochem. Soc. **157**, G241 (2010). https://doi.org/10.1149/1.3491381
23. K.-E. Elers et al., Film uniformity in atomic layer deposition. Chem. Vap. Depos. **12**, 13–24 (2006)
24. B. Choi, H.-U. Kim, N. Jeon, Uniformity of HfO_2 thin films prepared on trench structures via plasma-enhanced atomic layer deposition. Nano **13**, 161 (2022)
25. E. Knehr et al., Wafer-level uniformity of atomic-layer-deposited niobium nitride thin films for quantum devices. J. Vac. Sci. Technol. A **39**, 052401 (2021)
26. C.T. Lennon et al., High-uniformity atomic layer deposition of superconducting niobium nitride thin films for quantum photonic integration. Mater. Quantum Technol. **3**, 045401 (2023)
27. V. Miikkulainen, M. Leskelä, M. Ritala, R.L. Puurunen, Crystallinity of inorganic films grown by atomic layer deposition: overview and general trends. J. Appl. Phys. **113** (2013)
28. E. Thimsen et al., Atomic layer deposition of the quaternary chalcogenide Cu_2 $ZnSnS_4$. Chem. Mater. **24**, 3188–3196 (2012)
29. H. Van Bui, F. Grillo, J.R. van Ommen, Atomic and molecular layer deposition: off the beaten track. Chem. Commun. **53**, 45–71 (2017)
30. M. Shahmohammadi, R. Mukherjee, C. Sukotjo, U. Diwekar, C. Takoudis, Recent advances in theoretical development of thermal atomic layer deposition: a review. Nano **12**, 831 (2022)
31. J. Kim, T.W. Kim, Initial surface reactions of atomic layer deposition. JOM **61**, 17–22 (2009)
32. J.R. Scheffe et al., Hydrogen production via chemical looping redox cycles using atomic layer deposition-synthesized iron oxide and cobalt ferrites. Chem. Mater. **23**, 2030–2038 (2011)
33. K. Knapas, M. Ritala, *In Situ* studies on reaction mechanisms in atomic layer deposition. Crit. Rev. Solid State Mater. Sci. **38**, 167–202 (2013)
34. M. Napari, *Low-Temperature Thermal and Plasma-Enhanced Atomic Layer Deposition of Metal Oxide Thin Films* (University of Jyväskylä, Jyväskylä, 2017)
35. D.J. Monsma, *Atomic Layer Deposition Plasma-Enhanced ALD (PE-ALD)* (Beneq) https://beneq.com/en/atomic-layer-deposition/peald/
36. J. Musschoot, *Advantages and Challenges of Plasma Enhanced Atomic Layer Deposition* (Ghent University, 2011)
37. H. Kim, I.K. Oh, Review of plasma-enhanced atomic layer deposition: Technical enabler of nanoscale device fabrication. Jpn. J. Appl. Phys. **53**, 03DA01 (2014), Preprint at https://doi.org/10.7567/JJAP.53.03DA01
38. R.W. Johnson, A. Hultqvist, S.F. Bent, A brief review of atomic layer deposition: From fundamentals to applications. *Mater. Today*. **17**, 236 (2014), Preprint at https://doi.org/10.1016/j.mattod.2014.04.026
39. J. Bachmann, *Atomic Layer Deposition in Energy Conversion Applications* (Wiley, Weinheim, 2017)
40. S. O'Donnell et al., Thermal and plasma enhanced atomic layer deposition of ultrathin TiO_2 on silicon from amide and alkoxide precursors: growth chemistry and photoelectrochemical performance. J. Phys. D. Appl. Phys. **55**, 085105 (2022)
41. D. Muñoz-Rojas, V. Huong Nguyen, C. .M. de la Huerta, C. Jiménez, D. Bellet, Spatial atomic layer deposition, in *Chemical Vapor Deposition for Nanotechnology*, (IntechOpen, 2019). https://doi.org/10.5772/intechopen.82439
42. D. Muñoz-Rojas, J. MacManus-Driscoll, Spatial atmospheric atomic layer deposition: a new laboratory and industrial tool for low-cost photovoltaics. Mater. Horiz. **1**, 314–320 (2014)
43. J.P. Vale et al., Can we rationally design and operate spatial atomic layer deposition systems for steering the growth regime of thin films? J. Phys. Chem. C **127**, 9425–9436 (2023)

44. Spatial Atomic Layer Deposition System. https://uwaterloo.ca/research/about-research/waterloo-commercialization-office-watco/spatial-atomic-layer-deposition-system
45. V. Miikkulainen et al., Photoassisted atomic layer deposition of oxides employing alkoxides as single-source precursors. J. Vac. Sci. Technol. A **37**, 060911 (2019)
46. P.N. Heys, et al., Methods of Atomic Layer Deposition Using Hafnium and Zirconium-Based Precursors (2011), Preprint at https://patents.google.com/patent/US8039062B2/en
47. B.M. Coffey, H.C. Nallan, J.R. Engstrom, C.H. Lam, J.G. Ekerdt, Vacuum ultraviolet-enhanced oxidation – a route to the atomic layer etching of palladium metal. Chem. Mater. **32**, 6035–6042 (2020)
48. N.E. Richey, C. de Paula, S.F. Bent, Understanding chemical and physical mechanisms in atomic layer deposition. J. Chem. Phys. **152**, 040902 (2020)
49. C.W. Wiegand et al., Understanding the growth mechanisms of multilayered systems in atomic layer deposition process. Chem. Mater. **30**, 1971–1979 (2018)
50. R.L. Puurunen, Growth per cycle in atomic layer deposition: a theoretical model. Chem. Vap. Depos. **9**, 249–257 (2003)
51. R.L. Puurunen, Growth per cycle in atomic layer deposition: real application examples of a theoretical model. Chem. Vap. Depos. **9**, 327–332 (2003)
52. J. Li, G. Chai, X. Wang, Atomic layer deposition of thin films: from a chemistry perspective. Int. J. Extreme Manuf. **5**, 032003 (2023)
53. J.J.M. Sanders. ALD plasma & materials processing, https://research.tue.nl/en/equipments/ald
54. S.D. Elliott, Atomic-scale simulation of ALD chemistry. Semicond. Sci. Technol. **27**, 074008 (2012). https://doi.org/10.1088/0268-1242/27/7/074008
55. Z. Hu, J. Shi, C. Heath Turner, Molecular dynamics simulation of the Al_2O_3 film structure during atomic layer deposition. Mol. Simul. **35**, 270–279 (2009)
56. M.R. Shaeri, T.C. Jen, C.Y. Yuan, Improving atomic layer deposition process through reactor scale simulation. Int. J. Heat Mass Transf. **78**, 1243 (2014). https://doi.org/10.1016/j.ijheatmasstransfer.2014.07.079
57. T. Reiter et al., Modeling the impact of incomplete conformality during atomic layer processing. Solid State Electron. **211**, 108816 (2024)
58. A. Heyman, C.B. Musgrave, A quantum chemical study of the atomic layer deposition of Al_2O_3 using $AlCl_3$ and H_2O as precursors. J. Phys. Chem. B **108**, 5718–5725 (2004)
59. D. Pan, T.-C. Jen, C. Yuan, Effects of gap size, temperature and pumping pressure on the fluid dynamics and chemical kinetics of in-line spatial atomic layer deposition of Al_2O_3. Int. J. Heat Mass Transf. **96**, 189–198 (2016)
60. H.C.M. Knoops, S.E. Potts, A.A. Bol, W.M.M. Kessels, Atomic layer deposition, in *Handbook of Crystal Growth: Thin Films and Epitaxy*, vol. 3, 2nd edn., (Elsevier, 2015), pp. 1101–1134
61. S.E. Potts, W.M.M. Kessels, Energy-enhanced atomic layer deposition for more process and precursor versatility. Coord. Chem. Rev. **257**, 3254–3270 (2013)
62. C. Marichy, M. Bechelany, N. Pinna, Atomic layer deposition of nanostructured materials for energy and environmental applications. Adv. Mater. **24**, 1017–1032 (2012)
63. A.S. Asundi, J.A. Raiford, S.F. Bent, Opportunities for atomic layer deposition in emerging energy technologies. ACS Energy Lett. **4**, 908–925 (2019)
64. B. Gupta et al., Recent advances in materials design using atomic layer deposition for energy applications. Adv. Funct. Mater. **32** (2022). https://onlinelibrary.wiley.com/doi/abs/10.1002/adfm.202109105
65. A. Basile, L. Di Paola, F.I. Hai, V. Piemonte, Membrane reactors for energy applications and basic chemical production, in *Membrane Reactors for Energy Applications and Basic Chemical Production*, (Elsevier, 2015). https://doi.org/10.1016/C2013-0-16489-6
66. T.A. Schmauss, S.A. Barnett, Atomic layer deposition for surface area determination of solid oxide electrodes. J. Mater. Chem. A Mater. **11**, 3695 (2023)
67. H. Liu et al., Nitrogen doped titania stabilized Pt/C catalyst via selective atomic layer deposition for fuel cell oxygen reduction. Chem. Eng. J. **463**, 142405 (2023)

68. H.W. Park et al., Atomic layer deposited platinum on tungsten oxide support as high performance hybrid catalysts for polymer electrolyte membrane fuel cells. Appl Catal B **337**, 122956 (2023)
69. Y.S. Lee, J.W. Shim, Y.S. Choi, B.G. Seo, J.H. Shim, Stabilization of platinum catalyst surface-treated by atomic layer deposition of cobalt for polymer electrolyte membrane fuel cells. Int. J. Energy Res. **46**, 15387 (2022)
70. S.H. Jeon et al., Metal-oxide nanocomposite catalyst simultaneously boosts the oxygen reduction reactivity and chemical stability of solid oxide fuel cell cathode. Chem. Eng. J. **455**, 140611 (2023)
71. J.S. Park et al., High-performance hydroxide exchange membrane fuel cell comprising an atomic layer-deposited silver cathode. Nano Lett. **23**, 7825 (2023)

Chapter 2
Technologies of Hydrogen Generation and Purification

2.1 Introduction

The molecular form of hydrogen can be extracted from a wide variety of resources, which include water, biomass, and fossil fuels among others [1]. In order to obtain hydrogen from these various sources, there needs to be an abundant supply of readily available energy in a nonstop state of availability. Therefore, sustainable hydrogen could be created by utilizing technologies that tap into the potential of renewable energy, such as the sun, wave sea, wind, etc. [2–4]. The pathway to the generation of environmentally friendly hydrogen is depicted in Fig. 2.1. This chapter focuses mainly on environmentally friendly methods of producing hydrogen. In the future, the primary technologies utilized to produce hydrogen will be determined by the type of raw material that is put into production: renewable resources or fossil fuels.

2.2 Hydrogen Generation from Water

Even though the vast majority of hydrogen that is generated today comes from hydrocarbons, there has been a growing interest in producing green hydrogen from renewable resources [6]. Hydrogen, with a minimal environmental impact footprint, has the potential to dramatically reduce emissions of carbon dioxide (CO_2) from energy production which can contribute to keeping the increase in global temperature below 2 °C [7]. In this context, "green hydrogen" refers to the hydrogen produced from water or compounds derived from biomass.

Water is the most readily available resource for the generation of hydrogen, and it is possible to separate its components into hydrogen and oxygen by applying sufficient energy to the reaction in a manner that does not release any toxic by-products [8, 9]. Water splitting is possible using a variety of technologies. As

P. O. Oviroh et al., *Emerging Atomic Layer Deposition for Hydrogen Energy*,
Springer Series in Materials Science 348,
https://doi.org/10.1007/978-3-031-67774-8_2

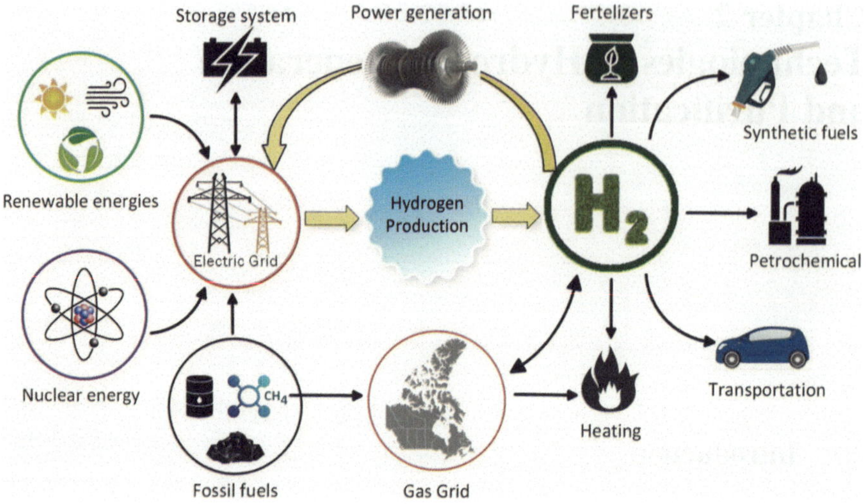

Fig. 2.1 Generation of environmentally friendly hydrogen pathway. (Reprinted with permission from Ref. [5])

shown in Fig. 2.2, depending on the amount of energy needed to split, these technologies can be divided into four broad categories. The simplest method of water splitting involves the use of an electrical current called electrolysis, flowing between two electrodes to separate the water molecules into oxygen and hydrogen [10]. Table 2.1 summarizes these and other features of this compound, including the fact that it can be divided using both thermal energy (thermolysis) and photonic energy (photoelectrolysis) as well as biophotolysis employing microorganisms [8, 11].

2.2.1 *Water Splitting by Electrolysis*

When it comes to the production of hydrogen from water, electrolysis is perhaps one of the most straightforward technologies now accessible. It is possible to summarize the process as the transformation of electrical power into chemical energy, which results in the production of oxygen and hydrogen as a by-product. There are two processes that take place in each electrode, which are referred to as the anode and the cathode [8, 15]:

anode

$$H_2O \rightarrow \frac{1}{2}O_2 + 2H^+ + 2e^- \tag{2.1}$$

cathode

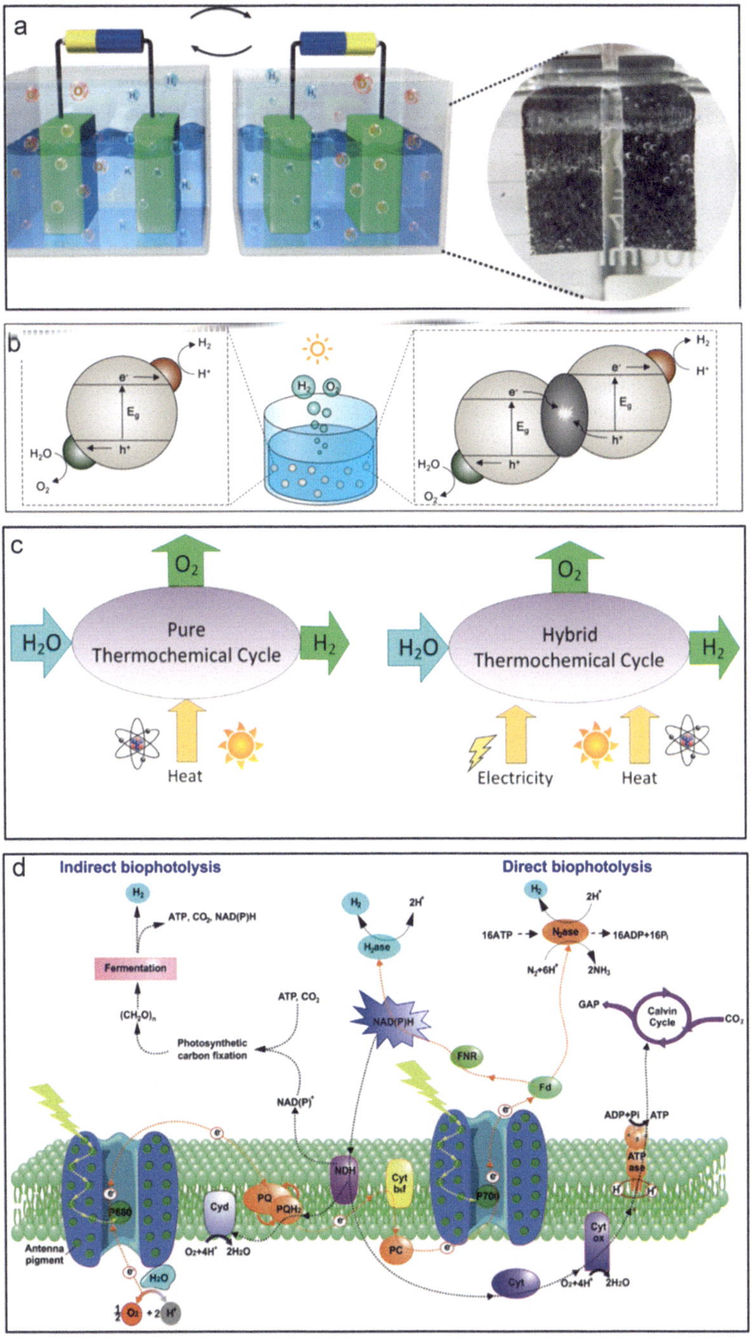

Fig. 2.2 Various H2 generation approaches: (**a**) electrolysis water splitting, (**b**) photocatalytic water splitting, (**c**) thermochemical water splitting, (**d**) biophotolysis H_2 production. [(**a**) Reprinted with permission from Ref. [12]. (**b**) Reprinted with permission from Ref. [13]. (**c, d**) Reprinted with permission from Ref. [14]]

Table 2.1 A review of water-based hydrogen production methods

Technology	Energy source	Operating conditions	Brief description	Maturity
Electrolysis	Electrical energy	Up to 30 bar; 50–900 °C (depending upon the method used)	Water decomposition into O_2 and H_2 by passing a direct current which drives electrochemical reactions	Commercial
Thermolysis	Thermal energy	Temperature of >2500 °C (<1000 °C for thermochemical cycles)	Steam is brought to temperatures of over 2500 K at which water molecule decomposes thermally	Research and development
Photoelectrolysis	Electrical + photonic	Ambient conditions	Photo-electrodes external source of electricity	Research and development
Biophotolysis	Photonic + biochemical	Ambient conditions	Biological systems based on cyanobacteria are used to generate hydrogen in a controlled manner	Research and development

$$2H^{+} + 2e^{-} \rightarrow H_2 \tag{2.2}$$

overall process

$$H_2O \rightarrow H_2 + \frac{1}{2}O_2 \tag{2.3}$$

Electrolysis of water can be accomplished using a variety of methods: the most common of which are proton-exchange membrane electrolysis, electrolysis using solid oxide, and alkaline water electrolysis [16, 17]. To stop the mixing of the gas products, the electrolysis of alkaline water needs to be accompanied by a gas separator. As an electrolyte, it makes use of highly concentrated lye, and its electrodes are composed of non-noble metals like nickel. Humidified polymer membranes act as the electrolyte in proton-exchange membrane electrolysis, while noble metals or their oxides, like iridium oxide or platinum, act as the electrocatalysts. Each of these technologies has a wide temperature range (50–80 °C) and pressure range (up to 30 bars) for operation. On the other hand, solid oxide electrolysis makes use of high temperatures (between 700 °C and 900 °C) to facilitate the transformation of water into hydrogen and oxygen, which in turn raises the requirement for thermal energy. Proton-exchange membrane electrolysis and alkaline water electrolysis are two technologies that are more desirable for general deployment because of their lower initial investment costs and longer lifetimes of the unit [16]. It is projected that water electrolysis driven by renewable energy sources (such as biomass [4], sea wave, and wind) will make it possible to increase the production of hydrogen (with a high purity of 99.9%) while simultaneously reducing the amount of carbon dioxide emissions. Since this is the case, it is now feasible to generate hydrogen on-site, eliminating the requirement for transportation [17]. However, the cost of H_2 that is created by electrolysis is still quite a bit greater than the cost of H_2 that is produced through the combustion of fossil fuels [18].

2.2.2 *Water Splitting by Thermolysis*

Water can be split into hydrogen and oxygen through a thermochemical process called thermolysis, which involves heating water to very high temperatures (Eq. 2.4):

$$2H_2O + \text{heat} \rightarrow H_2 + O_2 \tag{2.4}$$

Despite the apparent simplicity of this method, water breakdown really necessitates temperatures above 2500 °C [10, 19]. Since the thermolysis process can be reversed, it is imperative that the hydrogen and oxygen that are created be kept apart, as their recombination could result in an explosive mixture [20]. Materials that can function at the required temperatures are also hard to come by.

Thermochemical water-splitting cycles are an alternative to thermolysis that produces hydrogen and oxygen in separate phases at lower maximum operating temperatures (typically around 1000 °C). This prevents the two gases from recombining and eliminates the need for costly and energy-intensive gas separation downstream [19]. Although the experimental process becomes more difficult, the temperature needed to split water drops, as the number of cycles increases. Both low-temperature and high-temperature thermochemical cycles can be used to split water. Despite the high-temperature requirement, the sulfur iodine thermochemical cycle has received the greatest attention from researchers since it is seen as the most promising cycle. The opposite is true for low-temperature cycles, with reports of CuCl and $MgCl_2$ cycles being more common [21, 22]. Recent developments in these fields have centered on harnessing the power of renewable resources like the sun or nonfossil fuels like nuclear fission [8, 23].

2.2.3 *Water Splitting by Photoelectrolysis*

Photoelectrolysis is a method of water splitting that is comparable to electrolysis; however, it integrates the two processes into one unit, allowing for more consistent energy production over the long run. Consequently, photonic and electrical energies are converted into chemical energy in the form of hydrogen, which is the consequence of this process, which requires both solar and electrical power to function [24]. In order to perform water photoelectrolysis, semiconducting photoelectrodes must absorb photons with energy higher than their band gaps, which results in the generation of electrons and holes in the photoelectrochemical cell [25]. Electrons with energies that are higher than the bandgap of the semiconductor are utilized by the semiconductor (for example, titanium dioxide) in order to generate electron–hole pairs. These pairs are then separated by the electric field that is sent via the electrolyte [26]. The system's efficiency is determined by the photon-absorbing material's specifics, such as its surface characteristics, reactivity, corrosion resistance, and crystalline structure [24].

2.2.4 *Water Splitting by Biophotolysis*

Hydrogen is produced via a biochemical process called biophotolysis, which is a biochemical process (photonic-driven) [26]. Green microalgae and cyanobacteria, for example, use the hydrogenase enzyme's catalytic activity in anaerobic conditions [27] to divide water molecules into hydrogen and oxygen ions during photosynthesis. Hydrogen may be extracted from water in an aqueous environment at room temperature and pressure, which is a distinct advantage of this method. From the points of view of water conservation and CO_2 consumption, it may be a viable and environmentally friendly option [28, 29]. Due to the limited amount of hydrogen that is produced while using this approach, it is now necessary to have a wide surface area in order to capture adequate sunlight [10, 30]. Indirect biophotolysis, on the other hand, accumulates carbohydrates during the CO_2 fixation step, where oxygen is generated. The carbon source for the next stage of hydrogen production comes from the substrates made in earlier stages, reducing the need to add nutrients to the medium [27]. The process of biophotolysis can be broken down into two clear categories: direct bio-photolysis and indirect bio-photolysis. The first method is called the direct pathway, and it involves a photosynthetic reaction that is carried out by microalgae in the presence of light energy (Eq. 2.1) [31]:

$$2H_2O + \text{Light energy} \rightarrow O_2 + 2H_2 \tag{2.5}$$

The process of indirect bio-photolysis has two stages. The first stage involves photosynthesis (Eq. 2.6), and the second stage involves the production of hydrogen and carbon dioxide (Eq. 2.6) [31].

$$12H_2O + 6CO_2 + \text{Solar energy} \rightarrow 6O_2 + C_6H_{12}O_6 \tag{2.6}$$

$$C_6H_{12}O_6 + 12H_2O + \text{Solar energy} \rightarrow 6CO_2 + 12H_2 \tag{2.7}$$

2.3 Hydrogen Generation by Biomass

Biomass refers to plant and animal components [29] such as municipal solid trash, forest residues, crops, microalgae, and animal wastes that can be used to produce fuels and chemical feedstocks [32, 33]. The production of hydrogen from biomass and residual wastes is currently feasible in a number of developed countries, both from a technical and an economic standpoint [34]. It has been predicted that by 2050, biomass will provide enough energy to meet more than 25% of the total demand [35]. When compared to the fossil fuels usage, biomass-to-energy technologies results in a carbon-neutral scenario since they both reduce CO_2 emissions and absorb CO_2 from the environment [36]. Figure 2.3 depicts the several thermochemical conversion routes for biomass [32, 37]. The low cost of fossil fuels meant that technologies that convert biomass into energy through thermochemistry have been overlooked in the past. However, as the need for energy grows and people become more aware of the environmental impact of fossil fuels, these technologies are getting a second look. Biological and thermochemical processes are the two most common ways to turn biomass into hydrogen. Typically, the thermochemical process is significantly quicker and provides a higher hydrogen yield than the biological approach [28]. The processes of hydrothermal liquefaction (HTL), gasification, pyrolysis, and torrefaction, which are depicted in Fig. 2.3, are considered to be the most fundamental types of thermochemical conversion technologies.

Table 2.2 provides a high-level overview of the biomass type, operating conditions, and technological maturity associated with the two key technologies that will underpin the ongoing development of these processes.

2.3.1 *Biological Processes*

Recently, there has been a considerable increase in the hydrogen amount obtained through research on biological processes. This is mostly due to the growing focus on

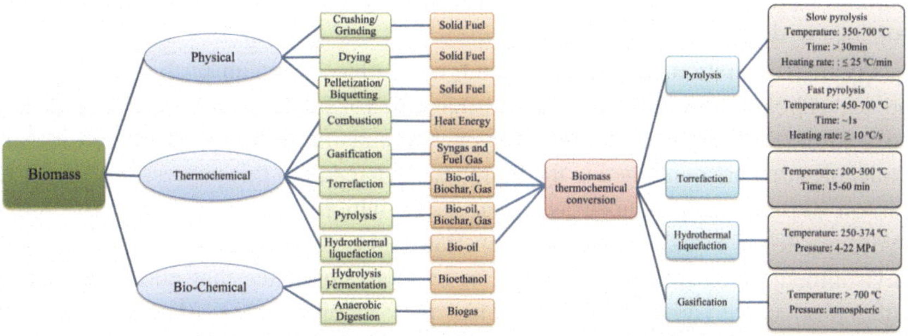

Fig. 2.3 Thermochemical conversion routes for biomass

Table 2.2 Brief overview of the technologies used to produce hydrogen from biomass

Technology	Energy source	Principle	Operating conditions	Maturity
Dark fermentative	Carbohydrate-rich substrates	Biological	Anoxic conditions	Research and development
Photo fermentative	Small organic molecules	Biological	Anaerobic conditions	Commercial
Pyrolysis	Dried biomass	Thermochemical	300–1000 °C in the absence of oxygen	Commercial
Gasification	Dried biomass	Thermochemical	800–900 °C	Research and development
Hydrothermal liquefaction[a]	Wet biomass	Thermochemical	250–370 °C	Commercial
Steam reforming	Biomass-derived liquids	Thermochemical	4–22 mpa 800–1000 °C	Research and development

[a]Requires a subsequent steam reforming process

sustainable development and the reduction of waste. The dark fermentative production of hydrogen and the photofermentative processes are the two primary types of biological procedures utilized in the hydrogen production process [10, 38].

In the process of dark fermentation, anaerobic bacteria are utilized on carbohydrate-rich substrates while being shielded from light and kept in anoxic conditions. This results in the conversion of the original biomass into hydrogen gas, organic acids, and carbon dioxide [28, 39, 40]. Because the presence of light is not necessary, the reaction can take place at any time. The synthesis of biohydrogen takes place through a process known as dark fermentation, which involves the use of enzymes in biochemical reactions that take place at room temperature and pressure [38, 40]. The procedure pathway biologically limits the highest hydrogen yield, even while the yield is strongly connected to the substrate type, inoculum, and operational factors (such as pH, temperature, and so on). For example, in the acetate dark fermentation process, one molecule of glucose should yield 12 mol of hydrogen based on the stoichiometric feasibility, but only 4 mol of hydrogen based on the thermodynamic perspective of glucose metabolism are recorded. Several advancements have been recorded in the scientific literature as a result of this, including the immobilization method [38] and the metal ions usage and nanoparticles oxide to enhance hydrogen production [41].

Photofermentation, on the other hand, takes place when anaerobic conditions are present. Photosynthetic bacteria are responsible for this process, which involves the utilization of sunlight as a source of energy and the digestion of minute organic molecules that are contained in the biomass. As a result of this process, hydrogen and carbon dioxide are produced as by-products, which makes it possible to produce hydrogen from a wide variety of substrates [41, 42]. A stringent control of the environmental conditions is required in order to guarantee that photofermentation will produce hydrogen in an effective manner [43]. Theoretically, the photofermentation process has a high hydrogen yield. A limit was placed on the economic viability of hydrogen production due to the nitrogenase metabolism, which is the enzyme responsible for producing hydrogen during the process of nitrogen fixation. Additionally, the light intensity was also a factor [44]. Generally speaking, the hydrogen production is smaller in the dark than it is in the light [28].

2.3.2 Thermochemical Processes

When it comes to extracting hydrogen-rich gases from biomass, the thermochemical procedures are among the most effective solutions [45, 46]. Pyrolysis, gasification, and hydrothermal liquefaction are the primary steps involved in these technological processes [6, 29, 32]. Gasification and pyrolysis are two thermochemical processes that can be utilized to convert dried biomass into a form that is thermochemically comparable to that of fossil fuels. Both of these methods result in the creation of carbon monoxide and carbon monoxide gas (CH_4 and CO), which, after being subjected to steam reforming and water-gas shift processes, can be used to boost the amount of hydrogen produced [28]. When dealing with moist biomass, the production of hydrogen should involve a combination of hydrothermal liquefaction and steam reforming.

Among the several thermochemical conversion technologies, pyrolysis is generally regarded as the first step in the process. The reason behind this is because it requires all chemical interactions to produce solid, liquid, and gas as its main by-products, all while avoiding the production of oxygen [47]. The use of biomass as a source of liquid fuels has proven to be the subject of substantial study in the last several years. Technologies such as pyrolysis, gasification, hydrothermal liquefaction, and the conversion of biomass into liquid are included in this research. Additionally, upgrading techniques are also being investigated [33]. The pyrolysis of dried biomass is typically carried out at temperatures ranging from 300 °C to 1000 °C [48, 49]. After heat decomposition, pyrolysis results in the production of bio-oil, biochar, and non-condensable gases such as H_2, CO_2, methane, and some other gaseous hydrocarbons. These are referred to as pyrolytic products. At a lower temperature less than 450 °C, the creation of biochar is encouraged, but when it is between 450 °C and 800 °C, bio-oil is the primary product. Product yields are dependent on the operational parameters [32]. The conditions of operation dictate whether the pyrolysis process is characterized by rapid or conventional pyrolysis. Fast pyrolysis, involving elevated temperatures and relatively short residence durations, is favored for the creation of hydrogen as opposed to conventional pyrolysis, which results in charcoal as the primary by-product [6]. In recent years, rapid pyrolysis has received a lot of attention as a potential method for the generation of liquid fuel. This is due to the fact that, in comparison to other processes, rapid pyrolysis boasts superior transport and storage capabilities, as well as lower investment costs and higher energy efficiencies. In the research that has been done, various types of pyrolysis reactors, such as fixed-bed, fluidized-bed, solar reactors, and microwave have been described. These reactors all have the benefit of employing a clean energy source for the heating process [50]. Although the majority of research associated with pyrolysis is concentrated on the generation of liquid biofuels. Directly producing hydrogen at elevated temperatures and residence times is feasible according to the reaction protocol provided in Eq. 2.8.

$$\text{biomasss} + \text{heat} \longrightarrow H_2 + CH_4 + CO + \text{other products} \tag{2.8}$$

The freshly generated CO_2 and H_2 can also undergo steam reforming and water-gas shift processes, respectively, to increase their hydrogen content. These reactions can take place in the presence of water. Catalytic pyrolysis utilizing catalysts based on metals (such as nickel or alkali metals) or nonmetals (such as activated carbon) could be used to improve the amount of hydrogen produced [51]. Both have shown that they are capable of increasing the amount of hydrogen that is produced. On the other hand, zeolites or fundamental materials are favored to increase the quality of bio-oil that is generated; expensive pre-upgradation processes do not have to be implemented [52].

In the process of biomass gasification, dry biomass is converted into a mixture of flammable gases at high temperatures (between 800 °C and 900 °C)) in order to enhance the amount of hydrogen produced [28]. It is a subtype of pyrolysis and, as a result, operates on the principle of partial oxidation [10]. At the conclusion of the gasification process, the majority of the biomass is transformed into carbon monoxide (CO) and hydrogen (H_2), along with a little amount of water, carbon dioxide (CO_2), and methane (CH_4). Gasification is impossible without oxidation agents like air, oxygen, or steam [32, 53]. Because of technological advancements, it is now feasible to produce a fuel that is considerably less polluting and more adaptable than the biomass it started from. There are two different kinds of gasifiers, which are known as fluidized-bed and fixed-bed reactors. To fluidized-bed reactors, fixed-bed reactors typically result in a lower concentration of dust and particles in the product gas. Additionally, fixed-bed reactors are generally more cost-effective than their fluidized-bed counterparts. These are the two primary advantages of fixed-bed reactors.

In contrast, the primary advantage of utilizing fluidized-bed reactors reside in their improved ability to transmit mass and heat between both gaseous and solid phases. This enables a temperature that is consistent throughout the reactor, which in turn improves carbon conversion while simultaneously reducing the production of tar [53, 54]. The literature reports that catalytic biomass gasification can be used to increase syngas output to its maximum potential. Adding a catalyst throughout the gasification process will make the cracking reaction more efficient while also lowering the activation energy required, which will result in a decrease in overall energy consumption. Ni-based catalysts stand out from other types of catalysts because of their costs and high reactivity, they are widely employed during biomass catalytic gasification. However, there are other categories of catalysts that can be explored for this purpose, such as mineral, alkali metals and, noble metal catalysts [55].

Hydrothermal liquefaction typically employs temperatures between 250 °C and 370 °C and pressures between 4 and 22 megapascals to decompose biomass's polymer structure [32, 53]. The liquid biocrude produced at the end of this process, with, a solid residue an aqueous phase, and a gaseous stream, are the four main products of this process [56]. To improve the amount of bio-oil produced, it is possible to feed the aqueous phase back into the hydrothermal unit. Additionally, it could be utilized in the production of hydrogen-rich syngas via steam reformation. The process of hydrothermal liquefaction may be broken down into three primary stages: the first stage involves the depolymerization of the biomass, the second stage involves the decomposition of the biomass into monomers, and the third stage involves the repolymerization and recombination of reactive fragments. Because the initial biomass was composed of a convoluted carbohydrates mixture, lignin, lipids, and proteins, the reaction chemistry that takes place during the biomass transformation into liquid products is notoriously difficult to accomplish [33]. Batch or continuous operation can be used for the reactor. Batch conditions have been the focus of the majority of research, despite the fact that continuous reactors achieve better feedstock conversion [57].

In addition, the choice of catalyst is a critical factor in maximizing the production of bio-oil while simultaneously reducing the amount of temperature and pressure that is necessary for this process. The choice of catalyst, whether it be alkaline, acid, metal-based, or mineral-based, is determined by the substrate that is being utilized. For instance, alkaline catalysts are recommended for use with lignocellulosic and microalgal substrates, whereas zeolites or acid catalysts provide the greatest results when working with alcohols [58]. Compared with pyrolysis, the primary benefits of hydrothermal liquefaction include a higher energy efficiency, a lower temperature at which the process is carried out, and a smaller yield of tar [33].

Processing the bio-oil that is created as a by-product of pyrolysis and hydrothermal liquefaction enables an increase in the amount of hydrogen that is produced. The energy density of the final liquid product is greater than that of the biomass from which it was produced. It is mostly made up of a broad variety of oxygenated compounds, including, but not limited to, phenols, aldehydes, ketones and acids, among others [32, 33, 59]. Because of its complex composition, which depends on the original biomass's composition [60, 61] bio-oil has several unfavorable properties, such as corrosiveness, low heating value, and immiscibility with other different fuels.

This complicated composition is dependent upon the composition of the original biomass [60, 61], and it is responsible for many undesirable qualities of bio-oil, such as its low heating value, corrosiveness, or immiscibility with other fuels. These properties cause it to be deemed a fuel of low grade [61–63]. One of the bio-oil's fractions can be organic, while the other can be aqueous, and each fraction can be separated from the other [64]. The organic fraction is made up of nonpolar chemicals; as a result, it is possible to upgrade it through the process of hydrotreating in order to get fuels. However, there are a handful of oxygenated organic compounds present in the aqueous phase, with concentrations varying from 15 to 60 wt%, while the aqueous phase primarily comprises water and has a low value [65]. As a result, it is possible to give it a new value by employing a process called catalytic reforming, which results in the production of additional hydrogen [66]. The hydrogen yield that may be achieved through this process is, however, limited by the thermodynamic constraints of the water-gas shift reaction as well as other subsequent processes, such as methanation and coking.

However, the thermodynamic limits of the water-gas shift reaction and other secondary reactions, which include methanation and coking, affect the hydrogen yield that can be achieved as a result of this process. This results in a decrease in the amount of hydrogen that can be produced [67]. Therefore, the development of appropriate catalysts takes on a significant amount of significance in order to achieve high activity, low deactivation (mostly as a consequence of coke deposition), and high selectivity toward hydrogen. The catalyst is also accountable for breaking O – H bonds in addition to C – C and C – H bonds. This task is one that it accomplishes independently. In this regard, catalysts based on noble metals have been characterized as stable and active, but they are also very expensive. However, catalysts based on transition metals, such as nickel and cobalt, are more cost-effective and have shown appropriate activity in steam reforming. Researchers have begun to focus on them more extensively in recent years for these reasons [67–69].

Table 2.3 lists the main pros and cons, energy efficiency, hydrogen yields, and operating costs of the different renewable resource-based hydrogen systems. While water-based hydrogen production methods have the ability to produce large amounts of hydrogen, their low energy efficiency makes them economically incomparable to other strategies. Photoelectrolysis, biophotolysis, and thermolysis are three examples of processes with low cost-effectiveness and low energy efficiencies. The electrolysis of water has a reasonably good energy efficiency, but it has the lowest cost-effectiveness of all water-based hydrogen production systems. As a consequence of this, in order to lessen the amount of energy required, researchers are concentrating their efforts on finding suitable catalysts and merging these kinds of technologies with renewable energies in order to lessen the amount of energy required and make them economically competitive.

2.4 Hydrogen Generation from Fossil Fuels

Since the costs of production are directly associated with the prices of fuel, which are currently kept at levels that are acceptable, fossil fuels continue to hold a prominent position in the world's hydrogen supply. There are currently a number of mature methods that can manufacture hydrogen from fossil fuels. The most common of these processes are hydrocarbon reforming and pyrolysis. These methods almost make it possible to produce the actual amount of hydrogen that is required [29]. To be more specific, in 2015, natural gas accounted for 48% of hydrogen production, while petroleum accounted for 30%, and coal contributed 18% [7]. Table 2.4 is a summary of the most important aspects of each fossil fuel-based technology that will be discussed in the subsequent subsections. These characteristics include the feedstock that is used, the maturity level, and operating conditions.

Table 2.3 Renewable energy-based hydrogen generation technologies [70]

Technology		Advantages	Drawbacks	Energy efficiency (%)	H_2 yield (g/kg feedstock)	Cost ($/kg of h_2)	References
Water	Electrolysis	Simplicity Low temperatures Zero carbon Emissions O_2 as a by-product Integration with fuel cells	There is a need for high pressure. Challenges with energy storage Low system efficiencies High capital costs	55–80	111^{α}	4.15–10.30	[24, 29, 71]
	Thermolysis	Clean and sustainable Zero carbon emissions O_2 as a by-product	The separation stage is necessary in order to prevent the recombination that would occur in an explosive mixture. High capital costs	20–50	111^{α}	7.98–8.40	[24, 29, 71]
	Photoelectrolysis	Contributes to the sustainability of the energy supply Photonic and electrical energies can be converted to chemical energy Low operating temperature and pressure	Low efficiency Requires a significant surface Photocatalytic material is required	0.06–14	111^{α}	4.98–10.36	[24, 29, 71, 72]
	Biophotolysis	Can produce hydrogen at ambient conditions consumes CO_2	Requires a significant surface area to collect enough sunlight Difficult operations to control the different bacteria Large reactor volume	10–15	111^{α}	1.42–2.13	[24, 29, 71]
Biomass	Dark fermentation	Can produce hydrogen at any time because light is not required Ambient conditions	Hydrogen yield is metabolically restricted High by-product generation Low efficiency Large reactor volume	60–80	4–44	1.68–2.57	[29, 71, 73]

(continued)

Table 2.3 (continued)

Technology		Advantages	Drawbacks	Energy efficiency (%)	H_2 yield (g/kg feedstock)	Cost (\$/kg of h_2)	References
		Simple reactor design Waste recycling					
	Photofermentation	Allows for hydrogen Production from a wide range of substrates, including waste streams High removal efficiency of the chemical oxygen demand Waste recycling	Strict control of environmental conditions is required Requires a significant surface area to collect enough sunlight Nitrogenase metabolism restricts economic viability of hydrogen production Large reactor volume	0.1–12	9–49	2.57–2.83	[29, 71, 73]
	Pyrolysis	Developed technology Abundant and cheap feedstock Solid, liquid, and gas product streams Carbon-neutral emissions	Hydrogen yield depends upon the feedstock tar formation	35–50	25–65	1.59–2.20	[29, 71, 73]
	Gasification	Abundant and cheap feedstocks Carbon-neutral emissions	Requires oxidation agents Hydrogen yield depends upon the feedstock tar formation	30–60	40–190	1.77–2.05	[29, 71, 73]
	Hydrothermal liquefaction	Abundant and cheap feedstocks Drying step is not required High energy efficiency solid, liquid, and gas product streams	Hydrogen yield depends upon the feedstock Presence of nitrogenated compounds	85–90	0.3–2	0.54–1.26	[33, 74, 75]
	Steam reforming	Developed technology Avoid the costly upgrading of the bio-oil	Produce carbon coproducts	74–85	40–130	1.83–2.35	[71, 73]

[a]Calculated from the water-splitting reaction stoichiometry

Table 2.4 Fossil fuel-based hydrogen production technologies [70]

Technology		Feedstock	Operating conditions	Maturity
Reforming technologies	Steam reforming	Light hydrocarbons, which are produced from liquefied petroleum gas and naphtha less commonly than other hydrocarbons	900–1000 °C	Commercial
	Partial oxidation	Hydrocarbons, heavy fuel oil, and coal	Temperature of >1000 °C	Commercial
	Autothermal reforming	Light hydrocarbons, which are produced from liquefied petroleum gas and naphtha less commonly than other hydrocarbons	Temperature of >1000 °C	Early commercial
Pyrolysis		Hydrocarbons	500–800 °C in the absence of oxygen	Commercial
Gasification		Coal	700–1200 °C	Commercial

2.4.1 *Hydrogen Generation by Fossil Hydrocarbon Reforming*

The way of reforming hydrogen that has seen the most advancement is the process of reforming hydrocarbons into hydrogen using a chemical reaction. In addition to hydrocarbon, this process also needs other reactants, and those reactants might be oxygen or steam. Depending on which one is used, the process could be referred to as the steam reforming reaction or partial oxidation reaction, respectively. The process is referred to as autothermal reforming when both processes are brought together at the same time, resulting in an unchanged net reaction enthalpy [76]. Figure 2.4 presents a conceptual diagram of reforming technologies and the progression in catalysts development.

2.4.1.1 Steam Reforming Reaction

Reforming is a reaction that can occur when a mixture of steam and hydrocarbons is subjected to high temperatures. This reaction is characterized by the creation of hydrogen and carbon oxides as a result of the reaction. Through a process known as steam reforming, natural gas and, to a much lesser extent, liquefied petroleum gas and naphtha are both capable of contributing to the creation of hydrogen inside the system [1]. As was said earlier, the steam methane reforming procedure, which starts with natural gas or light hydrocarbons and ends with reformed hydrocarbons, is the one that is utilized most frequently in hydrocarbon reformation. First, in this process, CO_2 is created with H_2, which results in the formation of synthesis gas ($CH_4 + H_2O \rightarrow CO + 3H_2$); then, CO_2 is transformed to carbon dioxide and

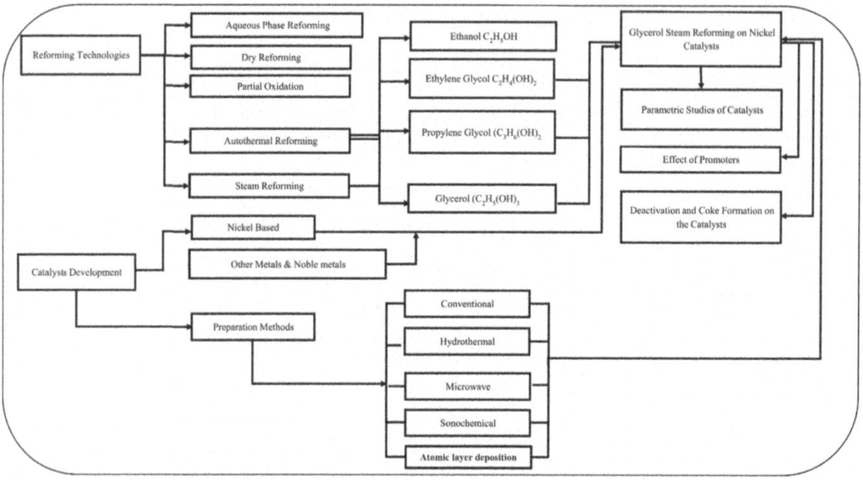

Fig. 2.4 The overview of reforming and catalysts development technologies

additional hydrogen ($CO + H_2O \rightarrow CO_2 + H_2$)- via the water-gas shift reaction; finally, the process concludes with the formation of CO_2 [77]. The entire reaction that occurs when methane is converted into water vapor by steam can be expressed as follows:

$$CH_4 + 2H_2O \rightarrow CO_2 + 4H_2 \tag{2.9}$$

The process of reforming is extremely endothermic, and hence, a significant quantity of heat is essential. Because of this, the temperatures at which these reactions are normally carried out range from 800 °C to 1000 °C [78]. Because of the high temperatures that must be reached in order to transform methane into hydrogen, the reformer must be built out of expensive materials in order to be able to survive the thermal pressures that are placed on it (for example, high alloy nickel-chromium steel) [79]. Other potential negatives include the creation of coke and the appearance of temperature profiles inside the catalyst bed, both of which would need to be taken into consideration [80]. To maximize hydrogen production and avoid the limitations of methane steam reforming, high-performance catalysts, usually based on nickel, are required. These catalysts are typically supported on ceramic oxides or oxides that have been stabilized by hydraulic cement [81]. In spite of this, once these factors are taken into account and reduced to a minimum, methane steam reforming generates a gas that is abundant in hydrogen while simultaneously reducing the amounts of methane, CO, and CO_2 produced. A phase of separation is essential for the purification process of hydrogen because it is typically found in a mixed state with other chemicals. The pressure swing adsorption system, which is used in hydrogen plants, is responsible for the purification process. This system is capable of producing hydrogen that is up to 99.999% pure, and it has a recovery rate of

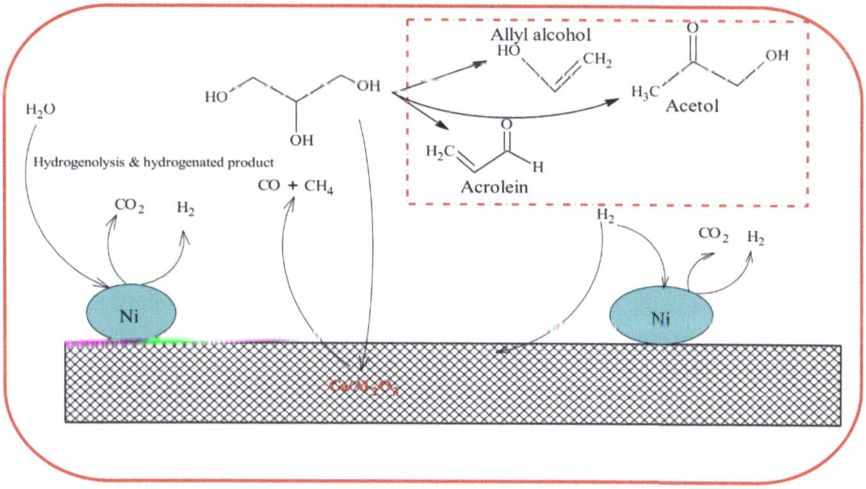

Fig. 2.5 Surface activity of Ca-doped nickel alumina in steam reforming process. (Reprinted with permission from Ref. [83])

between 70% and 95% [82]. A general illustration of the active metal sites on the Ni-Ca/Al_2O_3 catalyst used in steam reforming reactions is shown in Fig. 2.5.

2.4.1.2 Partial Oxidation

The steam reforming reactions can also be done by using an alternative method called partial oxidation. This method has the potential to work with a variety of feedstocks, including coal, heavy fuel oil, and even natural gas [1, 29]. When it comes to extracting hydrogen from coal and heavy fuel oil, the process known as partial oxidation is by far the superior option [76]. The exothermic process known as partial oxidation is used in the transformation of hydrocarbon fuels to give a components consists of carbon monoxide, hydrogen and other species that have been partially oxidized [84]. One of the benefits of this method is that it does not require the use of any additional sources of energy because the reactions that take place with oxygen are very exothermic [85]. Because the C/O ratio to determine the distribution of product into partial oxidation processes and because high reaction temperatures (more than 1000 °C) constitute a constraint on the distribution of products, partial oxidation reactions require heterogeneous catalysts at reduced temperatures. For partial oxidation reactions, catalysts based on transition metals are ideal because of their capacity to change oxidation states and to adsorb reactants and intermediates onto their surfaces. Transition metal-based catalysts are also less expensive than noble metals [86]. In its entirety, the reaction of partial oxidation can be characterized as follows:

$$C_nH_m\ 1/2_nO_2 \rightarrow nCO1/2mH_2 \quad (2.10)$$

Above 550 °C, the thermodynamically most important products of this process are H_2 and CO. CO is a precursor to coke, and its removal can be accomplished either through its oxidation into CO_2 or via the reaction of water-gas shift that results in an increase in the production of H_2. The operation of partial oxidation is straightforward, it has a reduced overall energy requirement, and it allows for a more adaptable feedstock. These are just some of the many benefits. However, there are still obstacles that need to be overcome before it can be used commercially. One of these obstacles is the relatively limited useful life of the unit, which is caused by the high reaction exothermicity. This exothermicity might generate hot spots, which would then result in the deactivation of the catalyst due to sintering [85, 87, 88].

2.4.1.3 Autothermal Reforming

Autothermal reforming relies on an exothermic partial oxidation with oxygen to power the endothermic steam reforming reactions [89]. When using a reformer, it is essential to feed both steam and oxygen simultaneously. This results in the oxidation and reforming reactions taking place simultaneously, which is necessary in order to achieve a thermodynamically neutral reaction [90]. The choice of the catalyst has an important impact on the performance of the process as a whole, just as it does in steam reforming or partial oxidation. Nickel-based catalysts are the most popular choice since they are both efficient and inexpensive. In light of the fact that this process has a high thermal efficiency, its energy requirements are significantly lower than those of partial oxidation or steam reforming [91]. However, the hydrogen yield produced by autothermal reforming is higher than that produced by partial oxidation but less than that produced by the steam reforming technique.

2.4.2 Hydrogen Generation by Fossil Hydro Carbon Pyrolysis

Thermal decomposition process that takes place in the absence of O_2 is called pyrolysis [92]. This process converts several light liquid hydrocarbons to C and H_2 in accordance with Eq. 2.11. When the procedure of thermal decomposition is performed on heavy residual fractions that have a boiling point that is higher than 350 °C, it is permissible to produce hydrogen using a two-step method. This is because the boiling point of these heavy residual fractions is higher than 350 °C. Hydrogasification (Eq. 2.12) and afterwards cracking of methane (Eq. 2.13) are necessary operations because heavy residual fractions comprise high amounts of sulphur and metals. This is because, in the direct breakdown, these elements will be transported into the coke, rendering it unusable for any further applications. This technique allows for the concurrent generation of high-value by-products, e.g., sulfur or metals, even though the composition of hydrogen generated is lower in comparison to the methodologies that utilize light hydrocarbon fractions. However, this process does allow for the generation of hydrogen [93]. There is no production of carbon dioxide during the pyrolysis of methane because all of the carbon is recovered in the solid state. This process has been discussed extensively in the scientific literature. The kinetic enthalpy for the generation of hydrogen by methane pyrolysis, also known as methane decarbonization, is 37 kJ/mol of hydrogen; this value is comparable to that for the synthesis of hydrogen by steam reforming, which is 41 KJ/mol of H_2, as long as the energy required to vaporize water is disregarded [94]. Mostly, natural gas serves as a feedstock for industrial operations rather than methane. As a result, additional molecules, such as carbon dioxide or water, are typically combined with methane in these cases. These compounds become reactive during the process of pyrolysis, which has a significant impact on both the selectivity of the products and their conversion. There have been reports of a variety of different catalysts being used for this process. These catalysts range from metallic catalysts to carbon-based catalysts. There has been a lot of application for Ni-based catalysts as metal-based catalysts due to their activity and ability to produce filamentous carbon at moderate temperatures. However, carbonaceous catalysts such as activated carbon and carbon black have also been employed [95].

$$C_nH_m \longrightarrow mC + \tfrac{1}{2}mH_2 \tag{2.11}$$

$$C_nH_m + (2n - m/2)H_2 \longrightarrow nCH_4 \tag{2.12}$$

$$CH_4 \longrightarrow C + 2H_2 \tag{2.13}$$

2.4.2.1 Coal Gasification

Thermochemical conversion of coal into gaseous products like hydrogen and carbon monoxide is known as "coal gasification" [95–97]. This process aspires to provide an alternate to burning coal that increases the fuel's energy density and decreases harmful emissions. In reality, steam, oxygen, or air facilitate the conversion of coal to synthesis gas at extraordinarily high pressures and temperatures [24, 98, 99]. In addition to the primary reactions, there can be secondary reactions when coal reacts with products of other reactions, like carbon dioxide, rather than oxygen or steam, leading to the production of more carbon monoxide via the Boudouard reaction. Coal may undergo further carbon monoxide production in secondary reactions that do not include steam or oxygen but rather other products of reaction, such as carbon dioxide, as a result of the Boudouard reaction. These reactions occur in addition to the primary reactions. These secondary reactions can also take place [95]. Given the high level of carbon content, the most significant drawback connected with producing hydrogen through the gasification of coal as opposed to utilizing other technologies that make use of various feedstocks is the increased level of CO_2 emissions. For this reason, advantages are now being explored by marrying various

Table 2.5 NETL fossil fuel-based hydrogen plant cases [1]

Case	Plant type	Feedstock	Technology type	CO_2 capture (%)	Hydrogen production capacity	Levelized cost of hydrogen
1	Reforming	Natural gas	Steam methane reforming		483,000 kg/day	$1.06/kg
2				96.2		$1.54/kg
3			Autothermal reforming	94.5	660,000 kg/day	$1.51/kg
4	Gasification	Coal	Shell/air products-type oxygen-blown, entrained low gasification			$2.58/kg
5				92.5		$2.92/kg
6		Coal/ Torrefied Woody biomass		92.6	133,000 kg/day	$3.44/kg

technologies based on carbon capture with the gasification of coal. Since coal gasification uses less expensive feedstock than other fossil fuel technologies, it stands out from an economic perspective; nonetheless, the unit's capital expenditures are higher [35].

An evaluation of the performance of several hydrogen production plants that use fossil fuel resources as main feedstocks was recently conducted by the National Energy Technology Laboratory (NETL) in a baseline study [1]. Natural gas, steam methane reforming, coal gasification, coal/biomass co-gasification, and autothermal reforming of natural gas are all processes that fall under this category of plants. The study was conducted independently and in a transparent manner. Table 2.5 provides a summary of the many scenarios that were considered, all of which were based on the assumption that commercially available technologies would be used. These scenarios considered both circumstances in which CO_2 would be captured and those in which it would not be captured in order to evaluate the potential performance and financial consequences of producing clean hydrogen utilizing these technologies.

2.5 Hydrogen Purification/Separation

Hydrocarbon or water electrolysis or gasification yields a crude hydrogen-based gas product that also contains other reactants and by-products such as carbon dioxide, oxygen, water vapor, carbon monoxide, and methane. Therefore, ultra-high pure hydrogen requires separation from by-products, coproducts, or residual reactants present to a certain level according to the intended application. The term "separation" is used for the first-stage operation to increase hydrogen concentration, while "purification" is used for the second-stage process to upgrade hydrogen.

There are three primary methods used in the purification of hydrogen, which are as follows [100]:

- Physical Absorption and Chemical Absorption
- Membrane Separation
- Pressure Swing Adsorption (PSA)

2.5.1 *Physical and Chemical Method*

The absorption method is the process of separating and purifying hydrogen using a solvent or solution. In this procedure, the soluble contaminant is washed into a liquid absorbent. Absorption can be classified as either physical or chemical. The physical absorption method uses the difference in solubility between H_2 as well as other components, whereas the chemical absorption approach uses chemical interactions between the impurity component and the absorbent.

Hydrogen is separated from feed gases using the physical absorption process, which involves determining the solubility of gaseous components. This is done at high impurity concentrations and high pressures in order to remove any remaining impurities.

There are three main ways in physical absorptions to remove carbon dioxide, methane, and nitrogen from the rich hydrogen-based gases. These include scrubbing pre-purified hydrogen gas at a low temperature and separating light hydrocarbon from gases rich in hydrogen by using heavy hydrocarbon such as gasoline, oil, or methylcyclohexane as a solvent.

Chemical absorption is capable of completely removing specified contaminants from crude hydrogen and recovering hydrogen from it at low pressure or concentration. The separation of acidic chemicals such as carbon dioxide, hydrogen sulfide, carbonyl sulfides, and hydrogen cyanide is mainly accomplished through chemical absorption. Hot alkaline or amine solutions are used as sorbents to eliminate acidic contaminants. An example is a cryogenic separation and metal hydride.

2.5.1.1 Cryogenic Separation

Cryogenic separation is a process used to separate individual species of gases from a mixture because of the varying gas constituents' boiling points. It is undertaken through numerous compression, cooling, and expansion steps. Since hydrogen has the lowest boiling point of any mixture (−252.9 °C), it remains in the form of a gas while the impurities gases condense to become a liquid phase.

The impurity gases condense to their liquid phase while hydrogen remains in its gaseous state since it has the lowest boiling point of a mixture (−252.9 °C). The entire process takes place at low temperatures; therefore, it is quite energy-intensive [101]. One significant drawback is the extracted hydrogen's low purity, which is just 99%. However, sustaining cryogenic temperatures is expensive because of gas compression in order to keep cold conditions. Hence, hydrogen collections are stored in liquid form instead (Fig. 2.6a) [102].

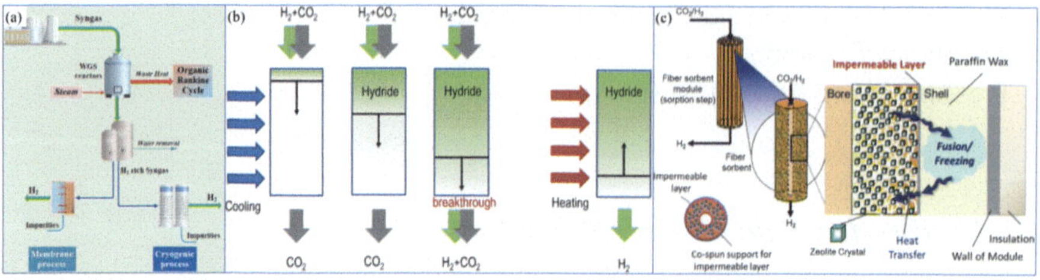

Fig. 2.6 Flow scheme of the (**a**) membrane and cryogenic distillation process, (**b**) metal hydride compressor, and (c) classical pressure swing adsorption (PSA). [(**a**) Reprinted with permission from Ref. [109]. (**b, c**) Reprinted with permission from Ref. [110]]

2.5.1.2 Metal Hydrides

Metal hydride separation and purification processes rely on hydrogen storage materials' selective sorption of hydrogen. The basic process involves physically adsorbed molecular hydrogen onto the metal surface, which then undergoes hydrolysis to produce individual hydrogen atoms. Low energy usage, easy and safe operation, and a high extraction rate of hydrogen are some of the benefits of the purification technique for metal hydride. In a nutshell, metal hydrides are created when molecular hydrogen physically adsorbs to a metal surface, then dissociates and diffuses into the metal lattice, creating MHx (where M is the metal) solid solution, and finally, the development of the MHy—metal hydride. Whether the MHx transformation in MHy can be reversed depends on the amount of hydrogen that comes into contact with the metal that forms hydrides as well as the reaction conditions [103].

To purify hydrogen, metal hydrides work on a following principle. Metal hydride is known to be a hydrogen-selective absorbent; this property allows it to remove impurities from gaseous hydrogen [104]. The gas mixture containing hydrogen is introduced into the reactor, which houses the metal or alloy capable of producing hydrides. Here, the metal starts to selectively absorb hydrogen through hydride production, and the coolant gets the heat that comes out of the reaction. When hydrogen starts to leave the metal on the other side of the reactor after passing through the hydride-forming reactor, the reaction is said to be complete. After subjecting the hydride to high temperatures, it can be desorption-treated to obtain hydrogen of high purity. One way to control the pressure of hydrogen at the output is to raise the temperature of the flowing liquid (Fig. 2.6b) [105].

The performance of metal hydride devices is impacted by contaminants in the feed gas to the reactor. The hydrogen storage properties can degrade due to a worsening of the sorption-poisoning reaction kinetics (e.g., by H_2S), reduction in hydride formation and rate of corrosion (e.g., by CO_2, moisture), the layer formation of inert gas (i.e., by N_2), and so on, depends on the alloys or metals selected for this task that can generate hydrides. Reducing the hydride material's hazardous qualities requires additional procedures, such as changing the alloy's composition, treating the surface, and using the redox method [103, 106].

2.5.2 *Pressure Swing Adsorption*

The adsorption separation method uses the preferred adsorption of other component species other than hydrogen because of their dissimilarity in adsorption equilibrium. Adsorbents include materials such as molecular sieves, activated carbon, silica gel, and activated aluminum. Adsorption processes at cryogenic temperatures are indispensable for the final removal of residual impurities in making ultra-high-purity hydrogen. They are widely applied to remove water vapor, dust, tars, heavy hydrocarbons, mists, and acidic components. An example of the adsorption separation method is Pressure Swing Adsorption (PSA).

A separation method based on the ability of adsorbent materials to form a stronger physical connection with impurity molecules at higher partial pressures than at lower partial pressures is called pressure swing adsorption [107]. Some examples of adsorbents are carbon molecular sieve, aluminite, silica gel, activated carbon, and zeolite. There are five basic steps in the procedure:

- Adsorption
- Concurrent depressurization
- Countercurrent depressurization
- Purge
- Repressurization.

PSA unit consists of four to twelve value columns containing specific absorbents. The columns are connected through a valve to perform pressurization and depressurization by sophisticated charge and discharge of hydrogen. During adsorption in the PSA system, feed gases enter at the highest working pressure, impurities are retained inside the adsorbent material, and high purity hydrogen is extracted from the end product. In the second procedure, some hydrogen is trapped along with certain contaminants, and the adsorbent is depressurized. During countercurrent depressurization, the adsorber partly depressurizes, causing contaminants to be withdrawn. To further renew the adsorber, hydrogen from the second step process is used to purge it. Finally, the adsorber is repressurized with hydrogen from the second step and a portion of the end product, and when the required pressure is attained, the adsorption is ready to restart the cycles.

Higher adsorption pressure results in purer hydrogen discharge, but at the expense of greater energy consumption [100, 108]. It can recover hydrogen with a purity of 99.9% depending on factors such as purge gas pressure, intake pressure, impurity level at start-up, and hydrogen concentration in the feed gas. Figure 2.6c shows the flow diagram of a traditional PSA system used to obtain H_2 from coal with N_2 as a significant contaminant.

2.5.3 *Membrane*

A separation membrane is any physical barrier that allows a particular mass species to transport. It is frequently used for separation and purification in many industries. A membrane layer's significant role is to regulate components' movement between two neighboring fluid phases. Its ability to effectively separate a mixture is deterring parameters—it is mined by permeability and selectivity. H_2 separation by permeation-selective membranes requires applying a pressured mixture of gases that are enriched in H_2 to one side of the membrane [111]. For the purpose of separating and purifying H_2, several different types of membranes have been created, such as electrochemical membranes, inorganic, palladium-based, ionic-liquid-based, carbon molecular sieves, and mixed-matrix membranes [112]. Research demonstrated a recovery rate of >95% and a purity of 99.5% for H_2 produced at fueling stations from ammonia when comparing hybrid multi-configuration schemes and membrane-based stand-alone with PSA. In comparison to the stand-alone setups, the hybrid schemes produced much better and more promising results. Table 2.6 compares information from the available literature on the technique used, feed treated, and benchmark features of hydrogen purity and recovery, among other helpful things.

Hydrogen separation membrane devices are gaining popularity as a means of producing hydrogen in the industrial sector. Metal membranes, in particular, look promising since they can withstand the stresses of a normal hydrogen purification process [127]. The advantages and disadvantages of membrane separation are summarized in Table 2.7. Hydrogen selectivity and permeability are required for the commercial application of hydrogen separation membranes, and most are very high [128].

There are primarily two types of separation membranes: non-porous (dense metallic) and nanoporous. In contrast to dense membranes, whose separation process relies on solution diffusion, porous membranes often use a molecular sieve mechanism [127]. Membrane separation can also be classified into a different class, and it was summarized in Table 2.8.

The properties listed in Table 2.8 are among the crucial characteristics to take into account when choosing a membrane for hydrogen separation. The two parameters that are perhaps most frequently used to assess the viability of a proposed membrane are permeate flux and membrane selectivity, while chemical and mechanical stability are also important.

2.5.3.1 Nanoporous Membrane

Nanoporous membrane consists of metal, ceramic, and polymer with precisely regulated pore sizes as a common feature. The pore size is generally in the nanometers (nm) range and they are classified as microporous (pore radius <1 nm),

Table 2.6 Process systems engineering investigations into H_2 purification and separation

Technology	Process detail	Hydrogen purity	Hydrogen recovery	Feed	References
Cryogenic	Refrigeration cycles			Coke oven gas	[113]
Cryogenic	Distillation			Synthetic natural gas to produce LNG	[114]
Cryogenic	Desublimation of CO_2	100.00%	92.7%		[115]
Cryogenic	Cryo-coolers and membrane	99.99%			[116]
Cryogenic	N_2 and propane (C_3) refrigeration cycles	88.1%	25.1%		[117]
Cryogenic	Diffusion–absorption and N_2 refrigeration cycle	88%		Refinery gas stream	[118]
Hybrid	PSA-membrane	99.98%	91.71%	Coal-derived SG	[119]
Hybrid	PSA-membrane	>99.5%	>95%	Ammonia at refueling stations	[120]
Membrane	Carbon molecular sieve membrane	>90%	>90%	Coal and biomass-derived SG	[121]
Membrane	Pilot-scale multitubular Pd module	>98%	64%	Coal-derived SG	[122]
Membrane	Electrochemical membrane		88%		[123]
Membrane	Carbon molecular sieve membranes	99.95%	97%	With large recycle flows and high production cost	[108]
Membrane	Cellulose-based carbon hollow-fiber membrane	99.5%	90%		[124]
Modified PSA	Steam rinsing and purging beyond the dew point	99.98%	99.33%	Coal-derived SG	[125]
PSA	6-step and 2-bed system	99.87%	73.52%		[126]

mesoporous (1 nm < pore radius < 25 nm), and macroporous (pore radius > 25 nm). Examples of nanoporous are silica, carbon, zeolite, and Al_2O_3 [129].

These porous membranes can function at high temperatures and pressures and with gases such as CO and H_2S that are toxic to metallic membranes because they are chemically inert. The fact that the H_2 flux is exactly proportional to the pressure difference across the membrane is a benefit of porous membranes. In a porous membrane, five different types of diffusion mechanisms—solution diffusion, molecular sieving, Knudsen diffusion, surface diffusion, and capillary condensation, and

Table 2.7 Advantages and disadvantages of membrane separation

Advantage	Disadvantage
Low energy consumption	Fouling tendency
The capacity to carry out a continuous separation	Low membrane lifetime
Relative low cost both in operation and capital terms	Low selectivity or flux
Membranes can easily be combined with other separation technology	More or less linear scaling up factor
Easy to scale up and operate	
Extreme operating conditions are not required	
Membranes provide the option of delivering highly pure hydrogen with minimum loss.	

Table 2.8 Provides a concise overview of the several membranes' features, qualities, and problems for easy comparison [127]

	Dense polymer	Microporous ceramic	Dense metallic	Porous carbon	Dense ceramic
Temperature range (°C)	<100	200–600	300–600	500–900	600–900
H_2 selectivity	Low	5–139	>1000	4–20	>1000
H_2 flux ($\times 10^{-3}$molm^{-2}s^{-1}) at $\Delta P = 100$ KPa	Low	60–300	60–300	10–200	6–80
Stability issues	Swelling, compaction, mechanical strength	Stability in H_2O	Phase transition	Brittle, oxidizing	Stability in CO_2
Poisoning issues	HCl, SO_2, CO_2		H_2S, HCL, CO	Strong adsorbing vapors, organic carbon	Stability in CO_2
Materials	Polymers	Silica, alumina, zirconia, etc.	Pd, and its alloys, group 5 metals	Carbon	Proton conducting ceramic
Transportation mechanism	Solution/diffusion	Molecular sieving	Solution/diffusion	Surface diffusion, molecular sieving	Solution/diffusion

molecular sieving—can be used to achieve separation. Porous membranes have the advantage of an H_2 flux that is directly proportional to the pressure gradient across them. Molecular sieving, surface diffusion, Knudsen diffusion, and capillary condensation are the five distinct diffusion mechanisms that can be employed in a porous membrane to accomplish separation.

The molecular transport processes in polymers and other materials can be universally described by solution-diffusion models. In contrast, permeation models include molecules' adsorption from the permeate onto the feed side and their subsequent diffusion via permeate side across the membrane [130, 131]. The steps for transporting molecules of gas or liquid are as follows: First, molecules in the permeate solution are adsorbed and dissolved into the membrane matrix at the surface of the membrane. Second, molecules in the permeate solution diffuse across the membrane following chemical potential gradient. Finally, molecules in the permeate solution desorb from the plate located along membrane surface downstream [131, 132].

In the case of membranes made from novel, highly porous materials, the molecules that are able to pass through the membrane do so via both diffusion and pore channels [133]. Diffusion displays transport properties that go beyond what can be explained by a single solution diffusion model [134] since it subjects to an amount of parameters, including the molecules in the permeate, variations in pore size and shape, and the characteristics of the surface of the pores themselves. The transport process in organic–organic porous composite membranes with twisted accumulation pores and rigid frameworks, or in inorganic–organic hybrid porous composite membranes with a solid framework, relies on molecular sieving, surface diffusion, capillary condensation, Knudsen diffusion, and solution-diffusion transport. In organic–organic hybrid porous composite membranes with a solid framework or organic–inorganic porous composite membranes with twisted accumulation pores and rigid frameworks, the transport process depends on molecular sieving, surface diffusion, capillary condensation, Knudsen diffusion, and solution-diffusion transport (Fig. 2.7).

Accurate regulation of pore size is essential for molecular sieving mechanisms. In theory, an ideal molecular sieve can be created when the pore size falls within the range of the two penetrating molecules' molecular dynamics diameters. In this instance, the smaller molecules pass through while the bigger ones are held. Porous materials' variable pore size characteristics open the door to the prospect of creating highly selective pores with finely tuned dimensions. According to Knudsen [135], the primary model for transmission theory applies when the pore size falls somewhere in the middle of the permeating molecule's pore size and its average free range. The Knudsen diffusivity equation is given by Eq. (2.14).

$$D_{\mathrm{K}} = \frac{\varepsilon d_{\mathrm{P}}}{\tau} \left(\frac{8RT}{9\mu M} \right)^{\frac{1}{2}} \quad (2.14)$$

The variables in Eq. (2.14) are the porosity (ε), pore size (d_{P}), gas constant (R), temperature (T), channel curvature (τ), velocity (μ), and molecular weight (M) of the molecules that penetrate. Equation (2.14) shows that there is a negative correlation between molecular weight and the Knudsen diffusion coefficient; that is, a low diffusion coefficient is produced for a high molecular weight, leading to a change in flow due to diffusivity. In efficient molecular sieving, Knudsen diffusion is very constrained. For instance, under these conditions, the optimal selectivity limit for H_2/CO_2 is just 4.7 [130]. The jumping-type surface diffusion mechanism is the favored method when the adsorption connection between the permeate molecules and the

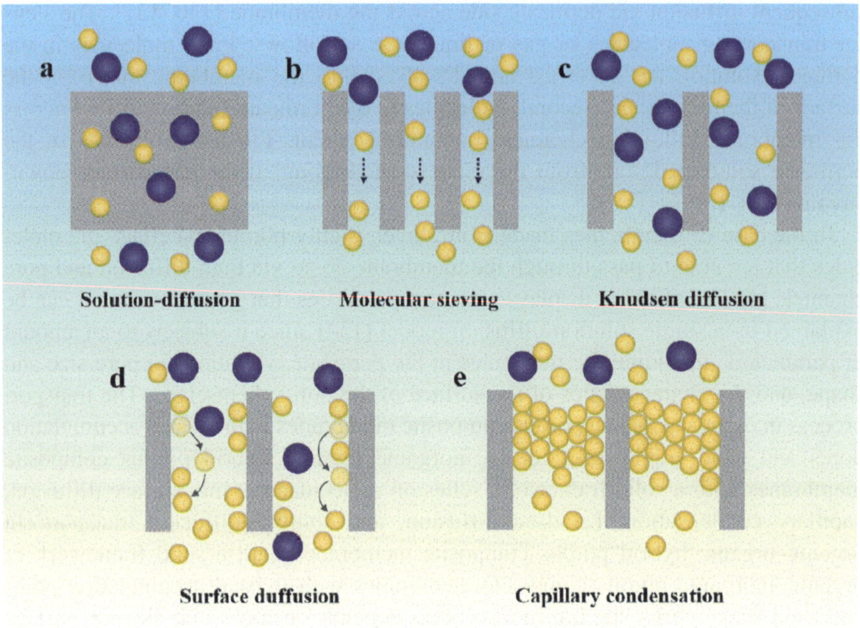

Fig. 2.7 Transport mechanism in pore membranes: (**a**) solution diffusion, (**b**) molecular sieving, (**c**) Knudsen diffusion, (**d**) surface diffusion, and (**e**) capillary condensation. (Reprinted with permission from Ref. [140])

pore channel is high. Because the suspension effect accelerates surface diffusion, it is significantly quicker than Knudsen diffusion [136]. If the adsorbed gas is able to condense, as is often the case when passing a mixture of gases through a capillary, then the gas-phase selectivity is increased or the inversion occurs. Furthermore, new studies have uncovered some encouraging diffusion-selective transport processes, including sieving mechanisms that require precise shape matching. It is possible to achieve the monoclinic diffusion mechanism of ordered one-dimensional pore arrangement in crystalline coordination solids by accurately designing one-dimensional channels [137]. Permeation mechanisms based on geometry are also a part of it. Nunes and colleaques [138] created organic solvent nanofiltration using POC membranes ultra-thin cyclodextrin-based that can differentiate between molecules with nearly identical molecular weights but distinct geometries. In order to separate ethylene and ethane by molecular cross-sectional size differentiation, Zhou et al. [139] created a MOF with serrated pores.

2.5.3.2 Dense Membrane

Dense membranes can be categorized into two groups due to the materials used for separation: Polymeric and Metallic Membrane.

Polymeric separation of hydrogen from substances such as nitrogen, carbon monoxide, and hydrocarbons is based on polyamide-imide block copolymer. They are inexpensive to manufacture and operate, can withstand high-pressure dips, and work at a low temperature.

The polymer can be categorized into a glassy (prepared at a temperature below the glass transition temperature) and a rubbery (prepared at a temperature above the glass transition temperature) polymeric membrane. Both glassy and rubbery varieties of the polymer exist, depending on whether they are made at temperatures below or above the glass transition point. Glassy polymer membranes have relative low flux and high selectivity, while membranes made of rubbery polymers have higher flux but worse selectivity, i.e., their selectivity and fluxes are both moderate.

Polymer membrane's advantages are the capacity to cope with a high-pressure drop, low cost, good scalability, and operating temperature limited to 90–100 °C. Its drawbacks are a limited level of resistance to certain chemicals (HCL, CO_2), low mechanical strength, welling high sensitivity and compaction, and a low level of chemical corrosion resistance [141, 142].

The solution-diffusion process is used to separate hydrogen across a polymeric membrane, similar to how dense metallic membranes operate for H_2 separation.

Dense metal membranes are the best-selected membranes used in industry to purify hydrogen because of the sensitive and selective permeation of hydrogen separation at a temperature (400–800 °C) [127]. In comparison to other materials, metal membranes are superior in separating H_2 to a high degree of purity and remaining stable at relatively elevated operating temperatures [127, 129]. However, metals appropriate for H_2 separation need to be inexpensive, have strong thermal stability at elevated temperatures (when hydrogen solubility and mobility are at their highest), and exhibit good solubility and bulk hydrogen diffusivity in addition to strong permeability. The dense metallic membrane can be made of pure metals like Palladium (Pd), Vanadium (V), Niobium (Ni), binary alloys (Pd-Cu, Pd-Ag, V-Cu), ternary alloys, and composite metals.

A metallic-based composite membrane combines two or more metals with properties that cannot be found in the parent metal, such as Palladium-Copper, Palladium-Silver, Vanadium-Silver, etc. Previous investigations have shown that metallic membranes may be used to purify H_2 since they can work at high temperatures [129]. Factors such as the temperature, membrane thickness, and hydrogen's solubility and diffusivity determine the hydrogen's permeability through a metal membrane.

The poor thermal stability of such composite membranes, which is brought on by interdiffusion between the coating and the primary membrane material, is their significant limitation [143]. Fabrication of composite membrane leads to improvement of the membrane performance, which is so dependent on the methods used. Different fabrication methods are used in a composite metallic membrane: Sol-gel, electroplating, Physical Vapor Deposition (PVD), Chemical Vapor Deposition (CVD), and Magnetic Sputtering. Hydrogen permeates through a dense metal membrane using the mechanism of the solution-diffusion technique.

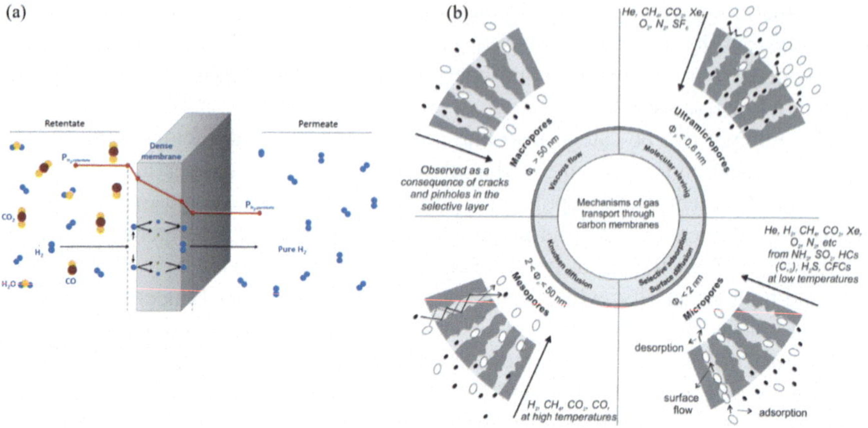

Fig. 2.8 Transport mechanism of (**a**) H_2 penetration through dense membrane from hydrocarbon and (**b**) carbon membrane during gas separation. [(**a**) Reprinted with permission from Ref. [145]. (**b**) Reprinted with permission from Ref. [110]]

The permeation of hydrogen through the membrane follows the solution-diffusion techniques, and hydrogen is moved through the metal membrane in several steps as follows:

- Adsorption of hydrogen on the membrane surface on the feed side.
- Dissociation into ions and electrons.
- Absorption or dissolution into the metal.
- Hydrogen atom diffusion by two transitional events through the metal lattice (surface-to-bulk and bulk-to-surface).
- Recombination of hydrogen atoms on the permeate side.

There are basically three main methods for hydrogen separation membranes, as shown in Fig. 2.8a.

Knudsen diffusion is the primary transport mechanism for mesoporous membranes, which are characterized by low selectivity. Microporous membrane separation mechanisms are more intricate and involve molecular sieve at high temperatures and selective adsorption at low temperatures, with a combination of the two processes occurring at intermediate conditions [144]. Although microporous and molecular sieve membranes are sometimes regarded to be distinct types of porous membranes, in reality, the latter refers exclusively to those with pores smaller than 0.6 nm (Fig. 2.8b).

The solubility and diffusion of H_2 in the metal, along with the thickness and temperature of the membrane, all influence the penetration of hydrogen across a metal membrane. Dense metallic membranes need to be able to conduct free electrons in order to successfully separate gases, and they also need particular catalysts to split hydrogen gas molecules into atoms on the feed side. The hydrogen ions and free electrons are reconnected upon leaving the membrane material to create hydrogen molecules. Gas separation using dense metallic membranes requires free-electron conducting capabilities and, on the feed side, certain catalysts for splitting hydrogen gas molecules into atoms. When hydrogen ions and free electrons exit the membrane material, they are rejoined to form molecules of hydrogen. The thick membrane structure prevents bigger gas molecules like CO, CO_2, O_2, and CH_4 from passing through it, leading to the high purity (99.99%) of hydrogen produced.

2.5.4 *Membrane Assembly*

When various porous materials are combined, it is called a membrane assembly. In order to manufacture high-performance porous composite membranes, it is necessary to make use of the structures and properties of various materials to construct more nanochannels, repair defects, improve mechanical properties, and so on [146, 147]. Recently, the idea of MOF-in-COF membranes for gas separation was proposed by Fan et al. [148]. As shown in Fig. 2.9a, these membranes entail the controlled expansion of MOFs within a supplementary COF layer. A membrane with a novel MOF-in-COF structure was created, and it showed very high H_2 permeability as well as great selectivity in the separation of H_2/CO_2 and H_2/CH_4 gas mixtures. The development of mixed-dimensional heterostructures and the assembly of various inorganic-based membranes porous materials are both limited by this approach, which may find use on organic substrates as well. When different nanosheets and porous framework materials are assembled, the concept of growth restriction is also incorporated. Li et al. [149] created composite membranes by growing MOFs between layers of PVDF hollow fibers and ammoniating reduced graphene oxide (rGO) that was evenly distributed across the fibers (Fig. 2.9b). The metal salt is dissolved into the enlarged interlayer space and reacts with the joint in situ to produce MOF, taking advantage of the oxygen polar groups and electronegativity of the hydrated rGO membrane. The hydrated rGO membrane's electronegativity and oxygen polar groups are used to create MOF by dissolving the metal salt into the expanded interlayer gap and reacting with the joint in situ. The crystalline region was contained to the interlayer space in order to produce MOFs with a nanosheet structure and a thickness of 10–15 nm. A variety of transmission pathways allow gas to be selectively permeated through MOF pores and the intrinsic flaws of rGO. According to Yang et al. [150], a dense interlocking COF membrane was created by first assembling 2D COF nanosheets with 1D cellulose nanofibers (CNFs) in combined dimensions. This allowed for the creation of porous framework materials with 0D, 1D, and 2D nanosheets. Because of this, materials with porous frameworks and 0D, 1D, or 2D nanosheets could be made (Fig. 2.9c). Furthermore, the two-dimensional CTF-1 nanosheets may be realized by stacking them on top of each other on a carrier with the help of GO. This allows for the adjustment of nanometer thickness and the related properties of continuous membrane separation [151].

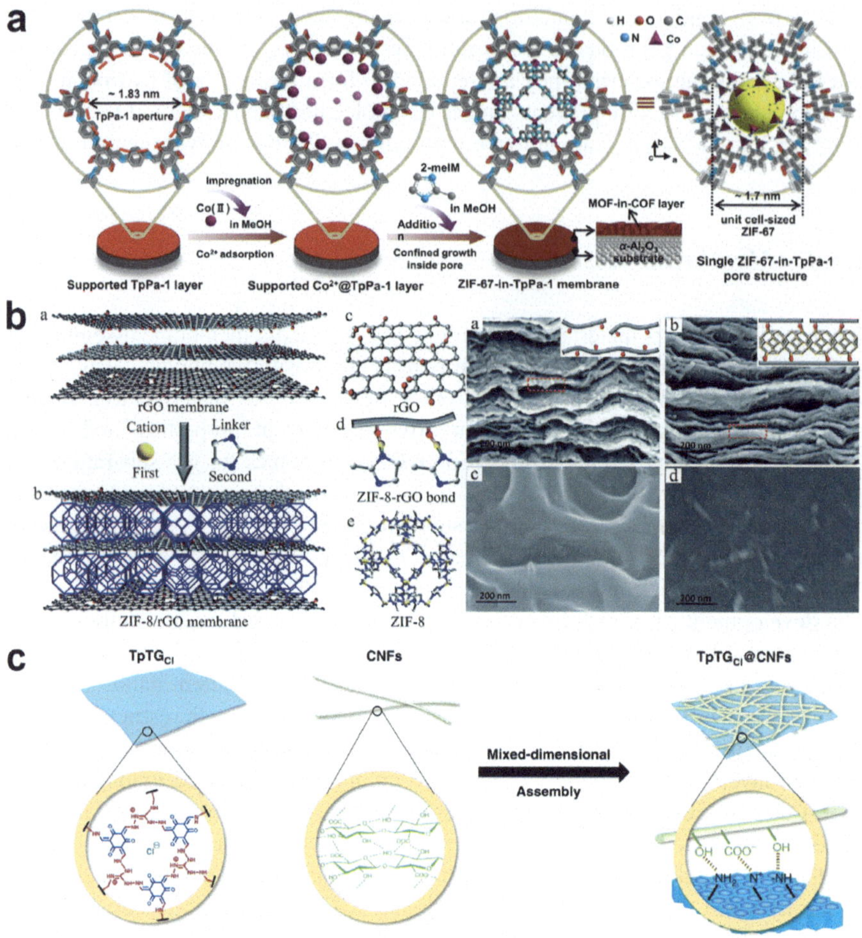

Fig. 2.9 (**a**) A molecular sieve membrane that is MOF-in-COF. (**b**) The ZIF-8/rGO membrane is created through organic crystallization in real time. (**c**) A COF membrane constructed from 2D COF and 1D CNF. [The original source. (**a**) Reprinted with permission from Ref. [148]]. (**b**) Reprinted with permission from Ref. [149]. (**c**) Reprinted with permission from Ref. [150]]

2.5.5 *Parameters Used to Determine the Efficiency of a Membrane*

A permeation cell can determine the performance of a metallic membrane using hydrogen flux, permeability, and solubility.

2.5.5.1 Hydrogen Solubility

Hydrogen solubility is hydrogen dissolution into a metallic membrane, which can either decrease or increase with temperature. This behavior is determined by the hydride metal's propensity to form solid solutions or pseudo-metallic ordered hydride phases. The hydride metal's tendency to create or pseudo-metallic solid solutions ordered hydride phases dictates this behavior. The transition metals of most interest for H_2 separation are exothermic occluders, i.e., the solubility decrease with increasing temperature. Hence, solubility is governed by Sievert's law.

$$C = KP^{0.5} \tag{2.15}$$

where K is the temperature-dependent Sievert's constant, $P^{0.5}$ is the relationship attributed to the breakdown of H_2 into ionic species resulting in the two tons being transmitted through each molecule of hydrogen.

2.5.5.2 Hydrogen Permeability

The rate of the hydrogen transferring through an alloy membrane is known as hydrogen permeability. This rate could be surface-limited, meaning that the rate of hydrogen dissociating at high pressure or reassociating at low pressure regulates the diffusion rate of atomic H through the metals. The rate-limiting process depends on the membrane structure's thickness. Narrow membranes can be constrained by the H_2 association and dissociation rate at each surface, whereas the rate of atomic H diffusion constrains thick membranes via the bulk alloy. Therefore, the functionality of "thin" and "thick" membranes is typically described in distinct ways [152]. The hydrogen permeability of dense metallic membranes is primarily determined by their lattice structure, the presence of lattice defects, and how they respond to H_2 or other feed stream gases. Structure, lattice defects, and response to feed stream gases, including hydrogen, are the primary determinants of dense metallic membranes' hydrogen permeability.

Hydrogen permeability (Q_H) through a metallic-based membrane is derived as follows:

$$Q_H = D_H \times S_H \tag{2.16}$$

where D_H is the hydrogen diffusivity through a membrane, S_H is the hydrogen solubility. Hence, high hydrogen solubilities and low activation energies of diffusion are desirable since they lead to a higher rate of permeation. As a result, for optimal permeation rates, it is preferable to have hydrogen solubilities that are high and diffusion activation energies that are low. Permeability is a temperature dependent and independent of thickness and H_2 partial pressure. Thus, it offers the best means of comparing the intrinsic transport properties of different bulk materials. Thickness and hydrogen partial pressure have no effect on permeability; rather, permeability is temperature dependent. In comparison with intrinsic transport qualities of various bulk materials, it provides the best means of comparison.

2.5.5.3 Hydrogen Diffusion

Hydrogen diffusivity is determined by diffusion coefficients. Numerous techniques can be used to determine H's diffusion coefficients across metals and alloys. The temperature and H/M ratio, the shape of the metal sample, the hydride metal's phase stability, and the grain-to-grain boundary ratio in the sample can all impact the method chosen. Multiple factors can influence the method of choice, including temperature, H/M ratio, sample shape, phase stability of the hydride metal, and grain-to-grain boundary ratio.

Hydrogen flux is the amount of H_2 in a mole that passes through a particular area over a time at a certain temperature and pressure differential, measured in units of (mol H_2) cm^{-2} s^{-1}. At any given temperature, the concentration gradient ($\frac{dC}{dx}$), and the hydrogen diffusion coefficient (D) are negative terms, which equals the flow of hydrogen (J) over a membrane, as stated in Fick's first law of diffusion.

$$J = -D\frac{dC}{dx} \tag{2.17}$$

This is recognized as Fick's first law of diffusion. Hydrogen Flux (J_{H_2}) can be derived from the equation below:

$$J_{H_2} = \frac{Q_H}{L}\left(P^{0.5}_{H_2\ ret} - P^{0.5}_{H_2\ perm}\right) \tag{2.18}$$

where L is the composite membrane's thickness $P^{0.5}_{H_2\ ret}$ and $P^{0.5}_{H_2\ perm}$ is the partial pressure at the reactant and permeate, respectively.

2.5.6 *Various Parameters That Affect the Permeation Rate of a Membrane*

Mass transfer in the Vanadium membrane is a complicated process, like Pd membrane, which can be affected by many factors. It is important to show the impact of different parameters on these membranes to validate their behavior under varied conditions. The main determinants covered in this part are the operating conditions, membrane thickness, permeation direction, and air pretreatment.

2.5.6.1 Operation Conditions

In the absence of chemical processes, Wei et al. recently showed that pressure enhances hydrogen flux but decreases selectivity, whereas temperature increases selectivity and hydrogen flux. Extreme temperatures, however, may reduce the membrane life from a durability perspective [153, 154]. It was also supported by Yocular et al. [155] using Pd/TiO_2 composite membrane to obtain the same result. Li et al. [156] and Didenko et al. [157] confirmed that sweep gas impacted hydrogen permeability. Sweep gas increases the transport rate by diluting the hydrogen, which decreases the permeation side's partial pressure. Despite the fact that sweep gas enhances membrane performance, its increased operating costs must be justified. While keeping a manageable operational cost, the minimum rate of sweep gas flow is expected to be sufficient to raise the average H_2 flux by at least 10% [156]. Thomas et al. [158] also agreed with Wei et al., stating that (using TiC/V) as the temperature increases, permeability increases, and as the pressure increases, the flux also increases.

2.5.6.2 The Thickness of the Membrane

Membranes, such as V and Pd membranes, are typically required to be very thin film in order to increase the rate of diffusion to produce high hydrogen separation flux [153]. However, the thickness of thin membrane should ensure that it has sufficient mechanical strength to stand on its own [155]. Hence, creating thin defect-free stand-alone V membrane is challenging. Numerous studies have focused on finding more efficient techniques for membrane preparation. As was already indicated, in order to create a thin layer of Pd or V with adequate mechanical strength, it is advantageous for mechanical support to be used as substrate [159], and the surface roughness of the support must be kept to a minimum. Chemical Vapor Deposition (CVD) and Physical Vapor Deposition (PVD) methods are the preparation procedures with the most potential for creating a thin metallic-based membrane. The thinness or thickness of the membranes determines the grain sizes and grain boundary of a membrane which has an influence.

Grain size is proportional to grain boundary volume and morphology. A critical microstructure feature is expected to influence metal's hydrogen permeation and durability to embrittlement [127]. Still, diffusion along grain boundaries and large areas of grain boundaries with smaller grains were thought to be responsible for the hydrogen flux rate. Additionally, an increase in hydrogen permeation is related to grain boundary gaps.

2.5.6.3 The Hydrogen Permeation Direction

Goto et al. [160] investigate the impact of the direction of hydrogen diffusion on the penetration of a composite palladium membrane in two different approaches, namely "Ceramic to palladium (CP) mode" and "Palladium to ceramic (PC) mode." The hydrogen direction in the CP model is from the support (ceramic support) to the membrane film (a palladium film), whereas the hydrogen direction in the PC model is the opposite. They conclude that the CP model has higher hydrogen permeation than the PC model. According to Liang and Hughes' research, the feedstock significantly impacts the direction of hydrogen penetration. Pure hydrogen has a greater rate of permeability from the support through the metal layer. However, the results were the opposite for a hydrogen/nitrogen mixture [161].

2.5.6.4 Air Pretreatment

Diego et al. [162] conducted experimental studies to find out how air pretreatment affected membrane performance. It was found that the Pd-Ag surface is activated and carbonaceous surface contaminants are eliminated through oxidation processes during the oxygen pretreatment. This may immediately impact the hydrogen permeability.

2.5.6.5 Impurity

Impurities such as carbon, nitrogen, and oxygen in bcc lattice membranes can decrease their ductility and workability because they can occupy interstitial gaps in the lattice. Dolan et al. found that niobium and vanadium have an increase in hydrogen solubility and a decrease in hydrogen permeability when exposed to dissolved oxygen. Since diffusivity and solubility are the two variables that determine permeability, it stands to reason that filling the interstitial sites will reduce diffusivity. Problems may arise if hydrogen reacts with carbon or oxygen contaminants to produce pressure steam or methane that becomes trapped in the membrane. The membrane may sustain serious harm as a result of this. The group of Vb metals is where this phenomenon is most frequently seen [152].

2.5.7 Overview of Dense Composite Membrane Fabrication Methods

Several methods are used to prepare composite dense metal membrane, and the most common methods are electroless plating, spray pyrolysis, electroplating, sputter-deposition, Physical Vapor Deposition (PVD), Magnetic Sputtering, Chemical Vapor Deposition (CVD), and Atomic Layer Deposition (ALD).

2.5.7.1 Electroless Plating

Electroless plating is known as a regulated autocatalytic deposition of a continuous layer on a catalytic surface by reacting complex molecules and a chemical reducing agent without using electrical energy [163].

It has two significant steps, which are the sensitization step and metallization steps:

- The sensitization process is known as the activation step. This process involves plating the active catalytic metal that will be involved in the reduction/oxidation reaction onto the surface that is intended to be coated.
- Metallization steps: the autocatalysis reduction/oxidation reaction occurs on the metal particles present [164].

Some of the advantages are as follows: (1) It is an easy and affordable technique, and (2) it offers consistent, dense coatings both on nonconducting and conductive surfaces. Its drawbacks include the need for multiple pretreatment states, including activation and sensitization (before the desired metal plating can be achieved), the formation of hazardous liquid water, and the use of toxic chemicals. Electroless plating (ELP) is also more time-consuming and complicated as a result.

2.5.7.2 Physical Vapor Deposition (PVD)

Physical Vapor Deposition is defined as a process in which thin, uniform metal and metal alloy films are deposited on a substrate source material. This process involves several steps:

- The very high-energy transfer that occurs when the source material is heated in a filament by bombardment of electrons or ions.
- In high vacuum states, the vaporization of the source material from energy is transferred through sublimation or evaporation into an atom.
- The vaporized source material is guided and transported toward the substrate surface.
- Process by which vapor reacts with reactor gas as it moves towards substrate surface.
- The process of evaporated material condensing on the substrate's surface.

Although PVD and CVD use the same processes, the precursors in PVD are often in the elemental state (as a pure metal), which prevents chemical decomposition reactions from happening on the surface. In contrast, the precursors in CVD are in the vaporization state (chemical compounds) [165]. Examples of PVD are thermal deposition, electron beam deposition, and Magnetron Sputtering. Since no liquid waste is created during the chemical reaction, this approach is environmentally friendly [154]. Its disadvantage is that it has low efficiency and is very expensive because of the costly high vacuum equipment required [166].

2.5.7.3 Chemical Vapor Deposition (CVD)

Chemical Vapor Deposition (CVD) is a flexible thin film deposition method of chemical reaction of gas-phase reactants over the substrate. This method can develop a thin film of simple and complex material at a time with anticipated morphology and chemical component. The following steps can be used in a procedure to create a thin film of solid materials:

- The precursor is injected into the reactor.
- Adsorption of the precursor onto the surface of the substrate triggers the initiation of the chemical reaction through heat or ionized plasma.
- In order to create the nucleation site and growing concern, the surface and decomposition of the precursor over the substrate advance at a given time for the appropriate thin film to form.
- The removal of the reactor's desorbed residue and by-product.

One of this method's significant advantages is the absence of wastewater production. The drawbacks include the high cost of the equipment, the incomplete deposition of all precursors on the support, and the difficulty in fabricating binary membranes [164]. Another limitation of this method is that CVD tends to give a lower flux and produce smaller H_2/N_2 selectivity values in comparison with electroless plating [144]. An experiment conducted by Kajiwara et al. [167] compared hydrogen permeation mechanism across supported platinum membrane prepared by CVD and electroless plating of platinum. They found that the defect structure created by CVD Pt layer increased H_2 flux than electroless plated Pt.

2.5.7.4 Atomic Layer Deposition (ALD)

Atomic Layer Deposition (ALD) is a chemical technique for thin-film deposition, where a series of self-limiting surface reactions are repeated discrete times [168, 169]. The reaction mechanism makes it possible to control the thickness at the atomic scale, and excellent uniformity and conformity are achieved at a low deposition. This will help eliminate intermetallic diffusion caused by other coating methods. There are two types of ALD which are Thermal ALD and Plasma-enhanced ALD. Plasma-Enhanced ALD (PEALD) utilizes oxygen plasma as an oxidizing agent to increase the reactivity at room temperature [169]. PEALD is a practical choice for film deposition, particularly when working with low-temperature substrates. Deposition of metal films has been challenging, but noble metals have overcome these challenges.

The procedure involves these four sequential steps:

- The initial precursor is applied to the substrate's surface, therefore saturating it;
- The reactor is evacuated of its surplus precursor and volatile by-products;
- The next step is to introduce the adsorbed layer so it can react with the second precursor;
- The second precursor's excess and volatile by-products are removed from the reactor.

2.6 Summary

A possible solution to the problem of global warming caused by the combustion of fossil fuels for power generation is the use of hydrogen as an energy carrier. The hydrogen fuel has found use in various industries, including the chemical, automotive, and aerospace sectors, among others. More than 95% of the hydrogen produced today comes from steam methane reforming, which is based on the reforming of fossil fuels. This is because fossil fuels have a high hydrogen concentration and are hence the most common source of hydrogen. Hydrogen production, however, is anticipated to progressively move towards environmentally friendly alternatives in the near future with the aim of reducing emissions of fossil CO_2. Although there is a wealth of literature on various advanced hydrogen production technologies, such as water splitting by electrolysis, photocatalytic water splitting, biological hydrogen generation, and high-temperature water splitting, there is ample opportunity to delve deeper into numerous scientific aspects that lack definitive experimental and theoretical answers.

Methods such as membrane separation, pressure swing adsorption, cryogenic separation, metal hydride usage, and catalytic purification are common in the hydrogen purification technology. To meet the global standard for hydrogen used in fuel cells for vehicles and to ensure the safety of a hydrogen production system based on electrolysis when coupled with renewable energy, a minimum of two hydrocarbon purification techniques are needed.

There are benefits and drawbacks to the various hydrogen separation technologies, which are determined by the needed hydrogen purity. The majority of the hydrogen will likely be utilized in fuel cells with solid polymer membranes. Because of its sensitivity, the low-temperature fuel cell can be catalytically poisoned by impurities like CO and H_2S. One well-established method for separating gases is the PSA technique. This method works well for mass productions. Incorporating metal hydrides into the purification process speeds up the process and increases the purity of the produced hydrogen by combining PSA with them. Even though the amount of hydrogen extracted from the gas mixture decreases, it is still feasible to get hydrogen of very high purity. One viable option for producing incredible amounts of ultra-high pure hydrogen is PSA adsorption. Using cryogenic technologies is the most effective way to decrease the amount of CO_2 in hydrogen. Only processes with high hydrogen fluxes are amenable to cryogenic technologies, which necessitate massive quantities of energy to attain low temperatures and high pressures. It is possible to separate hydrogen from a gas mixture using the cryogenic purification process, although this approach does not permit the achievement of high hydrogen purity.

When compared to other methods of hydrogen separation, the membrane cleaning technique shows the most promise. A variable amount of hydrogen can be produced by modifying the membranes. Dense palladium-based membranes are ideal for hydrogen purification. Although palladium is useful for the precise purification of very small amounts of hydrogen, its high cost prevents its scalability in this purification process. Dense membranes based on Group V metals are the technology that is now available. A less expensive catalyst or thin layers of Pd can be utilized. For hydrogen purification, polymeric proton exchange membranes work well. Electrochemical cell catalysts rapidly lose their efficacy in harsh environments because of the low operating temperature. Hydrogen purification is poor for porous membranes made of materials like zirconium oxide or aluminum ceramics. It is possible to mix porous membranes with other forms of cleaning. Gas mixtures containing carbon oxides and sulphur compounds do not penetrate ceramic barriers. To make catalyst-based cleaning procedures last longer, they work well as a pretreatment. More commonly, thin catalytic layers on porous membranes are employed. The presence of water vapor can cause ceramic membranes to degrade. The drawbacks of carbon membranes are identical. Some vapors require very high temperatures for operation because of their intense sorption. Making and using them is also a challenge.

References

1. International Energy Agency (IEA), in: Glob. Hydrog. Rev. IEA, Paris, France, 2022. https://doi.org/10.1787/a15b8442-en.
2. H. Boujjat, S. Rodat, S. Abanades, Techno-economic assessment of solar-driven steam gasification of biomass for large-scale hydrogen production. Processes. **9**, 1–20 (2021). https://doi.org/10.3390/pr9030462
3. M. Handwerker, J. Wellnitz, H. Marzbani, Comparison of Hydrogen Powertrains with the Battery Powered Electric Vehicle and Investigation of Small-Scale Local Hydrogen Production Using Renewable Energy. Hydrogen **2**, 76–100 (2021). https://doi.org/10.3390/hydrogen2010005
4. V. Franzitta, D. Curto, D. Milone, M. Trapanese, Energy saving in public transport using renewable energy. Sustain. **9** (2017). https://doi.org/10.3390/su9010106
5. E.B. Agyekum, C. Nutakor, A.M. Agwa, S. Kamel, A critical review of renewable hydrogen production methods: factors affecting their scale-up and its role in future energy generation. Membranes (Basel) **12** (2022). https://doi.org/10.3390/membranes12020173
6. S.E. Hosseini, M.A. Wahid, Hydrogen production from renewable and sustainable energy resources: Promising green energy carrier for clean development. Renew. Sustain. Energy Rev. **57**, 850–866 (2016). https://doi.org/10.1016/j.rser.2015.12.112
7. International Energy Agency (IEA), in: Technol. Roadmap Hydrog. Fuel Cells; IEA, Paris, France, 2015.

8. F. Safari, I. Dincer, A review and comparative evaluation of thermochemical water splitting cycles for hydrogen production. Energy Convers. Manag. **205**, 112182 (2020). https://doi.org/10.1016/j.enconman.2019.112182
9. A. Bolt, I. Dincer, M. Agelin-Chaab, A review of unique aluminum-water based hydrogen production options. Energy and Fuels. **35**, 1024–1040 (2021). https://doi.org/10.1021/acs.energyfuels.0c03674
10. J.D. Holladay, J. Hu, D.L. King, Y. Wang, An overview of hydrogen production technologies. Catal. Today. **139**, 244–260 (2009). https://doi.org/10.1016/j.cattod.2008.08.039
11. I. Dincer, C. Zamfirescu, Sustainable hydrogen production options and the role of IAHE. Int. J. Hydrogen Energy. **37**, 16266–16286 (2012). https://doi.org/10.1016/j.ijhydene.2012.02.133
12. Z. Zhuang, Y. Wang, C.Q. Xu, S. Liu, C. Chen, Q. Peng, Z. Zhuang, H. Xiao, Y. Pan, S. Lu, R. Yu, W.C. Cheong, X. Cao, K. Wu, K. Sun, Y. Wang, D. Wang, J. Li, Y. Li, Three-dimensional open nano-netcage electrocatalysts for efficient pH-universal overall water splitting. Nat. Commun. **10**, 1–10 (2019). https://doi.org/10.1038/s41467-019-12885-0
13. H. Song, S. Luo, H. Huang, B. Deng, J. Ye, Solar-Driven Hydrogen Production: Recent Advances. Challenges, and Future Perspectives, ACS Energy Lett. **7**, 1043–1065 (2022). https://doi.org/10.1021/acsenergylett.1c02591
14. M. Martino, C. Ruocco, E. Meloni, P. Pullumbi, V. Palma, Main hydrogen production processes: an overview. Catalysts. **11** (2021). https://doi.org/10.3390/catal11050547
15. O. Schmidt, A. Gambhir, I. Staffell, A. Hawkes, J. Nelson, S. Few, Future cost and performance of water electrolysis: An expert elicitation study. Int. J. Hydrogen Energy. **42**, 30470–30492 (2017). https://doi.org/10.1016/j.ijhydene.2017.10.045
16. J. Brauns, T. Turek, Alkaline water electrolysis powered by renewable energy: A review. Processes. **8** (2020). https://doi.org/10.3390/pr8020248
17. J. Chi, H. Yu, Water electrolysis based on renewable energy for hydrogen production, Cuihua Xuebao/Chinese. J. Catal. **39**, 390–394 (2018). https://doi.org/10.1016/S1872-2067(17)62949-8
18. J. Jia, L.C. Seitz, J.D. Benck, Y. Huo, Y. Chen, J.W.D. Ng, T. Bilir, J.S. Harris, T.F. Jaramillo, Solar water splitting by photovoltaic-electrolysis with a solar-to-hydrogen efficiency over 30%. Nat. Commun. **7**, 1–6 (2016). https://doi.org/10.1038/ncomms13237
19. S. Abanades, Metal oxides applied to thermochemical water-splitting for hydrogen production using concentrated solar energy. Chem. Eng **3**, 1–28 (2019). https://doi.org/10.3390/chemengineering3030063
20. C. Acar, Y. Bicer, M.E. Demir, I. Dincer, Transition to a new era with light-based hydrogen production for a carbon-free society: an overview. Int. J. Hydrogen Energy. **44**, 25347–25,364 (2019). https://doi.org/10.1016/j.ijhydene.2019.08.010
21. O. Oruc, I. Dincer, Assessing the potential of thermochemical water splitting cycles: A bridge towards for clean and sustainable hydrogen generation. Fuel. **286**, 119325 (2021). https://doi.org/10.1016/j.fuel.2020.119325
22. R.S. El-Emam, H. Ozcan, C. Zamfirescu, Updates on promising thermochemical cycles for clean hydrogen production using nuclear energy. J. Clean. Prod. **262** (2020). https://doi.org/10.1016/j.jclepro.2020.121424
23. M.A. Rosen, Advances in hydrogen production by thermochemical water decomposition: A review. Energy. **35**, 1068–1076 (2010). https://doi.org/10.1016/j.energy.2009.06.018
24. I. Dincer, C. Acar, Review and evaluation of hydrogen production methods for better sustainability. Int. J. Hydrogen Energy. **40**, 11094–11111 (2014). https://doi.org/10.1016/j.ijhydene.2014.12.035
25. Z. Chen, T.F. Jaramillo, T.G. Deutsch, A. Kleiman-Shwarsctein, A.J. Forman, N. Gaillard, R. Garland, K. Takanabe, C. Heske, M. Sunkara, E.W. McFarland, K. Domen, E.L. Milled, H.N. Dinh, Accelerating materials development for photoelectrochemical hydrogen production: Standards for methods, definitions, and reporting protocols. J. Mater. Res. **25**, 3–16 (2010). https://doi.org/10.1557/jmr.2010.0020

26. I. Dincer, Green methods for hydrogen production. Int. J. Hydrogen Energy. **37**, 1954–1971 (2012). https://doi.org/10.1016/j.ijhydene.2011.03.173
27. N. Akhlaghi, G. Najafpour-Darzi, A comprehensive review on biological hydrogen production. Int. J. Hydrogen Energy. **45**, 22492–22,512 (2020). https://doi.org/10.1016/j.ijhydene.2020.06.182
28. Z. Abdin, A. Zafaranloo, A. Rafiee, W. Mérida, W. Lipiński, K.R. Khalilpour, Hydrogen as an energy vector. Renew. Sustain. Energy Rev. **120** (2020). https://doi.org/10.1016/j.rser.2019.109620
29. P. Nikolaidis, A. Poullikkas, A comparative overview of hydrogen production processes. Renew. Sustain. Energy Rev. **67**, 597–611 (2017). https://doi.org/10.1016/j.rser.2016.09.044
30. V. Franzitta, D. Curto, D. Rao, A. Viola, Hydrogen production from sea wave for alternative energy vehicles for public transport in Trapani (Italy). Energies **9** (2016). https://doi.org/10.3390/en9100850
31. R. Kumar, A. Kumar, A. Pal, An overview of conventional and non-conventional hydrogen production methods. Mater. Today Proc. **46**, 5353–5359 (2020). https://doi.org/10.1016/j.matpr.2020.08.793
32. H.C. Ong, W.H. Chen, A. Farooq, Y.Y. Gan, K.T. Lee, V. Ashokkumar, Catalytic thermochemical conversion of biomass for biofuel production: A comprehensive review. Renew. Sustain. Energy Rev. **113**, 109266 (2019). https://doi.org/10.1016/j.rser.2019.109266
33. A.R.K. Gollakota, N. Kishore, S. Gu, A review on hydrothermal liquefaction of biomass. Renew. Sustain. Energy Rev. **81**, 1378–1392 (2018). https://doi.org/10.1016/j.rser.2017.05.178
34. M. Shahabuddin, B.B. Krishna, T. Bhaskar, G. Perkins, Advances in the thermo-chemical production of hydrogen from biomass and residual wastes: Summary of recent techno-economic analyses. Bioresour. Technol. **299**, 122557 (2020). https://doi.org/10.1016/j.biortech.2019.122557
35. C. Acar, I. Dincer, Comparative assessment of hydrogen production methods from renewable and non-renewable sources. Int. J. Hydrogen Energy. **39**, 1–12 (2014). https://doi.org/10.1016/j.ijhydene.2013.10.060
36. M.N. Uddin, K. Techato, J. Taweekun, M.M. Rahman, M.G. Rasul, T.M.I. Mahlia, S.M. Ashrafur, An overview of recent developments in biomass pyrolysis technologies. Energies. **11** (2018). https://doi.org/10.3390/en11113115
37. W.H. Chen, B.J. Lin, M.Y. Huang, J.S. Chang, Thermochemical conversion of microalgal biomass into biofuels: A review. Bioresour. Technol. **184**, 314–327 (2015). https://doi.org/10.1016/j.biortech.2014.11.050
38. C. Zhang, X. Kang, N. Liang, A. Abdullah, Improvement of biohydrogen production from dark fermentation by cocultures and activated carbon immobilization. Energy and Fuels **31**, 12217–12,222 (2017). https://doi.org/10.1021/acs.energyfuels.7b02035
39. C. Acar, I. Dincer, Review and evaluation of hydrogen production options for better environment. J. Clean. Prod. **218**, 835–849 (2019). https://doi.org/10.1016/j.jclepro.2019.02.046
40. Z.M.A. Bundhoo, Potential of bio-hydrogen production from dark fermentation of crop residues: A review. Int. J. Hydrogen Energy. **44**, 17346–17362 (2019). https://doi.org/10.1016/j.ijhydene.2018.11.098
41. P. Mishra, S. Krishnan, S. Rana, L. Singh, M. Sakinah, Z. Ab Wahid, Outlook of fermentative hydrogen production techniques: an overview of dark, photo and integrated dark-photo fermentative approach to biomass. Energy Strateg. Rev. **24**, 27–37 (2019). https://doi.org/10.1016/j.esr.2019.01.001
42. E. Eroglu, A. Melis, Photobiological hydrogen production: Recent advances and state of the art. Bioresour. Technol. **102**, 8403–8413 (2011). https://doi.org/10.1016/j.biortech.2011.03.026

43. H. Argun, F. Kargi, Bio-hydrogen production by different operational modes of dark and photo-fermentation: An overview. Int. J. Hydrogen Energy. **36**, 7443–7459 (2011). https://doi.org/10.1016/j.ijhydene.2011.03.116
44. H.S. Lee, W.F.J. Vermaas, B.E. Rittmann, Biological hydrogen production: Prospects and challenges. Trends Biotechnol. **28**, 262–271 (2010). https://doi.org/10.1016/j.tibtech.2010.01.007
45. Y. Kalinci, A. Hepbasli, I. Dincer, Biomass-based hydrogen production: A review and analysis. Int. J. Hydrogen Energy. **34**, 8799–8817 (2009). https://doi.org/10.1016/j.ijhydene.2009.08.078
46. M.A. Hossain, J. Jewaratnam, P. Ganesan, Prospect of hydrogen production from oil palm biomass by thermochemical process – a review, Int. J. Hydrogen Energy. 41 (2016) 16637–16,655. doi:https://doi.org/10.1016/j.ijhydene.2016.07.104.
47. M. Patel, X. Zhang, A. Kumar, Techno-economic and life cycle assessment on lignocellulosic biomass thermochemical conversion technologies: A review. Renew. Sustain. Energy Rev. **53**, 1486–1499 (2016). https://doi.org/10.1016/j.rser.2015.09.070
48. S. Mutsengerere, C.H. Chihobo, D. Musademba, I. Nhapi, A review of operating parameters affecting bio-oil yield in microwave pyrolysis of lignocellulosic biomass. Renew. Sustain. Energy Rev. **104**, 328–336 (2019). https://doi.org/10.1016/j.rser.2019.01.030
49. H. Zhang, Z. Gao, W. Ao, J. Li, G. Liu, J. Fu, C. Ran, X. Mao, Q. Kang, Y. Liu, J. Dai, Microwave-assisted pyrolysis of textile dyeing sludge using different additives. J. Anal. Appl. Pyrolysis. **127**, 140–149 (2017). https://doi.org/10.1016/j.jaap.2017.08.014
50. R.E. Guedes, A.S. Luna, A.R. Torres, Operating parameters for bio-oil production in biomass pyrolysis: A review. J. Anal. Appl. Pyrolysis. **129**, 134–149 (2018). https://doi.org/10.1016/j.jaap.2017.11.019
51. T. Suprianto, W. Winarto, I.N.G. Wijayanti, Wardana, Synergistic effect of curcumin and activated carbon catalyst enhancing hydrogen production from biomass pyrolysis. Int. J. Hydrogen Energy. **46**, 7147–7164 (2021). https://doi.org/10.1016/j.ijhydene.2020.11.211
52. H.B. Goyal, D. Seal, R.C. Saxena, Bio-fuels from thermochemical conversion of renewable resources: A review. Renew. Sustain. Energy Rev. **12**, 504–517 (2008). https://doi.org/10.1016/j.rser.2006.07.014
53. Y. Zhang, L. Wan, J. Guan, Q. Xiong, S. Zhang, X. Jin, A review on biomass gasification: effect of main parameters on char generation and reaction. Energy and Fuels. **34**, 13438–13,455 (2020). https://doi.org/10.1021/acs.energyfuels.0c02900
54. S. Safarian, R. Unnþórsson, C. Richter, A review of biomass gasification modelling. Renew. Sustain. Energy Rev. **110**, 378–391 (2019). https://doi.org/10.1016/j.rser.2019.05.003
55. J. Ren, J.P. Cao, X.Y. Zhao, F.L. Yang, X.Y. Wei, Recent advances in syngas production from biomass catalytic gasification: A critical review on reactors, catalysts, catalytic mechanisms and mathematical models. Renew. Sustain. Energy Rev. **116**, 109426 (2019). https://doi.org/10.1016/j.rser.2019.109426
56. J.S. Rowbotham, P.W. Dyer, H.C. Greenwell, M.K. Theodorou, Thermochemical processing of macroalgae: a late bloomer in the development of third-generation biofuels? Biofuels. **3**, 441–461 (2012). https://doi.org/10.4155/BFS.12.29
57. D.C. Elliott, P. Biller, A.B. Ross, A.J. Schmidt, S.B. Jones, Hydrothermal liquefaction of biomass: Developments from batch to continuous process. Bioresour. Technol. **178**, 147–156 (2015). https://doi.org/10.1016/j.biortech.2014.09.132
58. I.A. Basar, H. Liu, H. Carrere, E. Trably, C. Eskicioglu, A review on key design and operational parameters to optimize and develop hydrothermal liquefaction of biomass for biorefinery applications. Green Chem. **23**, 1404–1446 (2021). https://doi.org/10.1039/d0gc04092d
59. K.G. Kalogiannis, S.D. Stefanidis, A.A. Lappas, Catalyst deactivation, ash accumulation and bio-oil deoxygenation during ex situ catalytic fast pyrolysis of biomass in a cascade

thermal-catalytic reactor system. Fuel Process. Technol. **186**, 99–109 (2019). https://doi.org/10.1016/j.fuproc.2018.12.008
60. G. Esteban-Díez, M.V. Gil, C. Pevida, D. Chen, F. Rubiera, Effect of operating conditions on the sorption enhanced steam reforming of blends of acetic acid and acetone as bio-oil model compounds. Appl. Energy. **177**, 579–590 (2016). https://doi.org/10.1016/j.apenergy.2016.05.149
61. N. Gao, C. Quan, Z. Ma, C. Wu, Thermal Characteristics of Biomass Pyrolysis Oil and Potential Hydrogen Production by Catalytic Steam Reforming. Energy and Fuels. **32**, 5234–5243 (2018). https://doi.org/10.1021/acs.energyfuels.8b00365
62. A.V. Bridgwater, Review of fast pyrolysis of biomass and product upgrading. Biomass and Bioenergy. **38**, 68–94 (2012). https://doi.org/10.1016/j.biombioe.2011.01.048
63. S. Xiu, A. Shahbazi, Bio-oil production and upgrading research: a review. Renew. Sustain. Energy Rev. **16**, 4406–4414 (2012). https://doi.org/10.1016/j.rser.2012.04.028
64. R.V.S. Silva, V.B. Pereira, K.T. Stelzer, T.A. Almeida, G.A. Romeiro, D.A. Azevedo, Comprehensive study of the liquid products from slow pyrolysis of crambe seeds: Bio-oil and organic compounds of the aqueous phase. Biomass and Bioenergy. **123**, 78–88 (2019). https://doi.org/10.1016/j.biombioe.2019.02.014
65. P.J. Megía, M. Cortese, C. Ruocco, A.J. Vizcaíno, J.A. Calles, A. Carrero, V. Palma, Catalytic behavior of co-based catalysts in the kinetic study of acetic acid steam reforming. Ind. Eng. Chem. Res. **59**, 19531–19,538 (2020). https://doi.org/10.1021/acs.iecr.0c03599
66. P.J. Megía, A. Carrero, J.A. Calles, A.J. Vizcaíno, Hydrogen production from steam reforming of acetic acid as a model compound of the aqueous fraction of microalgae HTL using Co-M/SBA-15 (M: Cu, Ag, Ce, Cr) catalysts. Catalysts **9** (2019). https://doi.org/10.3390/catal9121013
67. J. Chen, J. Sun, Y. Wang, Catalysts for steam reforming of bio-oil: a review. Ind. Eng. Chem. Res. **56**, 4627–4637 (2017). https://doi.org/10.1021/acs.iecr.7b00600
68. V. Palma, C. Ruocco, E. Meloni, A. Ricca, Renewable hydrogen from ethanol reforming over CeO_2-SiO_2 based catalysts. Catalysts **7** (2017). https://doi.org/10.3390/catal7080226
69. M. Olazar, L. Santamaria, G. Lopez, E. Fernandez, M. Cortazar, A. Arregi, J. Bilbao, Progress on catalyst development for the steam reforming of biomass and waste plastics pyrolysis volatiles: a review. Energy and Fuels. **35**, 17051–17,084 (2021). https://doi.org/10.1021/acs.energyfuels.1c01666
70. P.J. Megia, A.J. Vizcaino, J.A. Calles, A. Carrero, Hydrogen production technologies: from fossil fuels toward renewable sources. A mini review. Energy and Fuels. **35**, 16403–16,415 (2021). https://doi.org/10.1021/acs.energyfuels.1c02501
71. S. Shiva Kumar, V. Himabindu, Hydrogen production by PEM water electrolysis – a review. Mater. Sci. Energy Technol. **2**, 442–454 (2019). https://doi.org/10.1016/j.mset.2019.03.002
72. P. Parthasarathy, K.S. Narayanan, Hydrogen production from steam gasification of biomass: influence of process parameters on hydrogen yield – a review. Renew. Energy. **66**, 570–579 (2014). https://doi.org/10.1016/j.renene.2013.12.025
73. T. Lepage, M. Kammoun, Q. Schmetz, A. Richel, Biomass-to-hydrogen: a review of main routes production, processes evaluation and techno-economical assessment. Biomass and Bioenergy. **144**, 105920 (2021). https://doi.org/10.1016/j.biombioe.2020.105920
74. Y. Zhu, M.J. Biddy, S.B. Jones, D.C. Elliott, A.J. Schmidt, Techno-economic analysis of liquid fuel production from woody biomass via hydrothermal liquefaction (HTL) and upgrading. Appl. Energy. **129**, 384–394 (2014). https://doi.org/10.1016/j.apenergy.2014.03.053
75. A. Cascioli, M. Baratieri, Enhanced thermodynamic modelling for hydrothermal liquefaction. Fuel. **298**, 120796 (2021). https://doi.org/10.1016/j.fuel.2021.120796
76. H.L. Chen, H.M. Lee, S.H. Chen, Y. Chao, M.B. Chang, Review of plasma catalysis on hydrocarbon reforming for hydrogen production – interaction, integration, and prospects. Appl. Catal. B Environ. **85**, 1–9 (2008). https://doi.org/10.1016/j.apcatb.2008.06.021

77. T.E. Lipman, Hydrogen production science and technology. Fuel Cells Hydrog. Prod., 783–798 (2019). https://doi.org/10.1007/978-1-4939-7789-5_755
78. L. Barelli, G. Bidini, F. Gallorini, S. Servili, Hydrogen production through sorption-enhanced steam methane reforming and membrane technology: a review. Energy. **33**, 554–570 (2008). https://doi.org/10.1016/j.energy.2007.10.018
79. T.Y. Amiri, K. Ghasemzageh, A. Iulianelli, Membrane reactors for sustainable hydrogen production through steam reforming of hydrocarbons: a review. Chem. Eng. Process. – Process Intensif. **157**, 108148 (2020). https://doi.org/10.1016/j.cep.2020.108148
80. M. Mbodji, J.M. Commenge, L. Falk, D. Di Marco, F. Rossignol, L. Prost, S. Valentin, R. Joly, P. Del-Gallo, Steam methane reforming reaction process intensification by using a millistructured reactor: experimental setup and model validation for global kinetic reaction rate estimation. Chem. Eng. J. **207–208**, 871–884 (2012). https://doi.org/10.1016/j.cej.2012.07.117
81. E. Meloni, M. Martino, V. Palma, A short review on Ni based catalysts and related engineering issues for methane steam reforming. Catalysts. **10** (2020). https://doi.org/10.3390/catal10030352
82. International Energy Agency Greenhouse Gas R&D Programme (IEAGHG), in Techno-Economic Eval. SMR Based Standalone Hydrog. Plant with CCS (IEAGHG: Cheltenham, U.K.). Technical report 2017/2, 2017.
83. M.T. Azizan, A. Aqsha, M. Ameen, A. Syuhada, H. Klaus, S.Z. Abidin, F. Sher, Catalytic reforming of oxygenated hydrocarbons for the hydrogen production: an outlook. Biomass Convers. Biorefinery. (2020). https://doi.org/10.1007/s13399-020-01081-6
84. D. Hariharan, R. Yang, Y. Zhou, B. Gainey, S. Mamalis, R.E. Smith, M.A. Lugo-Pimentel, M.J. Castaldi, R. Gill, A. Davis, D. Modroukas, B. Lawler, Catalytic partial oxidation reformation of diesel, gasoline, and natural gas for use in low temperature combustion engines. Fuel. **246**, 295–307 (2019). https://doi.org/10.1016/j.fuel.2019.02.003
85. R. Ma, B. Xu, X. Zhang, Catalytic partial oxidation (CPOX) of natural gas and renewable hydrocarbons/oxygenated hydrocarbons – a review. Catal. Today. **338**, 18–30 (2019). https://doi.org/10.1016/j.cattod.2019.06.025
86. P. Arku, B. Regmi, A. Dutta, A review of catalytic partial oxidation of fossil fuels and biofuels: Recent advances in catalyst development and kinetic modelling. Chem. Eng. Res. Des. **136**, 385–402 (2018). https://doi.org/10.1016/j.cherd.2018.05.044
87. B. Sawatmongkhon, K. Theinnoi, T. Wongchang, C. Haoharn, C. Wongkhorsub, E. Sukjit, Modeling of hydrogen production from catalytic partial oxidation of ethanol over a platinum-rhodium-supported catalyst. Energy and Fuels. **35**, 4404–4417 (2021). https://doi.org/10.1021/acs.energyfuels.0c04125
88. C. Pirez, W. Fang, M. Capron, S. Paul, H. Jobic, F. Dumeignil, L. Jalowiecki-Duhamel, Steam reforming, partial oxidation and oxidative steam reforming for hydrogen production from ethanol over cerium nickel based oxyhydride catalyst. Appl. Catal. A Gen. **518**, 78–86 (2016). https://doi.org/10.1016/j.apcata.2015.10.035
89. M. Luneau, E. Gianotti, F.C. Meunier, C. Mirodatos, E. Puzenat, Y. Schuurman, N. Guilhaume, Deactivation mechanism of Ni supported on Mg-Al spinel during autothermal reforming of model biogas. Appl. Catal. B Environ. **203**, 289–299 (2017). https://doi.org/10.1016/j.apcatb.2016.10.023
90. R. Baruah, M. Dixit, P. Basarkar, D. Parikh, A. Bhargav, Advances in ethanol autothermal reforming. Renew. Sustain. Energy Rev. **51**, 1345–1353 (2015). https://doi.org/10.1016/j.rser.2015.07.060
91. R. Tariq, F. Maqbool, S.Z. Abbas, Small-scale production of hydrogen via auto-thermal reforming in an adiabatic packed bed reactor: parametric study and reactor's optimization through response surface methodology. Comput. Chem. Eng. **145**, 107192 (2021). https://doi.org/10.1016/j.compchemeng.2020.107192

92. S. Wang, G. Dai, H. Yang, Z. Luo, Lignocellulosic biomass pyrolysis mechanism: a state-of-the-art review. Prog. Energy Combust. Sci. **62**, 33–86 (2017). https://doi.org/10.1016/j.pecs.2017.05.004
93. N.Z. Muradov, How to produce hydrogen from fossil fuels without CO_2 emission. Int. J. Hydrogen Energy. **18**, 211–215 (1993). https://doi.org/10.1016/0360-3199(93)90021-2
94. S. Schneider, S. Bajohr, F. Graf, T. Kolb, State of the art of hydrogen production via pyrolysis of natural gas. Chemie-Ingenieur-Technik. **92**, 1023–1032 (2020). https://doi.org/10.1002/cite.202000021
95. K. Stańczyk, K. Kapusta, M. Wiatowski, J. Świdrowski, A. Smoliński, J. Rogut, A. Kotyrba, Experimental simulation of hard coal underground gasification for hydrogen production. Fuel. **91**, 40–50 (2012). https://doi.org/10.1016/j.fuel.2011.08.024
96. T. Sutardi, L. Wang, N. Karimi, M.C. Paul, Utilization of H_2O and CO_2 in coal particle gasification with an impact of temperature and particle size. Energy Fuels. **34**, 12841–12,852 (2020). https://doi.org/10.1021/acs.energyfuels.0c02280
97. L. Emami Taba, M.F. Irfan, W.A.M. Wan Daud, M.H. Chakrabarti, The effect of temperature on various parameters in coal, biomass and CO-gasification: a review. Renew. Sustain. Energy Rev. **16**, 5584–5596 (2012). https://doi.org/10.1016/j.rser.2012.06.015
98. G.J. Stiegel, M. Ramezan, Hydrogen from coal gasification: an economical pathway to a sustainable energy future. Int. J. Coal Geol. **65**, 173–190 (2006). https://doi.org/10.1016/j.coal.2005.05.002
99. J. Mularski, H. Pawlak-Kruczek, N. Modlinski, A review of recent studies of the CFD modelling of coal gasification in entrained flow gasifiers, covering devolatilization, gas-phase reactions, surface reactions, models and kinetics. Fuel. **271**, 117620 (2020). https://doi.org/10.1016/j.fuel.2020.117620
100. S. Adhikari, S. Fernando, Hydrogen membrane separation techniques. Ind. Eng. Chem. Res. **45**, 875–881 (2006). https://doi.org/10.1021/ie050644l
101. G. Xu, L. Li, Y. Yang, L. Tian, T. Liu, K. Zhang, A novel CO_2 cryogenic liquefaction and separation system. Energy **42**, 522–529 (2012). https://doi.org/10.1016/j.energy.2012.02.048
102. A.B. Hinchliffe, K.E. Porter. A comparison of membrane separation and distillation. Chem Eng Res Des **78**(2), 255–268 (2000). https://www.sciencedirect.com/science/article/abs/pii/S0263876200718880
103. X.Y. Chen, L.X. Wei, L. Deng, F.S. Yang, Z.X. Zhang, A review on the metal hydride based hydrogen purification and separation technology. Appl. Mech. Mater. **448–453**, 3027–3036 (2014). https://doi.org/10.4028/www.scientific.net/AMM.448-453.3027
104. A. Fujisawa, S. Miura, Y. Mitsutake, M. Monde, Simulation study of hydrogen purification using metal hydride. J. Alloys Compd. **580**, S423–S426 (2013). https://doi.org/10.1016/j.jallcom.2013.03.262
105. D. Dunikov, V. Borzenko, D. Blinov, A. Kazakov, C.Y. Lin, S.Y. Wu, C.Y. Chu, Biohydrogen purification using metal hydride technologies. Int. J. Hydrogen Energy. **41**, 21787–21,794 (2016). https://doi.org/10.1016/j.ijhydene.2016.08.190
106. K.D. Modibane, M. Williams, M. Lototskyy, M.W. Davids, Y. Klochko, B.G. Pollet, Poisoning-tolerant metal hydride materials and their application for hydrogen separation from CO_2/CO containing gas mixtures. Int. J. Hydrogen Energy. **38**, 9800–9810 (2013). https://doi.org/10.1016/j.ijhydene.2013.05.102
107. M. Nordio, S.A. Wassie, M. Van Sint Annaland, D.A. Pacheco Tanaka, J.L. Viviente Sole, F. Gallucci, Techno-economic evaluation on a hybrid technology for low hydrogen concentration separation and purification from natural gas grid. Int. J. Hydrogen Energy. **46**, 23417–23,435 (2021). https://doi.org/10.1016/j.ijhydene.2020.05.009
108. L. Lei, A. Lindbråthen, M. Hillestad, X. He, Carbon molecular sieve membranes for hydrogen purification from a steam methane reforming process. J. Memb. Sci. **627**, 119241 (2021). https://doi.org/10.1016/j.memsci.2021.119241
109. A. Naquash, M.A. Qyyum, Y.D. Chaniago, A. Riaz, F. Yehia, H. Lim, M. Lee, Separation and purification of syngas-derived hydrogen: A comparative evaluation of membrane- and cryogenic-assisted approaches. Chemosphere. **313**, 137420 (2023). https://doi.org/10.1016/j.chemosphere.2022.137420

110. A. Lider, V. Kudiiarov, N. Kurdyumov, J. Lyu, M. Koptsev, N. Travitzky, D. Hotza, Materials and techniques for hydrogen separation from methane-containing gas mixtures. Int. J. Hydrogen Energy. **48**, 28390–28,411 (2023). https://doi.org/10.1016/j.ijhydene.2023.03.345
111. H. Yin, A.C.K. Yip, A review on the production and purification of membrane technologies. Catalysts **7**(10), 297 (2017). https://doi.org/10.3390/catal7100297
112. G. Bernardo, T. Araújo, T. da Silva Lopes, J. Sousa, A. Mendes, Recent advances in membrane technologies for hydrogen purification. Int. J. Hydrogen Energy. **45**, 7313–7338 (2020). https://doi.org/10.1016/j.ijhydene.2019.06.162
113. K. Chang, Q. Li, Q. Li, Refrigeration cycle for cryogenic separation of hydrogen from coke oven gas. Front. Energy Power Eng. China. **2**, 484–488 (2008). https://doi.org/10.1007/s11708-008-0096-0
114. W. Lin, J. Xu, L. Zhang, A. Gu, Synthetic natural gas (SNG) liquefaction processes with hydrogen separation. Int. J. Hydrogen Energy. **42**, 18417–18,424 (2017). https://doi.org/10.1016/j.ijhydene.2017.04.141
115. T. Yurata, H. Lei, L. Tang, M. Lu, J. Patel, S. Lim, P. Piumsomboon, B. Chalermsinsuwan, C. Li, Feasibility and sustainability analyses of carbon dioxide – hydrogen separation via de-sublimation process in comparison with other processes. Int. J. Hydrogen Energy. **44**, 23120–23134 (2019). https://doi.org/10.1016/j.ijhydene.2019.07.032
116. B. Ghorbani, M. Mehrpooya, M. Sadeghzadeh, Process development of a solar-assisted multi-production plant: power, cooling, and hydrogen. Int. J. Hydrogen Energy. **45**, 30056–30,079 (2020). https://doi.org/10.1016/j.ijhydene.2020.08.018
117. M. Aasadnia, M. Mehrpooya, B. Ghorbani, A novel integrated structure for hydrogen purification using the cryogenic method. J. Clean. Prod. **278**, 123872 (2021). https://doi.org/10.1016/j.jclepro.2020.123872
118. M. Mehrpooya, S.A. Mousavi, M. Asadnia, A. Zaitsev, R. Sanavbarov, Conceptual design and evaluation of an innovative hydrogen purification process applying diffusion-absorption refrigeration cycle (exergoeconomic and exergy analyses). J. Clean. Prod. **316**, 128271 (2021). https://doi.org/10.1016/j.jclepro.2021.128271
119. B. Li, G. He, X. Jiang, Y. Dai, X. Ruan, Pressure swing adsorption/membrane hybrid processes for hydrogen purification with a high recovery. Front. Chem. Sci. Eng. **10**, 255–264 (2016). https://doi.org/10.1007/s11705-016-1567-1
120. L. Lin, Y. Tian, W. Su, Y. Luo, C. Chen, L. Jiang, Techno-economic analysis and comprehensive optimization of anon-sitehydrogen refuelling station system using ammonia: hybrid hydrogen purification with both high H2purity and high recovery. Sustain. Energy Fuels. **4**, 3006–3017 (2020). https://doi.org/10.1039/c9se01111k
121. D. Parsley, R.J. Ciora, D.L. Flowers, J. Laukaitaus, A. Chen, P.K.T. Liu, J. Yu, M. Sahimi, A. Bonsu, T.T. Tsotsis, Field evaluation of carbon molecular sieve membranes for the separation and purification of hydrogen from coal- and biomass-derived syngas. J. Memb. Sci. **450**, 81–92 (2014). https://doi.org/10.1016/j.memsci.2013.08.008
122. B. Castro-Dominguez, I.P. Mardilovich, R. Ma, N.K. Kazantzis, A.G. Dixon, Y.H. Ma, Performance of a pilot-scale multitube membrane module under coal-derived syngas for hydrogen production and separation. J. Memb. Sci. **523**, 515–523 (2017). https://doi.org/10.1016/j.memsci.2016.10.024
123. L. Schorer, S. Schmitz, A. Weber, Membrane based purification of hydrogen system (MEMPHYS). Int. J. Hydrogen Energy. **44**, 12708–12714 (2019). https://doi.org/10.1016/j.ijhydene.2019.01.108
124. X. He, L. Lei, Z. Dai, Green hydrogen enrichment with carbon membrane processes: techno-economic feasibility and sensitivity analysis. Sep. Purif. Technol. **276**, 119346 (2021). https://doi.org/10.1016/j.seppur.2021.119346
125. Z. Liu, P. Hao, S. Li, X. Zhu, Y. Shi, N. Cai, Simulation and energy consumption comparison of gas purification system based on elevated temperature pressure swing adsorption in

ammonia synthetic system. Adsorption. **26**, 1239–1252 (2020). https://doi.org/10.1007/s10450-020-00224-5

126. L. Tong, P. Bénard, Y. Zong, R. Chahine, K. Liu, J. Xiao, Artificial neural network based optimization of a six-step two-bed pressure swing adsorption system for hydrogen purification. Energy AI. **5**, 100075 (2021). https://doi.org/10.1016/j.egyai.2021.100075
127. J.W. Phair, R. Donelson, Developments and design of novel (non-palladium-based) metal membranes for hydrogen separation. Ind. Eng. Chem. Res. **45**, 5657–5674 (2006). https://doi.org/10.1021/ie051333d
128. J.Y. Han, C.H. Kim, S.H. Kim, D.W. Kim, Development of Pd alloy hydrogen separation membranes with dense/porous hybrid structure for high hydrogen perm-selectivity. Adv. Mater. Sci. Eng. **2014** (2014). https://doi.org/10.1155/2014/438216
129. J. Phair, S. Badwal, T.C. Scientific, Materials for separation membranes in hydrogen and oxygen production and future power generation (2006). https://doi.org/10.1016/j.stam.2006.11.005
130. D. Shi, X. Yu, W. Fan, V. Wee, D. Zhao, Polycrystalline zeolite and metal-organic framework membranes for molecular separations. Coord. Chem. Rev. **437**, 213794 (2021). https://doi.org/10.1016/j.ccr.2021.213794
131. R.G. Golledge, The solution-diffusion model: a review. J. Memb. Sci. **107**, 1–21 (1995). https://doi.org/10.1016/S0166-4115(08)60038-2
132. I. Pinnau, L.G. Toy, Transport of organic vapors through poly (1-trimethylsilyl-1-propyne). J. Memb. Sci. **116**, 199–209 (1996). https://doi.org/10.1016/0376-7388(96)00041-5
133. C. Li, S.M. Meckler, Z.P. Smith, J.E. Bachman, L. Maserati, J.R. Long, B.A. Helms, Engineered transport in microporous materials and membranes for clean energy technologies. Adv. Mater. **30**, 0–33 (2018). https://doi.org/10.1002/adma.201704953
134. M. Liu, P.A. Gurr, Q. Fu, P.A. Webley, G.G. Qiao, Two-dimensional nanosheet-based gas separation membranes. J. Mater. Chem. A. **6**, 23169–23196 (2018). https://doi.org/10.1039/C8TA09070J
135. A. Huang, Y. Chen, N. Wang, Z. Hu, J. Jiang, J. Caro, A highly permeable and selective zeolitic imidazolate framework ZIF-95 membrane for H2/CO2 separation. Chem. Commun. **48**, 10981–10983 (2012). https://doi.org/10.1039/c2cc35691k
136. Y. Liu, J.B. Friesen, E.M. Grzelak, Q. Fan, T. Tang, K. Durić, B.U. Jaki, J.B. McAlpine, S.G. Franzblau, S.N. Chen, G.F. Pauli, Sweet spot matching: a thin-layer chromatography-based countercurrent solvent system selection strategy. J. Chromatogr. A. **1504**, 46–54 (2017). https://doi.org/10.1016/j.chroma.2017.04.055
137. C. Lutz, M. Kollmann, C. Bechinger, Single-file diffusion of colloids in one-dimensional channels. Phys. Rev. Lett. **93**, 625–628 (2004). https://doi.org/10.1103/PhysRevLett.93.026001
138. T. Huang, T. Puspasari, S.P. Nunes, K.V. Peinemann, Ultrathin 2D-layered cyclodextrin membranes for high-performance organic solvent nanofiltration. Adv. Funct. Mater. **30**, 1–8 (2020). https://doi.org/10.1002/adfm.201906797
139. X. Zou, G. Zhu, Microporous organic materials for membrane-based gas separation. Adv. Mater. **30**, 1–13 (2018). https://doi.org/10.1002/adma.201700750
140. Z. Xu, Z. Fan, C. Shen, Q. Meng, G. Zhang, C. Gao, Porous composite membrane based on organic substrate for molecular sieving: Current status, opportunities and challenges. Adv. Membr. **2**, 100027 (2022). https://doi.org/10.1016/j.advmem.2022.100027
141. R. Xu, L. He, L. Li, M. Hou, Y. Wang, B. Zhang, C. Liang, T. Wang, Ultraselective carbon molecular sieve membrane for hydrogen purification. J. Energy Chem. **50**, 16–24 (2020). https://doi.org/10.1016/j.jechem.2020.03.008
142. N. Sazali, A review of the application of carbon-based membranes to hydrogen separation. J. Mater. Sci. **55**, 11052–11,070 (2020). https://doi.org/10.1007/s10853-020-04829-7

143. S. Bellini, G. Azzato, Y. Sun, F. Gallucci, A. Caravella, 14 – Metal membranes in hydrogen separation and purification, in *Current Trends and Future Developments on (Bio-) Membranes*, (Elsevier Inc., 2020). https://doi.org/10.1016/B978-0-12-817384-8.00014-5
144. M.R. Rahimpour, F. Samimi, A. Babapoor, T. Tohidian, S. Mohebi, Palladium membranes applications in reaction systems for hydrogen separation and purification: a review. Chem. Eng. Process. Process Intensif. **121**, 24–49 (2017). https://doi.org/10.1016/j.cep.2017.07.021
145. D. Alique, D. Martinez-Diaz, R. Sanz, J.A. Calles, Review of supported Pd-based membranes preparation by electroless plating for ultra-pure hydrogen production. Membranes (Basel). **8** (2018). https://doi.org/10.3390/membranes8010005
146. Y. Li, X. Zhang, A. Yang, C. Jiang, G. Zhang, J. Mao, Q. Meng, Polyphenol etched ZIF-8 modified graphene oxide nanofiltration membrane for efficient removal of salts and organic molecules. J. Memb. Sci. **635** (2021). https://doi.org/10.1016/j.memsci.2021.119521
147. F. Liao, Z. Xu, Z. Fan, Q. Meng, B. Lv, X. Ye, C. Shen, G. Zhang, Confined assembly of ultrathin dual-functionalized Z-MXene nanosheet intercalated GO nanofilms with controlled structure for size-selective permeation. J. Mater. Chem. A. **9**, 12236–12,243 (2021). https://doi.org/10.1039/d1ta01514a
148. H. Fan, M. Peng, I. Strauss, A. Mundstock, H. Meng, J. Caro, MOF-in-COF molecular sieving membrane for selective hydrogen separation. Nat. Commun. **12**, 1–10 (2021). https://doi.org/10.1038/s41467-020-20298-7
149. W. Li, Y. Zhang, P. Su, Z. Xu, G. Zhang, C. Shen, Q. Meng, Metal-organic framework channelled graphene composite membranes for H_2/CO_2 separation. J. Mater. Chem. A. **4**, 18747–18752 (2016). https://doi.org/10.1039/c6ta09362k
150. H. Yang, L. Yang, H. Wang, Z. Xu, Y. Zhao, Y. Luo, N. Nasir, Y. Song, H. Wu, F. Pan, Z. Jiang, Covalent organic framework membranes through a mixed-dimensional assembly for molecular separations. Nat. Commun. **10**, 1–10 (2019). https://doi.org/10.1038/s41467-019-10157-5
151. Y. Ying, D. Liu, J. Ma, M. Tong, W. Zhang, H. Huang, Q. Yang, C. Zhong, A GO-assisted method for the preparation of ultrathin covalent organic framework membranes for gas separation. J. Mater. Chem. A. **4**, 13444–13,449 (2016). https://doi.org/10.1039/c6ta04579k
152. M.D. Dolan, Non-Pd BCC alloy membranes for industrial hydrogen separation. J. Memb. Sci. **362**, 12–28 (2010). https://doi.org/10.1016/j.memsci.2010.06.068
153. L. Wei, J. Yu, Y. Huang, Silver coating on porous stainless steel substrate and preparation of H2-permeable palladium membranes. Int. J. Hydrogen Energy. **38**, 10833–10838 (2013). https://doi.org/10.1016/j.ijhydene.2013.02.137
154. S.A.R.K. Deshmukh, S. Heinrich, L. Mörl, M. van Sint Annaland, J.A.M. Kuipers, Membrane assisted fluidized bed reactors: Potentials and hurdles. Chem. Eng. Sci. **62**, 416–436 (2007). https://doi.org/10.1016/j.ces.2006.08.062
155. S. Yolcular, Hydrogen recovery from methylcyclohexane as a chemical hydrogen carrier using a palladium membrane reactor. Energy Sources Part A Recover. Util. Environ. Eff. **38**, 2148–2152 (2016). https://doi.org/10.1080/15567036.2015.1030476
156. X. Li, A. Li, C.J. Lim, J.R. Grace, Hydrogen permeation through Pd-based composite membranes: effects of porous substrate, diffusion barrier and sweep gas. J. Memb. Sci. **499**, 143–155 (2016). https://doi.org/10.1016/j.memsci.2015.10.037
157. L.P. Didenko, V.I. Savchenko, L.A. Sementsova, P.E. Chizov, Some peculiarities of dehydrogenation of propane and n-butane in a combined membrane reactor with the AOK-73-24 industrial alumina–chromia catalyst and a palladium membrane. Pet. Chem. **56**, 459–464 (2016). https://doi.org/10.1134/S0965544116040034
158. T.F. Fuerst, E.P. Petsalis, C.A. Wolden, J.D. Way, Application of TiC in Vanadium-based hydrogen membranes. Ind. Eng. Chem. Res. **57**, 16084–16,094 (2018). https://doi.org/10.1021/acs.iecr.8b02452
159. R. Checchetto, N. Bazzanella, B. Patton, A. Miotello, Palladium membranes prepared by r.f. magnetron sputtering for hydrogen purification. Surf. Coatings Technol. **177–178**, 73–79 (2004). https://doi.org/10.1016/j.surfcoat.2003.06.001

160. S. Goto, S. Assabumrungrat, T. Tagawa, P. Praserthdam, The effect of direction of hydrogen permeation on the rate through a composite palladium membrane. J. Memb. Sci. **175**, 19–24 (2000). https://doi.org/10.1016/S0376-7388(00)00405-1
161. W. Liang, R. Hughes, The effect of diffusion direction on the permeation rate of hydrogen in palladium composite membranes. Chem. Eng. J. **112**, 81–86 (2005). https://doi.org/10.1016/j.cej.2005.06.010
162. D. Pizzi, R. Worth, M. Giacinti Baschetti, G.C. Sarti, K.-i. Noda, Hydrogen permeability of 2.5 μm palladium-silver membranes deposited on ceramic supports. J. Memb. Sci. **325**, 446–453 (2008). https://doi.org/10.1016/j.memsci.2008.08.020
163. Y.S. Cheng, K.L. Yeung, Effects of electroless plating chemistry on the synthesis of palladium membranes. J. Memb. Sci. **182**, 195–203 (2001). https://doi.org/10.1016/S0376-7388(00)00563-9
164. M.J. den Exter, The use of electroless plating as a deposition technology in the fabrication of palladium-based membranes, in *Palladium Membrane Technology for Hydrogen Production, Carbon Capture and Other Applications*, (2014), pp. 43–67. https://doi.org/10.1533/9781782422419.1.43
165. G. Xomeritakis, Y.S. Lin, Fabrication of thin metallic membranes by MOCVD and sputtering. J. Memb. Sci. **133**, 217–230 (1997). https://doi.org/10.1016/S0376-7388(97)00084-7
166. K.L. Yeung, S.C. Christiansen, A. Varma, Palladium composite membranes by electroless plating technique. Relationships between plating kinetics, film microstructure and membrane performance. J. Memb. Sci. **159**, 107–122 (1999). https://doi.org/10.1016/S0376-7388(99)00041-1
167. M. Kajiwara, S. Uemiya, T. Kojima, E. Kikuchi, Hydrogen permeation properties through composite membranes of platinum supported on porous alumina. Catal. Today. **56**, 65–73 (2000). https://doi.org/10.1016/S0920-5861(99)00263-1
168. T. Muneshwar, M. Miao, E.R. Borujeny, K. Cadien, *Atomic Layer Deposition: Fundamentals, Practice, and Challenges* (Elsevier Inc., 2018). https://doi.org/10.1016/B978-0-12-812311-9.00011-6
169. M. Becker, M. Sierka, Atomistic simulations of plasma-enhanced atomic layer deposition. Materials (Basel). **12**, 1–11 (2019). https://doi.org/10.3390/ma12162605

Chapter 3
Recent Advance Development of ALD in Designing Catalytic Materials for Hydrogen Production Processes

3.1 Introduction

Nanotechnology plays a significant part in the ongoing research and development of catalytic materials for hydrogen production. Thus, the aims of this research is to increase the catalyst's performance and to raise the catalyst's intrinsic activity level and active site [1]. The research that has been done in this field has resulted in the development of various techniques for producing nanomaterials that can range anywhere from 1 to 100 nm in size. These include, for example, processes like sputter deposition, molecular beam epitaxy, electric arc deposition, and chemical vapor deposition [2]. Some processes are based on a chemical solution, such as hydrothermal synthesis, electrodeposition, sonochemical synthesis, and sol-gel processing. Wet chemistry fabrication processes can frequently be economically feasible at smaller scales because they generally do not require very complex instruments. This is because they produce less waste. Wet chemistry, on the other hand, might make it significantly more difficult to achieve precise control over the chemical composition of the catalyst and an evenly distributed coverage of 3D supports. This is due to the fact that wet chemistry typically involves the use of aqueous solutions.

In order to synthesize material-based catalyst thin films, atomic layer deposition (ALD) has been used as a nanofabrication tool due to its exceptional atomic-scale control and good-to-excellent coverage of three-dimensional structures. Because the reactions involved in ALD are self-limiting, the deposited amount can be digitally controlled, the material can be deposited uniformly into complex porous structures, the chemical composition of the material can be controlled, and the process is reproducible [3]. In addition, ALD's potential utility for benchmarking material-based catalysts is enhanced by the fact that material loading can be controlled, and results are reliably reproduced. The ALD process can be employed in a feedback loop to create catalytic materials with rational design (see Fig. 3.1). The first step is

P. O. Oviroh et al., *Emerging Atomic Layer Deposition for Hydrogen Energy*,
Springer Series in Materials Science 348,
https://doi.org/10.1007/978-3-031-67774-8_3

the synthesis of model catalysts with carefully regulated distributions of active sites, which may be utilized to better understand reaction processes, catalytic basics, and structure–property correlations. Once the fundamentals are understood, optimally tailored catalytic structures may be synthesized with the help of ALD due to the high degree of control it provides.

In recent years, it has been clear that ALD can be used for the self-limiting synthesis of catalysts in nanostructured environments. Examples of this include the use of ALD to inhibit metal growth on substrates [4], the creation of core-shell catalysts and alloy [4–6], and the development of functional and protective overcoating [7–9], to name just a few. These novel catalysts are more stable even under the high temperature and aqueous conditions required for biomass conversion, and they exhibit increased selectivity for individual catalytic reactions [10]. The use of ALD-derived overcoats has allowed the synthesis of base-metal nanoparticle

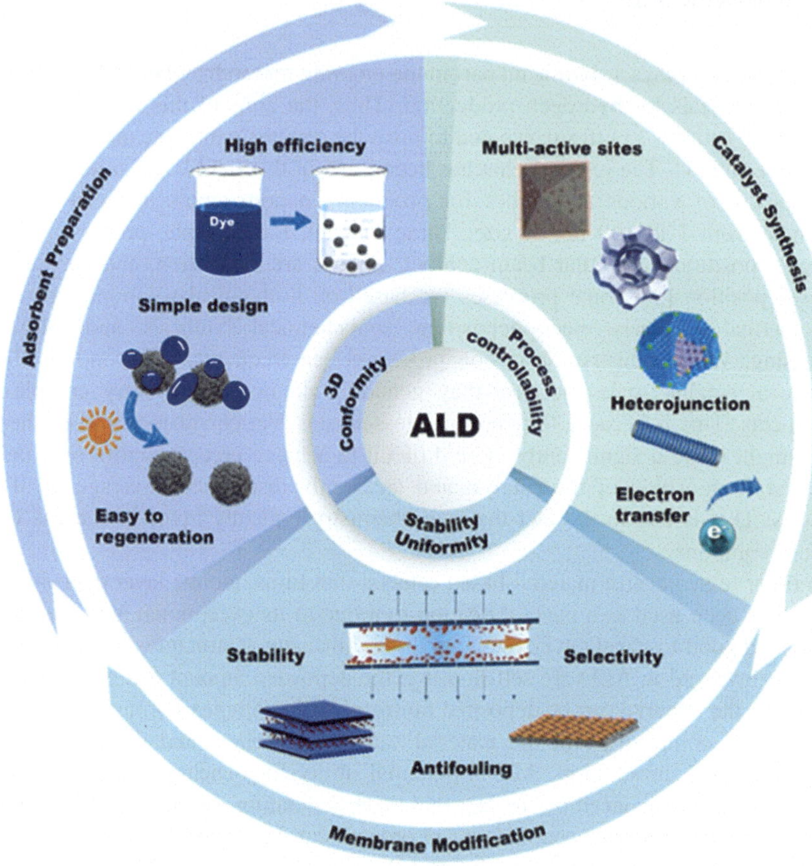

Fig. 3.1 Schematics of the ALD procedure for hydrogen-producing catalysts

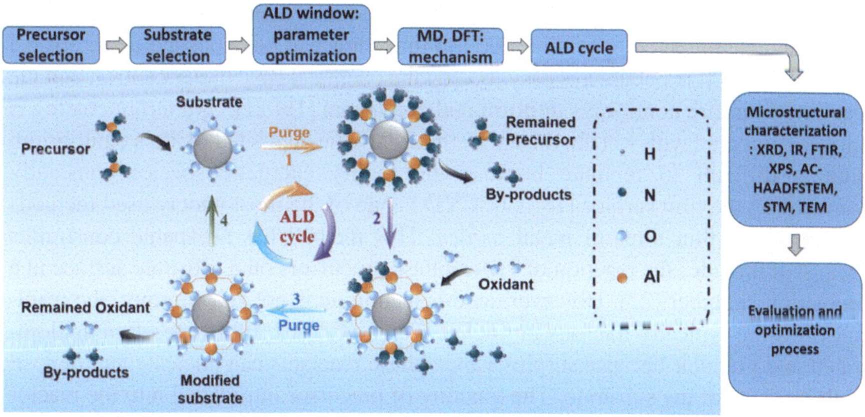

Fig. 3.1 (continued)

catalysts with enhanced stability and selectivity at a reduced cost [10–14]. Furthermore, ALD could be used to prepare extremely selective and active bimetallic catalysts in which the precise position of each metal is controlled. In addition, the atomically precise ALD synthesis combined with state-of-the-art ex situ and in situ spectroscopic techniques (SAXS, STM, TEM, EXAFS, XANES, etc.) allowed for a comprehensive structural and electronic characterization of the catalytic materials, laying the groundwork for the development of novel applications. As a result, ALD can potentially be a pivotal nanofabrication technique for both commercializing already existing catalysts for producing hydrogen and creating brand new ones [10]. The next several paragraphs will elaborate on the recent developments of ALD in hydrogen production technologies.

3.2 ALD in Solar Thermochemical Processes

In order to transform the energy from the sun into the chemical energy of hydrogen, scientists have developed solar-driven thermochemical water-splitting cycles that use a solar reactor system in conjunction with catalytic water dissociation processes. This technique is seen as promising to efficiently manufacture clean and renewable solar fuels. Two key components are necessary for thermochemical water-splitting cycles to be successful: the redox catalyst and the solar reactor system. The catalyst reduces the high temperatures required for direct water thermolysis, separates the processes of O_2 and H_2 evolution into distinct steps, and makes it easier to develop solar reactors. The reactor system drives the endothermic water splitting reaction, which provides the high thermal heating temperature necessary for doing so using concentrated solar energy. Plasma spray deposition, sputtering, Chemical vapor deposition (CVD), and different wet chemistry processes like sol-gel deposition are all used for thin film deposition [15–20]. However, these techniques have drawbacks that prevent them from providing fine-tuned control over film thickness, shape, and chemical composition. 100 nm thin films can be deposited using sol-gel techniques, but it is not always easy to keep the chemical state under control, and the resulting films are not always uniform and conformal [19–21]. Sputtering coating is thinner compared with wet chemistry methods, but the deposited oxide's equilibrium state is difficult to regulate because the highly energetic species constantly bombarding the film surface [18, 19]. CVD is one of the most widely used methods for depositing thin films of metal oxides. This method has no spatial constraints because it includes the reaction of vapor-phase precursors on a substrate surface in a deposition chamber [22]. However, as this technique is not self-limiting, the resultant films are typically microns in thickness or more [23]. Films are often agglomerated and granular because surplus vapor-phase reactants can nucleate and deposit on the surface of the substrate. The quantity of precursor introduced into the reactor for deposition is also sensitive to its vapor pressure [24], making it challenging to regulate film stoichiometry.

In order to produce hydrogen using thermochemical processes, ALD has been employed to manufacture thin films of metal oxide that can act as reactive intermediates. With ALD, you can deposit extremely conformal films one molecule at a time, as the method is self-limiting [25–28]. The basic idea behind deposition, as shown below, is to split a common CVD reaction called a binary reaction in half and expose the gas-phase precursors to a surface in opposite directions.

$$x \cdot \mathrm{M(L)}_n + y \cdot \mathrm{H_2O} \rightarrow \mathrm{M}_x\mathrm{O}_y + n \cdot (\mathrm{HL}) \tag{3.1}$$

$$[\mathrm{S}]: (\mathrm{OH})_y^* + y(\mathrm{ML}_n) \rightarrow [\mathrm{S}]: \mathrm{O}_y\mathrm{M}_x(\mathrm{L})_{n-y}^* + y(\mathrm{HL}) \tag{3.2}$$

$$[\mathrm{S}]: \mathrm{O}_y\mathrm{M}_x(\mathrm{L})_{n-y}^* + y \cdot \mathrm{H_2O} \rightarrow [\mathrm{S}]: \left(\mathrm{M}_x\mathrm{O}_y\right): (\mathrm{OH})_y^* + (n-y)(\mathrm{HL}) \tag{3.3}$$

The metal-to-oxide ratio is denoted by x:y, a surface site is denoted by S, the number of ligands and metal species is denoted by L and M, respectively, and active surface species is denoted by * [29]. The operating parameters are adjusted so that each half-reaction's precursors react exclusively with their counterparts. Since each half-reaction can only contribute one submonolayer of species, this guarantees atomic-layer control [30].

3.2.1 Deposition of Reactive Intermediate Redox Catalysts

In solar-driven of two-step thermochemical cycles of water splitting, redox catalysts play a significant role as one of the most crucial components. The redox catalysts that are utilized have a significant impact on the most important aspects of two-step thermochemical solar energy conversion methods. These aspects include the temperature at which thermal heating occurs, the design of solar reactors, the efficiency with which solar energy is converted into fuel, and large-scale practical applications.

Since the 1960s [31], researchers have been conducting screening tests and looking for suitable redox catalysts. In 1977, Nakamura reported the discovery of the first functional redox catalyst composed of Fe_3O_4 [32]. For the manufacture of thermochemical solar fuel, many different types of metal oxide-based redox catalysts have been created over the course of the past several decades. Currently, the redox catalysts that have been subjected to the most research can be broken down into two distinct groups: those that are based on volatile metal oxides, such as ZnO and SnO_2, and those that are based on nonvolatile metal oxides, such as ferrites, perovskites, hercynites, ceria, and doped ceria.

3.2.2 *Metal Oxide Redox Catalysts*

ALD has been the driving force behind the fast development of thermochemical water splitting cycles that use nonvolatile metal oxide-based redox catalysts in recent years. In contrast to volatile metal oxide-based redox catalysts, nonvolatile redox catalysts maintain their condensed state throughout the thermochemical water-splitting process. This allows them to sidestep the need for an additional step that involves quick quenching and circumvents an undesired recombination problem. The nonvolatile types of redox catalysts that have received the most research attention are hercynite, perovskite oxides, ferrites, Fe_3O_4, CeO_2, and materials based on CeO_2.

It has been established that ALD may be used to deposit iron oxide on a wide variety of crystalline surfaces by utilizing $Fe(acac)_3$ and $Fe(thd)_3$ as precursors [33, 34]. In recent years, oxygen and $Fe(C_5H_5)_2$ (ferrocene) have been utilized in the process of depositing iron oxide coatings onto anodic aluminum and Si(100) [35]. The growth rate of Si(100) was 0.14 nm per cycle, while anodic aluminum experienced a growth rate of 0.06 nm per cycle. When the temperatures for deposition were lower than 500 °C, a combination of hematite and another phase could not be characterized. Iron oxide ALD on zirconia nanoparticles was explored by Scheffe et al., employing ferrocene and oxygen. Oxygen and ferrocene were injected into the reactor in alternating fashion at temperatures ranging from 367 °C to 534 °C. Analysis with induced coupled plasma–atomic emission spectroscopy and in situ mass spectrometry indicated self-limiting chemical reactions in the sample. Through the use of high-resolution transmission electron microscopy, both the conformality and uniformity of the film were successfully examined, and its rate of growth was calculated as 0.15 per cycle. There is a lot of interest in using iron oxide and cobalt ferrites as intermediates and catalysts in chemical looping combustion processes and thermal redox cycles. It has been established that these materials are effective catalysts that facilitate the breakdown of nitrogen oxides (NO_x) [35, 36], different organic contaminants [37, 38], the biomass tars decomposition [39], and the formation of H_2 through water gas shift reaction [40, 41]. In addition, they have been utilized as support for catalysts in substantial environmental catalysis operations. This is done in order to capitalize on their magnetic properties, which in turn helps with the operations of separation units further downstream [42]. In these applications, the magnetic properties of the catalysts have been used in exercising advantage of the magnetic properties of the supports. Utilizing precursors such as $Co(thd)_2$ [43, 44], CoI_2 [45], and bis(N,N′-diisopropylacetamidinato)cobalt(II) [46], one is able to build cobalt oxide thin films on a wide variety of different surfaces. In addition, thin films iron oxide have been formed using ferrocene [45, 47], iron(III) tertbutoxide [48], $Fe(thd)_3$ [33], $Fe(thd)_2$ [49], $Fe(acac)_3$ [34], and bis(N,N′-79 diisopropylacetamidinato)iron(II) [46]. Alternating doses of selected organometallic precursors that are in the right proportions can be used to deposit various stoichiometries of $CoxFe_3\text{-}xO_4$, as stated in [50].

Scheffe et al. [51] used the ALD approach to deposit ferrites on the ZrO_2 support surface. Additionally, they used this process to produce Co-ferrite films on the ZrO_2 accurately supports. The surface area was increased by synthesizing nano-sized ZrO_2 particles that were approximately 50 nm in size (Fig. 3.2a, b, e, and f). Then, using the ALD technique, a very thin layer of Co-ferrite film with a thickness of 2 nm was successfully formed on the ZrO_2 support surface (Fig. 3.2c, d, g, and h). The robust ALD-Co-ferrite film that was produced showed sustained catalytic activity throughout the thermochemical water-splitting cycles. In light of these findings, Scheffe and colleagues proposed that the ALD approach be used for fabricating additional ferrite systems on porous substrates.

ALD was used by Scheffe et al. [52] to create thin films of cobalt ferrite ($Co_xFe_{3-x}O_4$) and iron oxide ($\gamma\text{-}Fe_2O_3$) on m-ZrO_2 substrates with a substantial surface area (50 m^2/g). Cobalt oxide and iron oxide were deposited in a certain order in order to produce the oxide layers. To achieve the desired stoichiometry, an adjustment was made to the ratio of cobalt oxide cycles to iron oxide cycles. A chemical reduction

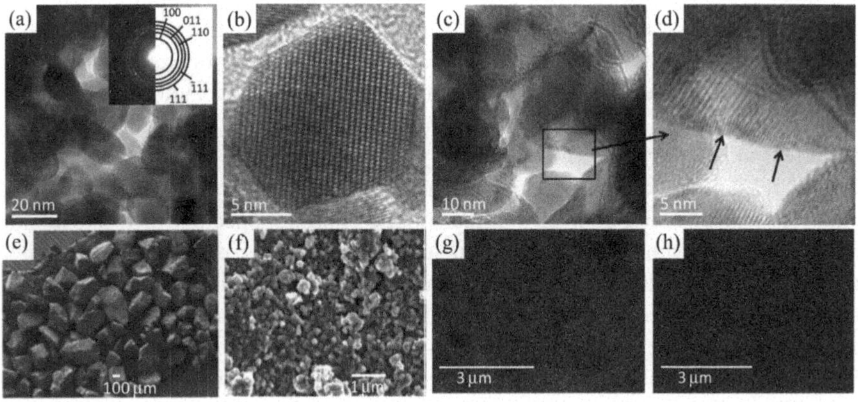

Fig. 3.2 Thin film deposition of ferrite materials. (**a–d**) showing the HRTEM images of (**a**) the ZrO_2 substrate and matching diffraction pattern, (**b**) ZrO_2 nanoparticle embedded in the substrate, (**c**) deposition of Co-ferrite, and (**d**) a crystalline Co-ferrite film deposited on m-ZrO_2. (**e–f**) showing the FESEM images of (**e**) low magnification of bulk m-ZrO_2 supports, (**f**) m-ZrO_2 nanoparticles inside the bulk support. (**g–h**) showing the EDS images of (**g**) Zr map after $Co_{0.85}Fe_{2.15}O_4$ deposition, (**h**) Fe map after $Co_{0.85}Fe_{2.15}O_4$ deposition [51]

procedure was performed on the samples while they were transported through in situ x-ray diffraction-equipped flow rector. They were also put through a reduction process and chemical reduction in a stagnation flow reactor as part of the process of determining whether or not the material's activity could be utilized in chemical looping cycles that were designed to produce hydrogen through the process of splitting water. This was done to determine whether or not the material could be used. When γ-Fe_2O_3 films were chemically reduced in mixtures of hydrogen, carbon monoxide, and carbon dioxide at 600 °C, they produced the phases Fe_3O_4 and FeO.

Additionally, the hydrogen production rates of the films decreased with time in a discernible manner. $Co_{0.85}Fe_{2.15}O_4$ films efficiently cycled without getting deactivated, and they generated hydrogen four times as much γ-Fe_2O_3 did. The formation of a CoFe alloy during the reduction process was primarily responsible for this result. A mixture of α-Fe_2O_3 and ZrO_2 powders that were physically milled could not maintain significant water splitting activity owing to grain formation and sintering. Iron loading was present in the same proportion in the mixture as it was in the thin films.

Scheffe et al. [52] also deposited $CoFe_2O_4$ on an alumina substrate using atomic layer deposition (Fig. 3.3a). The results of their research showed that there was a

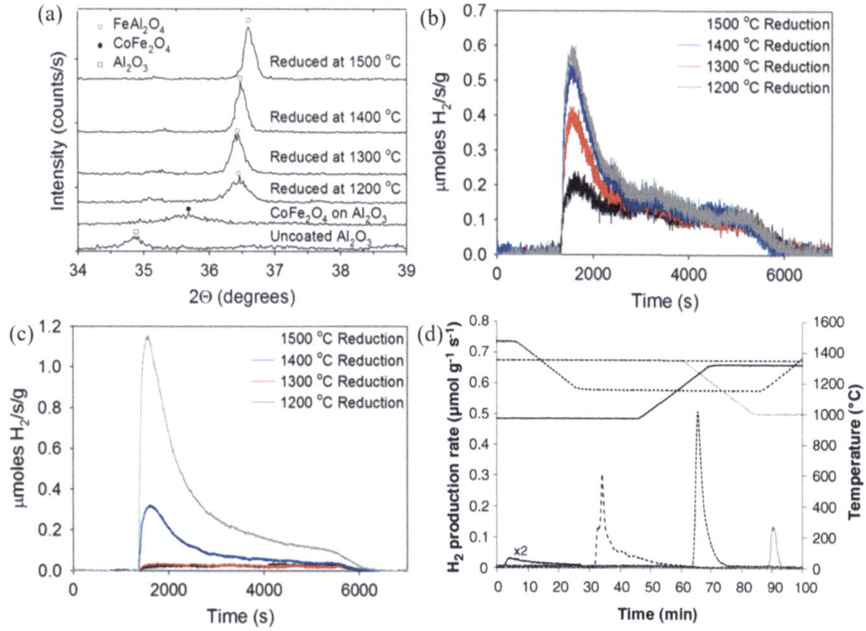

Fig. 3.3 (**a**) $CoFe_2O_4$ powder XRD data deposited on Al_2O_3 substrate at various reduction temperatures. (**b–d**) H_2 reaction rate as a function of temperature for (**a**) $CoFe_2O_4$ deposited on Al_2O_3, (**b**) $CoFe_2O_4/ZrO_2$, and (**c**) isothermal H_2O splitting with a hercynite catalyst (short dashed line) and conventional two-step thermochemical H_2O splitting with a hercynite (solid line) and CeO_2 catalyst (dotted line) [54]

reduction in temperature between 200 °C and 300 °C when $CoFe_2O_4$ was deposited on Al_2O_3. $CoFe_2O_4/Al_2O_3$ was predicted to produce nearly four times as much hydrogen as $CoFe_2O_4$ at 1200 °C/1000 °C, with no evidence of sintering or deactivation after having undergone many cycles (Fig. 3.3b, c). Muhich et al. [53, 54] created a screening strategy to determine whether the suggested redox materials are feasible. This approach was used to create the $FeAl_2O_4$, $CoAl_2O_4$, and doped-hercynite cycle. When compared to a stoichiometric procedure, it was demonstrated that the spinel materials belonging to the aluminate family conduct a redox reaction through an O-vacancy mechanism. In terms of H_2 generating capacity, it was hypothesized that cobalt aluminate and iron spinels would be organized as follows: $CoAl_2O_4 < Co_{0.5}Fe_{0.5}Al_2O_4 < FeAl_2O_4$, with approximate H_2 production capacity ratios of $2 \times 10^4 - 0.7 - 1.0$, respectively. This was anticipated because these spinels are made up of iron and cobalt aluminate. Un-doped hercynite ($FeAl_2O_4$) has a greater H_2 production capability than normal doped-hercynite materials and has been demonstrated to be a visible material for thermochemical water splitting. This was a huge step forward for the industry. Scheffe et al. postulated a novel reaction mechanism in which, during heat reduction, $CoAl_2O_4$ interacts with Al_2O_3 supports to generate a new kind of redox catalyst, $FeAl_2O_4$ (hercynite). In broad strokes, the redox process may be described as follows:

$$CoFe_2O_4 + 3Al_2O_3 + \text{heat} \rightarrow CoAl_2O_4 + 2FeAl_2O_4 + 1/2\ O_2 \quad (3.4)$$

$$CoAl_2O_4 + 2FeAl_2O_4 + H_2O \rightarrow CoFe_2O_4 + 3Al_2O_3 + H_2 \quad (3.5)$$

Figure 3.3d shows the results of isothermal water splitting using hercynite redox catalysts at a 1350 °C constant temperature. The yield of hydrogen gas (H_2) was found to be three times greater than that achieved with hercynite and 12 times higher than that achieved with CeO_2 in conventional two-step thermochemical water splitting cycles (at 1350 °C thermal reduction occurs and reoxidation occurs at 1000 °C). However, hercynites have poor catalytic durability and sluggish reaction kinetics because reactive particles aggregate during thermochemical water-splitting cycles, in contrast to ferrite-type catalysts. Because of these limitations, ferrite-type catalysts are preferred over hercynites. As a result, the key modifications that need to be done in hercynites with the goal for them to find further applications in the area of thermochemistry are to overcome the sluggish reaction kinetics of O_2 and H_2 generation and to increase the long-term catalytic stability.

3.3 ALD in Electrolytic Processes

There has been a lot of research on electrolytic catalysts for hydrogen generation in acidic or alkaline media. Hydrogen overpotential causes significant electrical energy consumption in the electrolytic processes used to generate hydrogen from water [55]. Therefore, efforts to reduce energy use must focus heavily on minimizing overpotential. Stable catalytic performance may be maintained if the breakdown of catalysts in electrodes is reduced or eliminated [56]. So far, much work has been done on developing and producing electrodes using innovative materials and architectures to improve electrocatalytic performance. There are two basic applications of atomic layer deposition (ALD), which is frequently utilized in electrolytic catalysts. The first application involves the deposition of a protective coating on the electrode to avoid catalyst degradation. The second application involves the direct coating of functional materials, e.g., single-atom, transition metals, and TMD catalysts. A relatively recent technique is ALD-oxide nanocatalysts, used as a seed/reactant layer for promoting the MOFs assembly for H_2 evolution. This section provides a concise overview of how the ALD method may be used to create protective layers and functional materials.

3.3.1 Deposition of Active Films

Electrode-active materials in current designs of electrolytic catalysts are severely degraded by electrolytic reactions, leading to a loss of efficiency [57]. As a result, the charge transfer and active materials that occur between the electrode and the electrolyte might potentially be conserved by the utilization of a coating layer that is deposited on electrode surface. The ALD approach has evolved to become the best one for achieving this goal.

In a standard photoelectrolytic H_2 evolution method, for example, catalysts such as diimine-dioxime cobalt complexes (CoC11Ps) and p-type silicon are used; however the performance of the active material becomes unstable due to corrosion caused by electrolytes. Chandrasekaran and colleagues [56] suggested a method for producing a modified electrode to solve this issue (Fig. 3.9a). In the method, a 15 nm ALD TiO_2 layer was used to provide protection for a p-type silicon surface. CoC11P was grafted onto TiO_2 using a covalent bonding technique, and then an extra of 0.2 nm ALD TiO_2 film was deposited for stabilization. Both the shielding and offering of loaded active materials are capabilities of ALD TiO_2 in this setting. Figure 3.4b–d shows a scanning electron micrograph of an electrode that has been treated with ALD TiO_2. This image indicates that the top ALD TiO_2 covering the electrode surface completely. Because of the protection provided by the two ALD TiO_2 layers on the electrode upper and lower surfaces, the newly designed electrode exhibits sequential changes as a result of the intermittent release of hydrogen bubbles from the electrode surface. The presence of ALD TiO_2 coatings on the electrode's upper and lower surfaces is responsible for this. A thick ultra-thin TiO_2 coating is created by ALD, which stabilizes the fixation of the molecular catalyst. This may explain why the ALD-modified electrode has a much improved property.

However, the quick electrodes disintegration in watery conditions or in humid air remains an issue; therefore, halide perovskite photocathodes are still considered an appropriate active material for hydrogen evolution electrodes [57]. To combat this issue, halide perovskite photocathodes have been protected by nanomembranes that

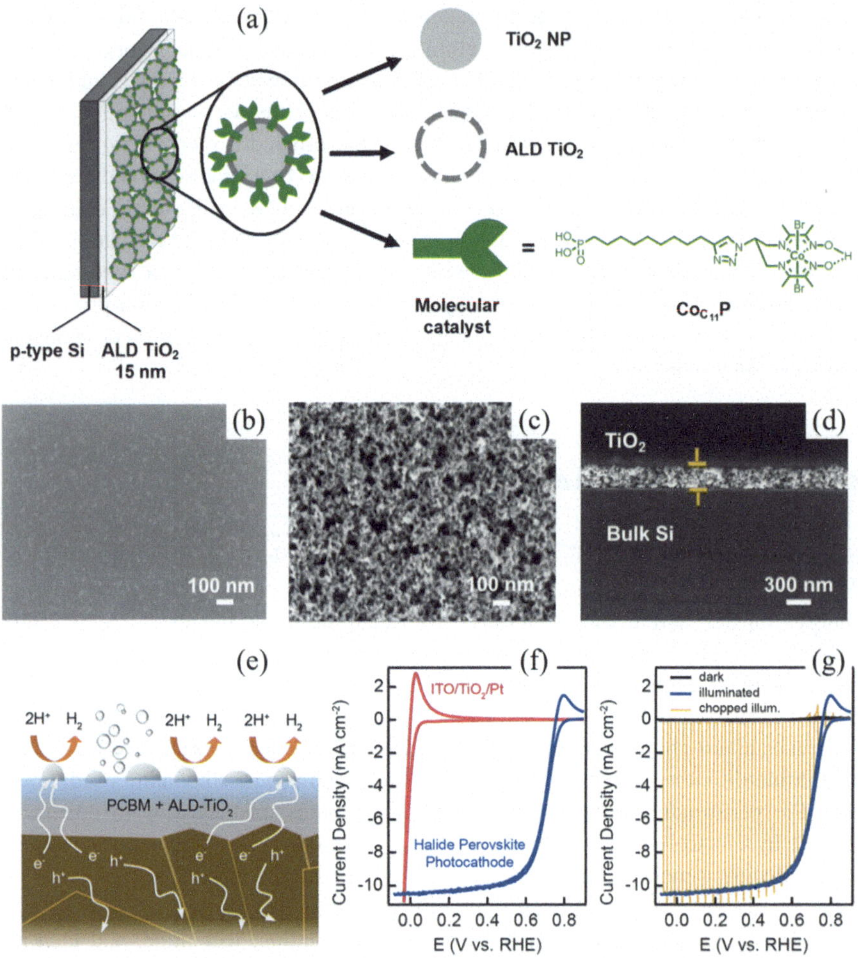

Fig. 3.4 (**a**) Atomically engineered ALD-TiO_2 On p-Si substrate. [(**a**) Reprinted with permission from Ref. [56]]. (**b–d**) SEM images of (**b**) a top view of TiO_2 layers deposited on p-Si: ALD-TiO_2, (**c**) a top view of ALD-TiO_2/SC-TiO_2, and (**d**) cross-sectional view of ALD-TiO_2. (**e–g**) A schematic representation of halide perovskite as ALD protective layer. (**e**) Electrons are transferred from the TiO_2 to the Pt catalysts through the conductivity of the electrons. The Pt catalysts then employ electrons to reduce protons to hydrogen. (**f**) Catalytic performance of the halide perovskite/TiO_2/Pt modified electrode. (**g**) Behavior of the electrode under dark, solar illumination and chopped illumination. [(**b–g**) are reprinted with permission from Ref. [57]]

were produced using ALD. By coating the electrode with a combination of PC61BM and ALD TiO_2 nanomembrane, as shown in Fig. 3.4e, Kim et al. [57] demonstrated a straightforward approach for creating a hybrid passive layer. By itself, the ALD TiO_2 nanomembrane facilitated a decrease in photo-assisted protons. As shown in Fig. 3.4f, g, the device had a filling factor of 65.5% and a current density short-circuit of 21.40 mA cm^{-2}. In addition, a photovoltaic average of 0.68 V was generated. The scientists found that ALD TiO_2's protective properties when the device contacted the electrolytes were responsible for the improved performance. This approach might be used with various hybrid organic light absorbers and semiconductors to create H_2 through photoelectrochemistry.

3.3.2 ALD-Based Active Materials for Electrolytic Processes

Various materials with a wide variety of different morphologies may be synthesized using ALD as active materials. For the evolution of hydrogen, for instance, a single-atom catalyst has garnered a great deal of interest owing to its very high catalytic efficiency. On the other hand, the technique of producing single-atom catalysts is too complicated, which is a barrier to their use in practical applications. Cheng et al. [58] recently developed a high-efficiency and uncomplicated approach to overcome this challenge. This approach may be used to synthesize stable and large-scale isolated single Pt atoms by using the ALD technology. Graphene nanosheets doped with nitrogen were employed as supports for single Pt atoms created using the ALD method. It is important to note that the density and size of Pt catalysts on graphene can be precisely regulated by altering the ALD growth cycles. In Fig. 3.10a, b, a schematic representation of the morphology of platinum that has been deposited on graphene substrate after 50 growth cycles is presented. The graphene supporter was covered with a large number of individual platinum atoms, which appeared as bright spots. By comparing the material to commercial Pt/C and Pt foil, the extended X-ray absorption fine structure (EXAFS) of the material is depicted in Fig. 3.5c. This comparison was done to further validate that the single atom of Pt was generated via ALD. Compared with the Pt foil, the ALD-prepared sample shows a considerable shift in the intensity of Pt–C or Pt–O bonds and a severe deviation in the Pt–Pt bond. These differences suggest the existence of single atoms of Pt. The HER activity of the Pt samples that ALD produced was analyzed using linear sweep voltammetry (LSV) scanning in a solution containing 0.5 M (1 M = 1 mol L^{-1}) H_2SO_4. Because of the single-atom catalyst's strong catalytic activity, the samples displayed remarkable hydrogen evolution catalytic features. For example, the samples had an extremely little overpotential of 0.05 V while operating at a 12.5 mA cm^{-2} current density (Fig. 3.5c). Catalysts containing a single atom of platinum that have been created via ALD have the potential to considerably cut costs associated with the industry's usage of precious metals in catalysts.

Hydrogen evolution often involves nanoparticles-like active materials. For hydrogen evolution catalysts in alkaline conditions, Zhang et al. [59] developed an ALD-based method of synthesizing nickel nanoparticles deposited onto Co-doped carbon particles (Fig. 3.5d, e). The uniform deposition of nickel nanoparticles by ALD techniques enables the formation of a suitable interface between nickel and carbon particles that have been doped with cobalt. This, in turn, results in enhanced electrochemical behavior and an overpotential of 157 mV (Fig. 3.5f), which is significantly lower than the overpotential of pure cobalt-doped carbon. Another

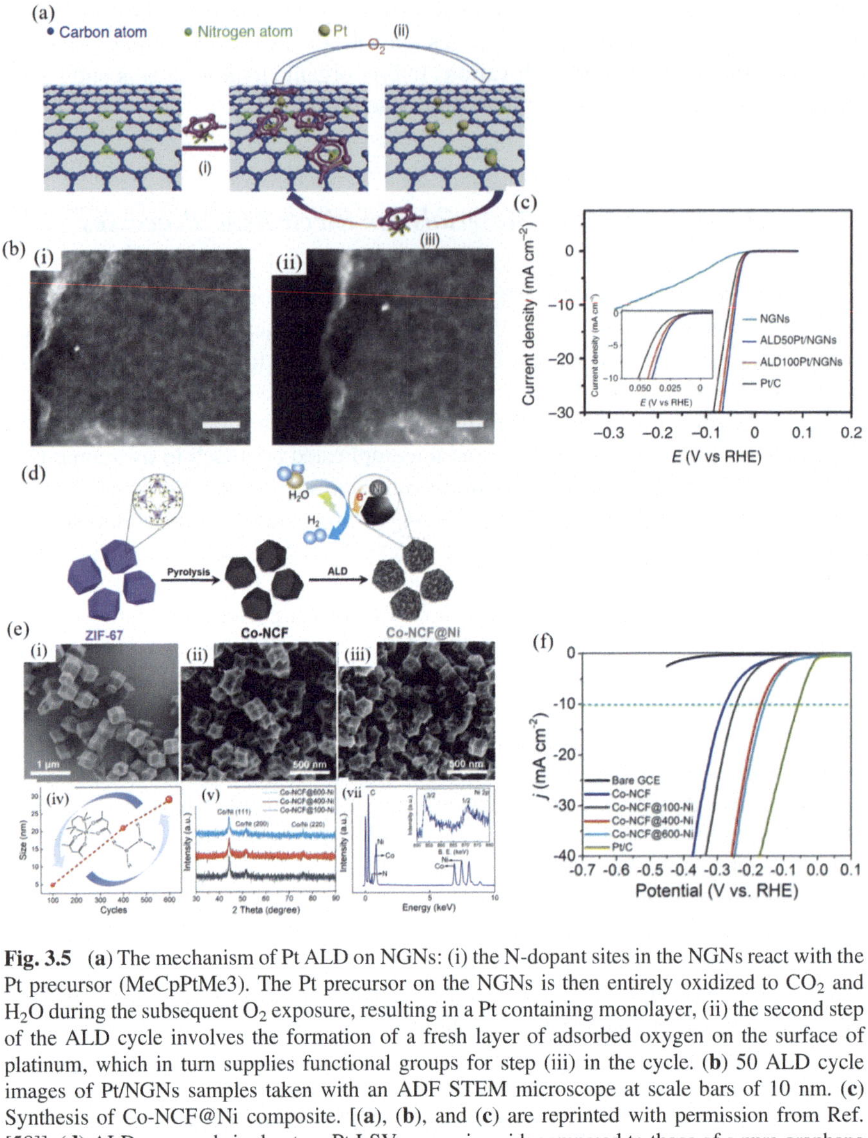

Fig. 3.5 (**a**) The mechanism of Pt ALD on NGNs: (i) the N-dopant sites in the NGNs react with the Pt precursor (MeCpPtMe3). The Pt precursor on the NGNs is then entirely oxidized to CO_2 and H_2O during the subsequent O_2 exposure, resulting in a Pt containing monolayer, (ii) the second step of the ALD cycle involves the formation of a fresh layer of adsorbed oxygen on the surface of platinum, which in turn supplies functional groups for step (iii) in the cycle. (**b**) 50 ALD cycle images of Pt/NGNs samples taken with an ADF STEM microscope at scale bars of 10 nm. (**c**) Synthesis of Co-NCF@Ni composite. [(**a**), (**b**), and (**c**) are reprinted with permission from Ref. [58]]. (**d**) ALD-prepared single-atom Pt LSV curves in acid, compared to those of a pure graphene and commercial Pt/C catalyst at 2 mVs^{-1} scan rate. (**e**) SEM images of (i) ZIF-67 and (ii) Co-NCF, (iii) Co-NCF@600-Ni. (iv) Variation in Ni particle size in response to ALD cycle count, (v) XRD patterns, and (vi) EDX spectrum of Co-NCF@600-Ni. (**f**) LSV curves of pure Co-doped and composite carbon particles. [(**d**), (**e**), and (**f**) are reprinted with permission from Ref. [59]]. (**g**) AS-ALD preparation steps for Co_3O_4 nanotrap-anchored Pt NPs deposited on Al_2O_3 supports. (**h**) The TEM and STEM images of Co_3O_4/Pt/Al_2O_3–600 and Pt/Al_2O_3–600. (**i**) Electrocatalytic reaction rate of Pt/Co_3O_4 nanocomposite. [(**g**), (**h**), and (**i**) are reprinted with permission from Ref. [65]]. (**j**) The effects of H-coverage and vdW interactions on the chemical adsorption of WX_2Y_2 on the WS_2 {010} surface. (**k**) HAADF-STEM images of ALD W2S grown by: (i) pure H_2S plasma and (ii) $H_2 + H_2S$plasma-enriched procedure. (**l**) The thickness-dependent LSV curves of ALD WS_2 electrocatalysts synthesized by H_2S or $H_2 + H_2S$ plasma boosted procedures. [(**j**), (**k**), and (**l**) are reprinted with permission from Ref. [68]]

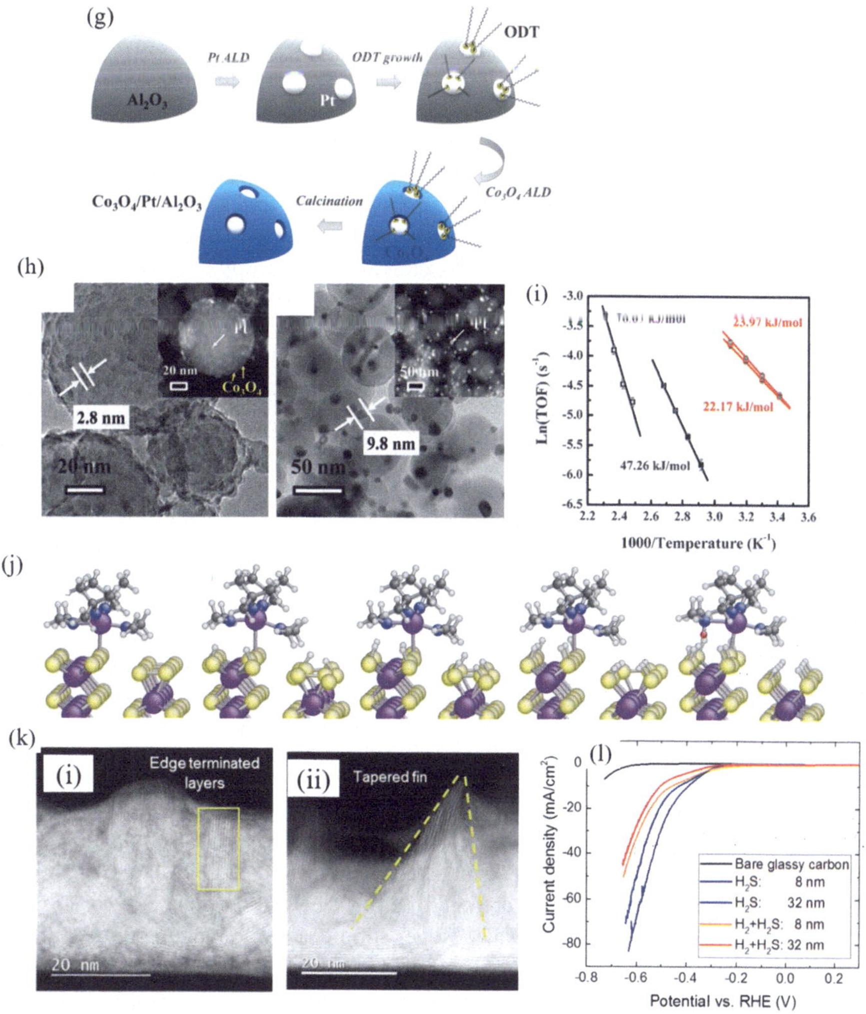

Fig. 3.5 (continued)

group, led by Zhang et al. [60], used ALD to produce 2D $Ti_3C_2T_x$ MXene composites coated with Pt nanoparticles. With ALD's excellent controllability, they were able to reduce Pt consumption to 0.98–3.10 wt.% (wt.% stands for weight percentage). With an overpotential of only 67.8 mV, the evenly distributed Pt particles deposited on MXenes demonstrated outstanding HER catalytic performance and stability. This performance was significant since it was relatively close to the Pt/C catalyst that is used in commercial applications (64.2 mV). Pt coating that was uniform and MXene supports high conductivity were found to be the causes of the high performance level.

Area-selective ALD can be descried as potential ALD approach for synthesizing active nanostructured materials by locally controlling the growth pattern of particular components (Fig. 3.5g). Chen et al. have put in a lot of work on this approach [61–65]. One such innovative structure was made by attaching Pt particles to metal oxide nanotubes [65]. In the first procedure, Al_2O_3 substrates were covered with a layer of Pt particles. The following deposition of Co_3O_4 onto Pt was prevented by using 1-octadecanethiol as a blocking agent. Then, Co_3O_4 was strategically deposited onto Al_2O_3 supports. After the inhibitor was taken away, the Pt particles were once again exposed. Figure 3.5h shows transmission electron microscopy (TEM) pictures that demonstrate an excellent Pt/Co_3O_4 interface and a uniform Pt dispersion with an average particle size of 2 nm. The resulting composites' impressive activation energy values are 22.17 kJ mol^{-1} (Fig. 3.5i).

TMDs are widely studied because of their promising electrochemical characteristics, strong catalytic activity, and ample active site exposure as 2D layered materials. Recent years have seen their widespread use in energy devices, particularly as catalysts for hydrogen production, suggesting that atomic-scale growth control might be crucial. The advantages of ALD as a onepot approach for directly depositing TMDs are low temperature, repeatability, self-limiting process, and the ability to regulate the thickness. Song et al. devised an innovative pressure-tuned stop-flow method using the ALD approach to coat a carbon nanotube array with ~5 nm of MoS_2 layer doped with 3D graphite foam for use as electrode in Li-O_2 batteries [66]. Incorporating ALD-prepared MoS_2, the overall design may reduce the energy barrier, and the one-of-a-kind 3D electrode showed off impressive catalytic activity in both the O_2 evolution and reduction reactions. Kwon et al. [67] devised a simple method of directly depositing the MoS_x film on porous carbon sheets to create high-performance catalysts for HER. Depositing the MoSx film required reactants dimethyldisulphide and molybdenum hexacarbonyl, which, when combined, create Mo and S, respectively, at 100 °C. Catalytic characteristics for the HER were dramatically improved by ALD-prepared MoS_x catalysts, which also displayed extremely low overpotential and Tafel slope, simplifying the manufacture of high-efficiency catalysts for hydrogen generation.

Furthermore, Balasubramanyam et al. [68] used the plasma-enhanced ALD method to increase the reactive edge sites density of TMDs, allowing for more efficient electrochemical hydrogen evolution events. By fine-tuning the H_2 and H_2S plasma atmosphere, they were able to perfectly regulate the shape, composition, and edge site density of the resulting WS_2 layer. Figure 3.5k shows that WS_2 was produced utilizing a pure H_2S process (Fig. 3.5k(i)), resulting in closely packed "nanoflakes" by individual lateral flake widths ranging from 10 to 20 nm. By changing the plasma environment to H_2 + H_2S, WS_2 changed its morphology to 2D stacking lamellar nanosheets (Fig. 3.5k(ii)). The electrochemical catalytic property of this open structure is shown in Fig. 3.5l. Comparing the LSV curves of WS_2 layers of varying thicknesses to those of a clean glassy carbon electrode reveals potential (200 mV) that is low at onset for HER. In addition, the standard current density of overpotential 10 mA cm^{-2} is lower for the sample created with pure H_2S plasma, indicating superior electrochemical catalytic performance for hydrogen generation. The results of this study confirmed that plasma-enhanced ALD is an effective method for producing TMDs to be used as electrochemical catalysts.

3.3.3 ALD-Based Catalysts as a Reactant for Electrolytic Processes

Metal-organic frameworks (MOFs) have received a lot of interest from scientists recently [69–74]. MOFs are three-dimensional (3D), long range organized assemblages of transition metal clusters and their matching organic frameworks [75–79]. Although MOFs have several potential uses in energy devices, their single particle form limits their use in many situations [79–81]. Microwave-facilitated approaches [82], suspension sprays [83], layer-by-layer deposition [84], room-temperature aging [85, 86], and electrodeposition [87] are only a few of the potential ways explored by researchers for producing MOF films. However, its widespread use is limited because of its poor adherence and uneven distribution [88]. Zhao et al. [89–93] came up with a unique approach to overcoming these difficulties by adopting ALD oxide thin membranes as seed/reactant layers to produce MOFs layers deposited targeted substrates. Outstanding homogeneity, conformal deposition, and firm attachment were accomplished with the MOFs, leading to good electrochemical characteristics when they were used in hydrogen evolution catalysts.

MOFs used in electrochemical energy applications are grown adopting oxide-induced growth ALD method depositing on conductive substrates like carbon fiber (Fig. 3.6a) [92]. To be more specific, a thin ZnO nanomembrane was formed as a reactant on carbon fiber surface before the formation of the ZIF-67 film could begin. The ALD ZnO-pretreated carbon fiber has a morphology (Fig. 3.6b) that is very close to that of untreated carbon fiber because of the uniform and conformal deposition. A phase transition may have happened with adding cobaltous nitrate hexahydrate, as seen in Fig. 3.6c, and a zinc/cobalt hydroxy double salt is considered [89]. A new phase formed as a result of a reaction between the organic linker and certain (Zn, Co) hydroxy double salts (Fig. 3.6d). The very active N^{3+} and NH^{4+} ions imbedded in hydroxy double salt inclines to generate fast anion exchanges, while the N^{3+} and NH^{4+} ions in the organic linkers offer active electrons that dissolve the crystal structure. Thus, the lamellar sheets begin to collapse upon exposure to organic linkers, NO^{3-} and OH^{-} are eliminated, and Co^{2+} is released; all of this was continually organized in situ together with 2-MI. In this way, crystals that resemble the elusive target particles in 3D are formed. When the aging process is complete, a ZIF-67 film is produced (Fig. 3.6e). The ZIF-67 layer on melamine foam was prepared similarly to that employed for forming MOF films deposited on carbon substrates [94]. The produced ZIF-67 film adhered uniformly and securely to the melamine foam, as shown in Fig. 3.6f of the SEM analysis. Figure 3.6g shows the X-ray diffraction (XRD) pattern associated with the Co-HDS and Zn layers that are fabricated through the induction process, whereas Fig. 3.6h shows the XRD pattern

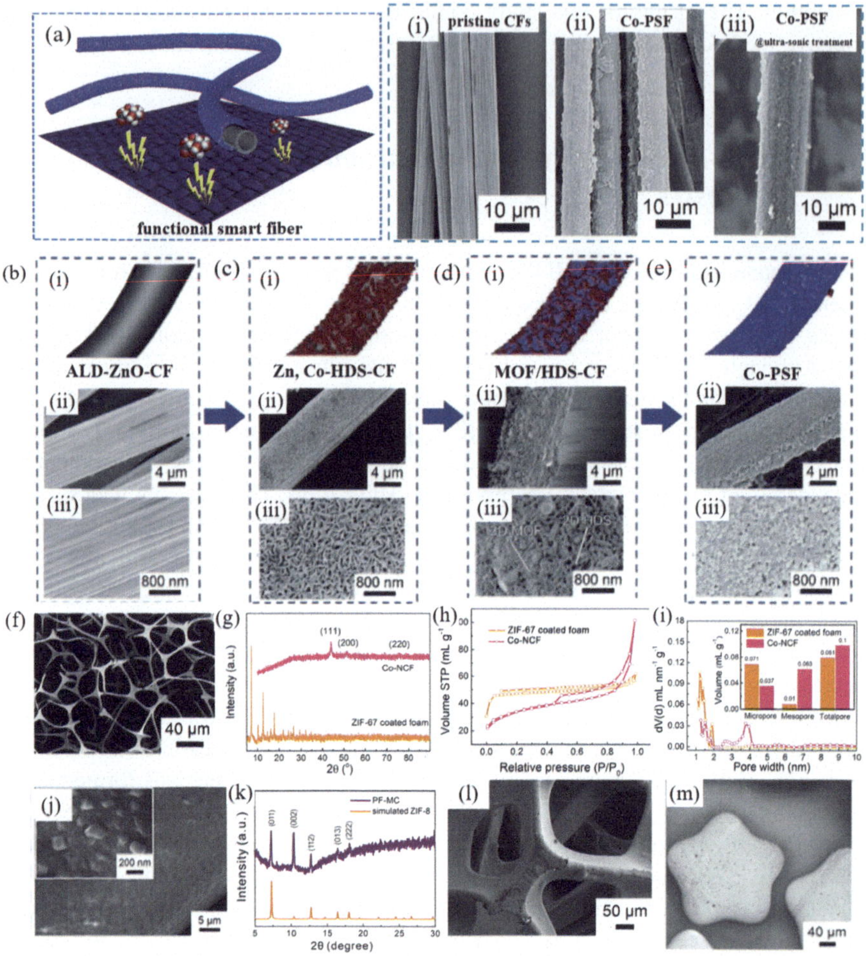

Fig. 3.6 (**a–e**) Schematic depictions of ALD oxide's role in the growth of MOF layers. The corresponding scheme, a low-magnification SEM image, and a high-magnification SEM image are shown in (i), (ii), and (iii), respectively. [(**a–e**) Reprinted with permission from Ref. [92]]. (**f**) A SEM showing the ZIF-67 layer growth on the melamine foam. (**g**) XRD patterns of Co-HDS, Zn, and ZIF-67 growth on melamine foam. (**h**) The adsorption isotherms of Nitrogen. (**i**) ZIF-67 layer growth on melamine foam and pyrolyzed ZIF-67 layer growth on melamine foam. Orange and pink represent the ZIF-67 layer and pyrolyzed ZIF-67 layer deposited on melamine foam. [(**f–i**) are adapted with permission from Ref. [94]]. SEM image (**j**) and XRD pattern (**k**) of ZIF-8 film growth on microtubular structure. [(**j–k**) are adapted with permission from Ref. [95]]. SEM image of the (**k**) PCN-333 pattern and (**l**) ZIF-8 pattern. [(**l–m**) are adapted with permission from Ref. [96]]

associated with the final ZIF-67 layer. Mesopores and micropores coexisting in the film are distinct from conventional MOF particles, and this innovative technique may build a hierarchically porous structure. The electrochemical property of the MOF film formed with the aid of ALD oxide is improved because more active sites are exposed, and ion transit is quicker, both of which may be ascribed to the film stack particles (Fig. 3.6h, i). MOF films may be accurately deposited on micro/nanostructures using ALD's high degree of control. The XRD pattern in Fig. 3.6k [95] confirms the crystal structure of ZIF-8, and the SEM imagine in Fig. 3.6j reveals a uniform ZIF-8 film adhering to the surface of a microtubular structure.

Furthermore, the MOF film may be formed by pattering oxide nanomembranes, since the MOF film development is critically reliant on ALD oxide. Using photolithography and acid etching, a patterned ZnO nanomembrane was created and then used in the experimental MOF film generation procedure. The pattern of PCN-333 (Fig. 3.6l) and the ZIF-8 (Fig. 3.6m) were both created using this method with great success [91]. The potential for on-chip energy devices made from patterned MOF films is high.

In order to investigate the potential of MOF generated through the induction of ALD oxide for use in electrochemical processes, the MOF film was deposited on carbon foam, subjected to high-temperature N_2 treatment. This was accomplished by utilizing a flexible substrate, a three-dimensional carbon foam framework, which provided multiple conductive pathways. Additionally, the ZIF-67 film that was produced (Fig. 3.7a) was carbonized into a hierarchically porous carbon layer (Fig. 3.7b) that was composed of spherical carbon particles, thereby maintaining a high pore volume and large surface area [89]. The ALD oxide layer also prevented Co-doped carbon particles from aggregating during pyrolysis [89], thanks to its tight bond between the active materials and substrates. This method of production, which makes use of ALD oxide layer induction effect, has been shown via experimentation to be capable of achieving an elevated mass loading of active materials by 40%. This excellent electrochemical catalytic performance may be traced back to the composite's high concentration of spherical carbon particles with large numbers of active sites and substantial amounts of Co-N. As shown in Fig. 3.7c, the resulting composite requires only an overpotential of 142 mV in an acid electrolyte to achieve a current density of 10 mA cm^{-2}. Electrodes are able to withstand multiple compressions and releases without losing their electrochemical stability (Fig. 3.7d) because they are supported by elastic and deformation-tolerant carbon foam. This groundbreaking study utilizes ALD to provide a fresh approach to the development of next-generation bendable electrodes for hydrogen production.

This above procedure has been used to prepare many MOFs so far. Using an ALD TiO_2 nanomembrane as one reactant, Zhao et al. [97] created UiO-66-NH_2-particle-doped polyamide-6 nanofibers. After a layer of highly loaded UiO-66-NH_2 particles was formed on top of the nanofibers, the resulting composite had an extremely high specific surface area of 205.9 m^2 g^{-1} to cultivate HKUST-1 MOF crystals. With the ALD oxide layer as the reactant, solvent-free MOF synthesis has also been achieved. ALD ZnO and 2-ethylimidazole powder were used by Stassen et al. [98] to establish a technique to produce ZIF-8 membranes on a silicon wafer. The gap-filling ultralow-k dielectric ZIF-67 was successfully synthesized by Krishtab et al. [99] by making use of ALD CoO and 2-methylimidazole powder. This was accomplished by the utilization of both of these components. In addition, Stassen et al. [100] were able to produce MAF-6 by utilizing 2-ethylimidazole and ALD ZnO in their

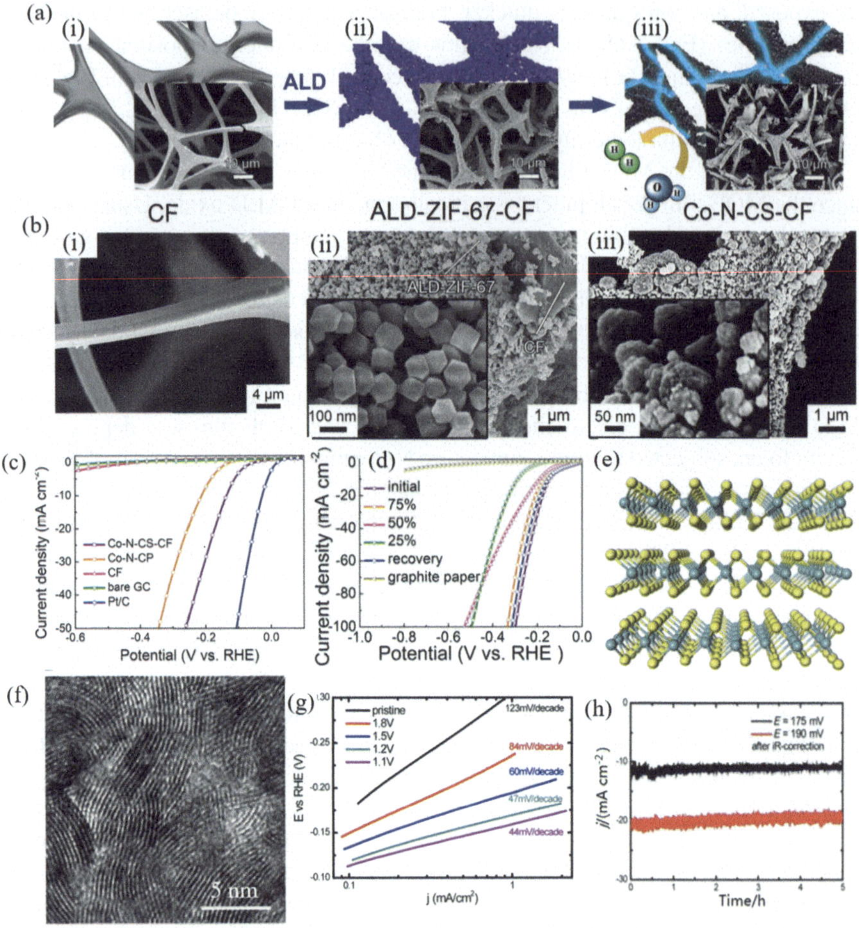

Fig. 3.7 (**a**) The synthesis of Co-N-CS-CF process. The insets show the SEM image of (i) CF, (ii) ALD-ZIF-67-CF, and (iii) Co-N-CS-CF. (**b**) SEM images of pyrolyzed (i) CF skeleton, (ii) ALD ZIF growth on CF, and (iii) Co-N-CS-CF. (**c**) LSV curves comparing different samples shown in (**b**). (**d**) Compression-induced changes in the LSV of the pyrolyzed ZIF-67-layer growth on carbon foam. [(**a–d**) Reprinted with permission from Ref. [89]]. (**e**) 2H-MoS_2 crystal structure. (**f**) TEM image of the developed pristine MoS_2 film. (**g**) Tafel curves of MoS_2 film. (**h**) Time-dependent electrocatalytic current of the developed MoS2 film. [(**e–f**) Reprinted with permission from Ref. [101]]

experiments. As a result of the remarkable positive correlation that exists between the thickness of the MOF layer and the thickness of the ALD metal oxide layer [99], the thickness of the MOF layer may be accurately adjusted by the number of ALD growth cycles. This phenomenon makes it possible to combine MOF with a greater variety of micro and nanostructures, which in turn enables a wider range of applications. Research on the use of MOF-related structures as a source of electrochemical energy is still in its infancy. The tactics have shown promise so far due to the widespread use of MOFs in related fields. It is anticipated that in the near future, additional MOF layers will be successfully manufactured, leading to devices with improved performance.

The extraordinary benefits that ALD provides have also led to its application in the process of depositing materials for reactions. Additionally, TMD has been produced through the process of performing posttreatment described below. For example, Wang et al. [101] used ALD to create a MoO_3 nanomembrane (Fig. 3.7e), which they then rapidly sulfurized to produce a MoS_2 film. The layered structure of MoS_2 is confirmed by the TEM image depicted in Fig. 3.7f. The high HER MoS_2 film catalytic activity is maintained despite the reduction in film thickness, as shown by the significantly enhanced 98 mV per decade of Tafel slope measured experimentally (Fig. 3.7g). No deterioration was seen even after a 5-hour stability test (Fig. 3.7h), and the cycling performance was excellent. This study highlights a novel approach to simply manufacturing potent electrochemical catalysts. Recent attempts to create electrochemical catalysts from functional materials via ALD are summarized in Table 3.1. More materials and architectures optimized for catalysis are likely to be developed in the near future.

3.4 ALD in Photoelectrolytic Water Splitting

It has become a hard situation to produce renewable energy supplies that are both earth-abundant and clean in order to balance for the energy problems that have been generated by the tremendous usage and degradation of fossil fuels [114]. It is going to be quite challenging to find a resource for alternative energy that can replace fossil fuels to resolve the present energy problem. Solar energy is often regarded as the most important source of energy, since it has the ability to pave the way for environmentally friendly and long-term energy solutions [115]. The process of converting solar energy into hydrogen fuel, known as photoelectrochemical (PEC) water splitting, has developed into a method that is both practical and environmentally friendly [116]. In contrast to primary and other more complex methods, PEC water splitting provides a number of benefits that are not available with other methods. In a word, this technique entails the incorporation of photoactive electrodes into an electrochemical cell, as well as the utilization of photons, in order to bring about electrochemical reactions [117–119]. Gerischer et al. [120], who worked in the late 1960s and established the basis for subsequent evolutionary investigations of semiconductor/electrolyte interfaces, are credited with having initially conceptualized this contemporary idea. A photoactive titanium oxide (TiO_2) catalyst was developed by researchers in recent years. This catalyst exhibited outstanding light-absorption capabilities and has been widely utilized in photovoltaics as well as the production of oxygen dioxide ever since [121]. A monolith photovoltaic-PEC water-splitting scheme was developed by Khaselev and Turner throughout the course of time and with the introduction of new technology both individually and collectively. Solar hydrogen is produced by the utilization of this technology, which

Table 3.1 Development of electrocatalysts as active materials by ALD

Reactant catalyst	Morphology	Substrate	Efficiency	Thickness/ ALD cycles	Ref.
Ni	Particle	Pyrolysis ZIF-67	157 mV at 10 mA cm^{-2}	600	[59]
MoS*x*	Film	Carbon fiber paper	55.8 $mVdec^{-1}$	100	[67]
WS_2	Nanosheet	Wafer	394 mV at 10 mA cm^{-2}	100	[68]
Pt	Single-atom	Graphene	0.05 V at 16 mA cm^{-2}	50	[58]
Pt	Particle	MoS_2	31 mV at 1 mA cm^{-2}	80	[102]
MoN*x*	Film	Porous carbon	236 mV at 10 mA cm^{-2}	700	[102]
RuO*x*	Film	CNTs	33 mV at 10 mA cm^{-2}	30	[103]
Pt	Particle	MXenes	67.8 mV at 10 mA cm^{-2}	40	[60]
MoS_2	Nanosheet	Wafer	98 $mVdec^{-1}$	1000	[101]
Pt	Film	Graphene	0.188 V at 1 mA cm^{-2}	200	[104]
CoS_y and FeS_y	Film	CNTs	294 mV at 10 mA cm^{-2}	11 nm	[104]
Pt	Film	Porous Ni	225 mV at 100 mA cm^{-2}	20	[105]
Pt and Ru	Single-atom	CNTs	28.9 $mVdec^{-1}$	/	[105]
MoS_2	Film	Wafer	200 mV at 15 mA cm^{-2}	9.4 nm	[106]
CoP	Film	Wafer	254 mV at 10 mA cm^{-2}	6.7 nm	[107]
CoS*x*	Particle	Graphite foil	0.12 V at 1 mA cm^{-2}	100	[108]
ZIF-67	Film	Carbon foam	142 mV at 10 mA cm^{-2}	300	[89]
Pt	Particle	Hollow fibers	190 mV at 10 mA cm^{-2}	250	[109]
Pt	Particle	Tungsten monocarbide	Nearly Pt/C	10	[110]
MoS_2	Film	SiO_2/Si	300 mV at 20 mA cm^{-2}	1000	[111]
Ni_3C	Film	CNT	132 mV at 10 mA cm^{-2}	300	[112]
Fe0.54Co0.46S0.92	Film	Carbon cloth	70 mV at 10 mA cm^{-2}	11 nm	[113]

combines the method of solar energy harvesting with the process of water electrolysis [122]. When it comes to the technical components of these systems, the semiconductor materials and the interface between the material and the electrolyte both play significant roles in determining the degree to which the units that are created perform effectively. In order to successfully produce hydrogen in PEC systems, photoelectrode materials (photoanode and photocathode) must overcome a number of obstacles, including (1) cost-effectiveness of device, (2) low overpotentials, (3) semiconductor designs with appropriate band-edge positions, (4) efficient use of photogenerated charge carriers, (5) high material stability towards photochemical reactions, and (6) absorption across the broad solar spectrum [117]

There have been tremendous strides in attempts to compete with natural photosynthetic processes since 1972, when Fujishima and Honda found artificial photosynthetic water splitting on TiO_2 using UV light [123]. Over the course of many decades, significant research efforts have centered on three main areas in order to develop PEC devices and attain the desired level of efficiency. One of the primary factors in establishing the electrolyte/semiconductor junction energetics is the specific composition/combination of materials used. Second, the electrode surface must be protected, passivated, and stabilized to allow for longer reactions [124]; the third is the electrode surface functionalization through coating, which allows for the absorption of light and subsequent separation/transport of the generated charge carriers. Since the photoelectrode is charged at a low voltage in a PEC, developing a PEC cell necessitates the use of photocatalytically active semiconductors at both photoelectrodes to facilitate charge split-up at the liquid/semiconductor-electrode boundary during the water-splitting process. Almost 1.23 eV of potential energy is sufficient for water splitting in theory; however, the energy actually needed is somewhat greater in practice, largely owing to the thermodynamics and kinetics of the phenomena participated in redox chemistry [125, 126]. Therefore, a 4-photon dual-absorbing (D4) tandem cell has substituted a single light-absorbing semiconductor in certain applications [127], allowing the electrode to absorb from a broader range of the solar spectrum. For rapid but steady charge transfer kinetics, it is important to have photochemical processes at the electrolyte/semiconductor and interconnecting semiconductor interfaces, and this is directly correlated to device efficiency.

3.4.1 Advanced Technology for the Development of PEC Devices

To be in compliance with the criteria, it is of the highest necessity to have a rational design or construction of photoelectrode that has control over the characteristics and qualities of thin films used for device integration. Among the current thin film technologies used in PEC water splitting devices is solution processing (electrochemical deposition [128], spin coating [129], drop casting [130], dip-coating [131], chemical bath deposition [132], solvothermal [133], and hydrothermal [134]), sputtering [135], and vapor phase processing (pulsed laser deposition (PLD) [136], chemical vapor deposition (CVD) [137], and physical vapor deposition (PVD) [138]) are occasionally unable to produce the desired thin films. From this vantage point, the evolution of chemical vapor deposition into the more sophisticated thin film technology known as "atomic layer deposition (ALD)" affords the chance to achieve atomic-level precise control with excellent conformality across both flat surfaces and structures with a high aspect ratio [139]. It is common knowledge that solution processing may be used to create photoelectrodes; however, the obstacles involved in the creation of nanostructured oxide photoelectrodes are seldom ever brought up [140]. Hence, solution processing by means of electrodeposition, solvothermal and hydrothermal is well recognized for its simplicity of procedure, cost-effectiveness, and scalability. In-depth investigations of the characteristics of PEC devices reveal that the performance of the system is particularly sensitive to the boundaries, which play a crucial role in charge carriers separation via shorter carriage lengths. This suggests that the interfaces are responsible for the majority of thc system's performance. The construction of such sensitive surfaces requires very precise control across the deposition process, as well as deposition thickness, accuracy, homogeneity, and repeatability, in order to achieve stable PEC responses throughout the whole surface of the photoelectrode. Although the formation of films by electrochemical deposition and solution processing is also conceivable, the quality and longevity of the films are greatly affected due to the unpredictable growth and nucleation process that occurs. Thermal and vapor phase deposition methods are also widely explored as active materials, in addition to being used in the photoelectrodes' surface treatment [141]. ALD is a technology that distinguishes out from the rest of the deposition methods because of the benefits it gives for the development of sophisticated interfaces, including control over the tunability of the material. ALD can make thin films of exceedingly high quality because the atomic thickness, highly dense layer growth, conformality, and stoichiometry can all be precisely controlled [142, 143]. The low-temperature processing combined with the self-limiting nature of the deposition creates a surface coating that is atomically smooth and is capable also of producing thin coatings over 3D objects that have good conformality. A superior level of adhesion to the substrate surface is shown by the thin films generated via ALD. Table 3.2 provides a summary of the several processes for the deposition of thin films, together with the corresponding obtained thicknesses and efficiencies. The purpose of this table is to shed light on the characteristics of different PEC devices. It is difficult to compare the data due to the wide range of materials and thicknesses of the photoelectrodes that were produced using the various approaches; however, the data appropriately highlight the utilization of diverse approaches to produce the host/core material, along with the requirement that the base material have a layer of protection or passivation, frequently referred to as the "interface layer."

The selection of the layer interface is largely dependent on the characteristics of the core materials since the material thickness and its interface play a significant role in determining the overall efficiency of the PEC. It is interesting to note that

Table 3.2 PEC materials created by a variety of deposition procedures (both wet and dry processes), together with their characteristic PEC properties [145]

Deposition technique	Active material	Substrate	Film thickness	Current density	Efficiency/stability	Ref.
Wet-chemical processing						
Hydrothermal	$MoS_2/BiVO_4$ photoanode	FTO (Fluorine doped Tin Oxide)	500 nm—$BiVO_4$ 30 nm—MoS_2	0.63 mA cm^{-2} at 1.23 V vs RHE	IPCE value of 3.25% at a wavelength of 450 nm	[146]
	Fe_2O_3/Fe_2TiO_5 photoanode	FTO	80 nm—Fe_2TiO_5	1.4 mA/cm^2 at 1.23 V vs RHE	Surface charge separation efficiency of 85%	[147]
	Ni:FeOOH-coated $WO_3/BiVO_4$ core/shell nanowire photoanode	FTO	5–10 nm thin Ni:FeOOH	4.5 mA/cm^2 at 1.23 V vs RHE	Increased charge-transfer efficiency by 91%	[134]
	Ferrihydrite (Fh)/$BiVO_4$ photoanode	FTO	10–20 nm Fh/$BiVO_4$	4.78 mA/cm^2 at 1.23 V vs RHE	ABPE value of 1.81% at 0.61 V	[148]
	$BiVO_4$ + CoP photoanode	FTO	–	4.0 mA/cm^2 at 1.23 V vs RHE	Low onset potential of 0.2 V	[149]
Solvothermal	ZnO/MoS_2 nanocable array	FTO	–	1.39 mA/cm^2 at 0.9 V vs. SCE	Low onset potential of 0.2 V vs. SCE	[133]
	$BiVO_4/Fe_2TiO_5$ photoanode	FTO	$BiVO_4$—1 μm Fe_2TiO_5—5 – 8 nm	3.2 mA/cm^2 at 1.23 vs RHE	Charge-transfer efficiency of 80% at 1.23 V	[150]
Chemical bath deposition	$CrFeWTiO_2$	Ti	$FeWTiO_2$—20–60 nm	0.3 mA/cm^2 at 0.6 V vs. Ag/AgCl	Chemical stability was evaluated for 1200 s at +0.60 V vs. Ag/AgCl	[151]
	TiO_2/Cu_2O core/shell NWAs	FTO	–	4.0 mA/cm^2 at 0.6 V vs. Ag/AgCl	Stable photocurrent of 2.3 mA/cm^2 for 5 hours	[132]

(continued)

Table 3.2 (continued)

Deposition technique	Active material	Substrate	Film thickness	Current density	Efficiency/stability	Ref.
Dip coating	$FeOOH/Fe_2O_3$ photoanode	FTO	2 nm FeOOH	1.21 mA/cm^2 at 1.23 V vs RHE	Current onset potential shifts in the cathodic direction by 0.12 V	[131]
	$GaInP_2$-TiO_2-cobaloxime photocathode	$GaInP_2$	35-nm TiO_2 ALD	9 mA/cm^2 at 0 V vs RHE	IPCE up to 80% through visible range (<690 nm)	[152]
Spin coating	$Pt/TiO_2/Sb_2Se_3$ photocathode	FTO	20 nm TiO_2 ALD	2 mA/cm^2 at 0 V vs RHE	Low overpotential for hydrogen evolution, showing near 30–40 mV at 10 mA/cm^2	[129]
	3-layered (<40 nm thick) MoS_2/Cu_2O photocathode	FTO	<40 nm thick MoS_2 Cu_2O film—2 μm	6.5 mA/cm^2 at 0.2 V vs. RHE	MoS_2 modifies the stability of Cu_2O maintaining even after 1 hours	[153]
	$BiVO_4$	FTO	300 nm $BiVO_4$	0.67 mA/cm^2 at 1.23 V vs RHE	Photo-conversion efficiency of 0.4% at 1.23 V vs RHE	[154]
	$Si/CoSe_2$	Si wafers	–	9 mA/cm^2 were achieved at 0 V vs. RHE	Strong shift of onset potential 0.18 V vs. RHE	[155]
	$Co_2N_{0.67}$-loaded $BiVO_4$	FTO	–	2.21 mA/cm^2 at 1.23 V vs RHE	Photocurrent increases by 80% than that of bare $BiVO_4$	[156]
Droplet casting	Co_2P-coated p-Si photocathode	p-Si	25 nm Co_2P	35.2 mA/cm^2 at 0 V vs RHE	PEC catalyzed HER above 30 mA/cm^2 at least 150 hours without notable degradation	[157]
	TiO_2 NAs/1T-MoS_2 photoanode	FTO	–	1.91 mA/cm^2 at 1.23 V vs RHE	12 hours PEC test with 92% retention of initial current	[130]
	Exfoliated LDH/α-Fe_2O_3	FTO	–	Increased photocurrent density by 50% at 1.0 V vs RHE	Onset potential shifted by 140 mV	[158]
	Textured $CuBi_2O_4$ photocathode	FTO	400–500 nm	−0.5 mA/cm^2 at −0.4 V vs Ag/AgCl	5-fold increment in photocurrent at −0.6 V vs RHE	[159]

Electroless Plating	TiO_2/Pt/n^+np^+-Si photocathodes	Si wafer	15 nm TiO_2 ALD	25 mA/cm^2 at 0.4 V vs RHE	11.5% efficiency and 168 hours stability under a continuous PEC reaction	[160]
	Si-NWs/M-B electrodes	p-Si wafer	–	19.5 mA/cm^2 in neutral buffer solution under simulated 1 sun illumination	Onset potential of cathodic photocurrent is positively shifted by 530–540 mV to 0.44–0.45 V vs RHE	[161]
Electrochemical deposition	Ni–Mo/Cu_2O and MoS_{2+x}–CE/Cu_2O photocathodes	FTO	MoS_{2+x}—30–50 nm and Ni-Mo film—40 nm	−6.3 mA/cm^2 at 0 V versus RHE	7.7% solar to hydrogen conversion efficiency	[162]
	Cu_2O/Ga_2O_3/TiO_2/RuO_x/NiMo photocathode	FTO	100 nm TiO_2 ALD	10 mA/cm^2 at 0 V vs RHE	Stable operation exceeding 100 hours	[128]
	Si-MWs@MoO_xS_y photocathodes	Si wafer	–	2.25 mA/cm^2 at 0 V vs RHE	0.75 mA/cm^2 for up to 5 hours	[163]
	Si/a-$CoMoS_x$ hybrid photocathode	Si wafer	a-$CoMoS_x$ film of ~40 nm	17.5 mA/cm^2 at 0 V vs RHE	3 hours showed relatively stable photocurrent	[164]
	3D p-Si/$NiCoSe_x$ core/shell nanopillar (NP) array photocathode	p-Si	–	−37.5 mA/cm^2 at 0 V vs RHE	Exhibited only 9% decrease after 2 hours of PEC test	[165]
	NiO_x/Ni/n-Si photoanodes	n-Si	–	14.7 mA/cm_2 at 1.23 V vs RHE	~100% Faradaic efficiency	[166]
	Co-Ci/H, 3% Mo:$BiVO_4$ photoanode	FTO	Co-Ci—4–10 nm	5.1 mA/cm^2 at 1.23 V vs RHE	IPCE of ~80% at 420 nm and light harvesting threshold of ~510 nm	[167]
	Fe_2O_3/Al_2O_3/$CoFeO_x$	FTO	0.5–0.8 nm—$CoFeO_x$ Al_2O_3	1.6 and 2.5 mA/cm^2 at 1.0 and 1.23 V vs RHE, respective	200 mV cathodic shift of the photocurrent onset potential	[168]

(continued)

Table 3.2 (continued)

Deposition technique	Active material	Substrate	Film thickness	Current density	Efficiency/stability	Ref.
	Co–La(MDH)/$BiVO_4$	FTO	–	1.323 and 2.02 mA/cm^2 at 0.6 and 1.23 V vs RHE	2.09 mA/cm^2, obtained by applying 0.8 V between the working and counter electrodes for 4000 s with 93% retention	[169]
Dry-chemical processing						
Magnetron sputtering	$LaFeO_3$	FTO	350 nm	–25 μA/cm^2 at 0 V versus RHE	$LaFeO_3$ photocathode biased at 0.1 V versus RHE during CA for 30 min	[170]
	Mo-doped $BiVO_4$	FTO	150 nm	1.2 mA/cm^2 at a potential of 1.23 VRHE	IPCE values dropped off rapidly for wavelengths higher than 460 nm	[135]
PVD	$MoSe_2$/n^+p-Si photocathode	Si wafer	–	Saturated photocurrent of 29.3 mA/cm^2	$MoSe_2$ layer can protect the n^+p-Si photocathode with a 120 hours stability	[171]
	Nano-MoS_2 layer on Al_2O_3/n^+p-Si photocathode	Si wafer	2 nm Al_2O_3	35.6 mA/cm^2 saturated photocurrent	120 hours continuous water splitting under 100 mA/cm^2 Xe lamp illumination	[172]
	MoS_2/$GaInP_2$	GaAS	200 nm $GaInP_2$	19 mA/cm^2 under 1 sun illumination	60 hours stability and 40% of the incident light with photon energies above the band gap of $GaInP_2$ and 500-fold increase in stability	[138]
	pn^+-Si/Ti/NiP_2 photocathodes	Si wafer	10 nm Ti, 10 nm Ni layer	12 mA/cm^2 at 0 V vs RHE	impressive operation durability for at least 6 hours in 0.5 M H_2SO_4	[173]
CVD	MoS_2 nanoflakes on the Ag@Si NSs	Ag film	Si shell 10 nm	33.3 mA/cm^2 at –0.4 V vs RHE	12 hours PEC performance	[174]

	$SiNWs/MoS_2$ photocathodes	p-Si wafer	–	16.5 mA/cm^2 at 0 V vs RHE	48 hours of PEC operation were achieved under a simulated 1 sun irradiation	[137]
	3D MoS_2/TiO_2/p-Si	p-Si	10 nm TiO_2	Onset potential was 0.2 V at a J_{sc} = 10 mA/cm^2 and a saturated current density of 33 mA/cm^2	181 hours without noticeable degradation	[175]
PLD	n-Si/NiO_x heterojunction	n-Si	105 nm NiO	photocurrent density as high as 26 mA/cm^2	Overpotential decreased by 35 mV at a current density of 3.0–5.0 mA/cm^2	[176]
	WO_3/Al_2O_3	FTO	3 nm Al_2O_3 18 μm WO_3	3.1 mA/cm^2 with 1 sun AM1.5 G illumination	Faradaic efficiency of 80% was reached without diminishing the photocurrent density	[136]
ALD	PEDOT:PSS/CdSe/Pt heterogeneous Photocathode	FTO	200 cycles Pt ALD CQDs 200 nm PEDOT:PSS 160 nm	−1.08 mA/cm^2 at 0 V vs RHE	Only 8.3% activity degradation after 6 hours, compared with 80% degradation when the Pt layer is deposited by electrodeposition	[177]
	p^+n-Si/Co_3O_4/$Co(OH)_2$ photoelectrodes	Si wafer	3.2 ± 0.2 nm PE-ALD CoO_x film	30.8 mA/cm^2 at 1.23 V vs RHE	Stable under illuminated operation for at least 72 hours	[178]
	NiO on $CoO_x/BiVO_4$	ITO	Ultra-thin NiOOH with CoO_x ITO 500 nm	2.5 mA/cm^2 at 0.6 V vs RHE	Activity maintained for 12 hours at the operation potential of 0.6 and 0.9 V vs RHE with continuous and stoichiometric H_2 and O_2 evolution	[179]
	GaAs NW/NiO_x	Patterned SiO_2	36 nm layer of nickel oxide (NiO_x)	Maximum oxidation photocurrent of 11.1 mA/cm^2 was measured	NW-patterned sample generated photocurrent density of 0.52 mA/cm^2 and the GaAs substrate generated 0.01 mA/cm^2	[180]

(continued)

Table 3.2 (continued)

Deposition technique	Active material	Substrate	Film thickness	Current density	Efficiency/stability	Ref.
	MoS_2 on Si photocathodes	p-Si	5–20 nm MoS_2	Photocurrent density of 31 mA/cm^2	Overpotential of about 630 mV compared to bare Si	[181]
	CoO_x-modified InGaN/GaN quantum well photoanode	Sapphire	CoO_x-12 nm	1.26 mA/cm^2 at 1.23 V vs RHE	Onset potential of −0.03 V vs RHE and IPCE is 16.8%	[182]
	p-Si wire arrays/TiN nanolayer	p-Si	TiN nano-layer 10 nm thick	2.3 mA/cm^2	Faradaic efficiency of nearly 100% at 0 V vs. RHE over 20 hours of reaction	[183]
	Fe_2O_3/CoO_x	FTO	2–3 nm ALD CoO_x overlayer	Photocurrent density as a function of time during the first 3 hours of operation at 1.23 VRHE	External quantum efficiency (IPCE) obtained on hematite increases by 66% at 1.23 VRHE	[184]

processes such as electrochemical deposition, droplet casting, spin-coating, and dip-coating are gaining popularity as a result of their lower costs in creating interfacial layers. Vapor phase methods, on the other hand, provide precise control over nanolayer thickness, allowing for the production of thin films to thousands nanometers from a few nanometers. Nearly every kind of PEC device requires that the films be crisp and homogeneous across wide surface areas, have great repeatability, and have precise thickness [144]. Therefore, ALD may be seen as a versatile tool that can be used in the process of creating stable and effective PEC devices by eliminating the challenges that are involved with the development of a layer interface on the photoelectrode. In the next section, we will discuss the ALD processing as well as the particular function that ALD plays in PEC devices.

3.4.2 ALD-Based Active Material for Photoelectrolytic Processes

In the following section, we present a brief summary of materials/catalysts and processes involving H_2 and O_2 evolution and materials utilized in photocathodes and photoanodes by ALD approach [185]. This is necessary because the materials for PEC cells might include in-depth talks about the design of the materials themselves. It is necessary to have an understanding of the bandgap of a semiconductor as well as its expected band-edge locations in order to know the practicability of material that is deposited by ALD. With the aim of improving the fabricated-electrode and sustainable PEC device systems' efficiency, we have provided a brief overview of the surface reactions of ALD-based catalysts, various nanostructures that impact efficiency, and ALD parameters [140]. To summarize the process that takes place in PEC systems, it is important to have an understanding of the following three points: (1) the surface area configuration of the photoelectrode (also known as interfacial reactivity), (2) the photoelectrode's response to photons for the purpose of beginning the reaction, and (3) the dynamics of charge carrier that take place in the device. Future possibilities are aided by material-centric modeling that has been done in the past [186]. When employing computational modeling and simulations, it is important to have a solid understanding of the fundamental dimensional connections that exist between structure and qualities. When it comes to studying the important properties of photoelectrode materials, advanced analytical methods like scanning photocurrent microscopy, electrochemical scanning tunneling microscopy, and atomic force microscopy are the way to go. These key features include photoconductivity, double-layer forces, interfacial charge transfer rates, special resolution, distributions of electric field, local photocurrents, carrier diffusion lengths evaluation, and other chemical and physical properties. Although integral restrictions make it difficult to correlate experimental data collected under specific operating circumstances, such methods are helpful for establishing viability at different scales for the different materials involved in real-time PEC applications [187]. The photoelectrode interfacial activity has a significant impact on the carrier dynamics, and this activity is relative to the photonic absorption of the electrodes. Before using ALD in PEC device systems, it is important to have a solid understanding of the changes that take place in the interfacial activity caused by the layer.

Materials that are appropriate for the creation of visible-light photocatalysts and boost semiconductor activity were introduced by Saraswat et al. [186]. These materials belong to both the platinum and non-platinum categories. They went on to say that a lot of studies have focused on changing the same active materials, so maybe future studies should look into how different active materials may be used in PEC systems to get better results. As a result of ALD, there is a broad range of materials that are capable of being deposited, beginning with oxides and moving on to sulphides. Doping, mixed architectures, and nanolaminates are also possible, as illustrated in Fig. 3.8a. In any case, the following is a list of common active materials, and they are classified as (1) Oxides [188]: Al_2O_3, WOx, NiO, $FexO_y$, CoO_x, ZnO,Y_2O_3, V_xO_y, TiO_2, Ta_2O_5, Ta_2O_5, Sc_2O_3, MgO, Nb_2O_5, HfO_2, La_2O_3, Ga_2O_3, HfO_2, CaO, CuO, and others; (2) Sulfides [189]: ZnS, GeS, Li_2S, SnS, GaSx, MoS_2, Sb_2S_2, TiS_2, WS_2, In_2S_2, Cu_xS, CdS, PbS, CaS, BaS, SrS, and others; (3) Nitrides [190, 191]: AlN, $TiSi_xN_y$, $TiAl_xN_y$, MoN_x, NbN, TiN_x, W_2N, TaN_x,

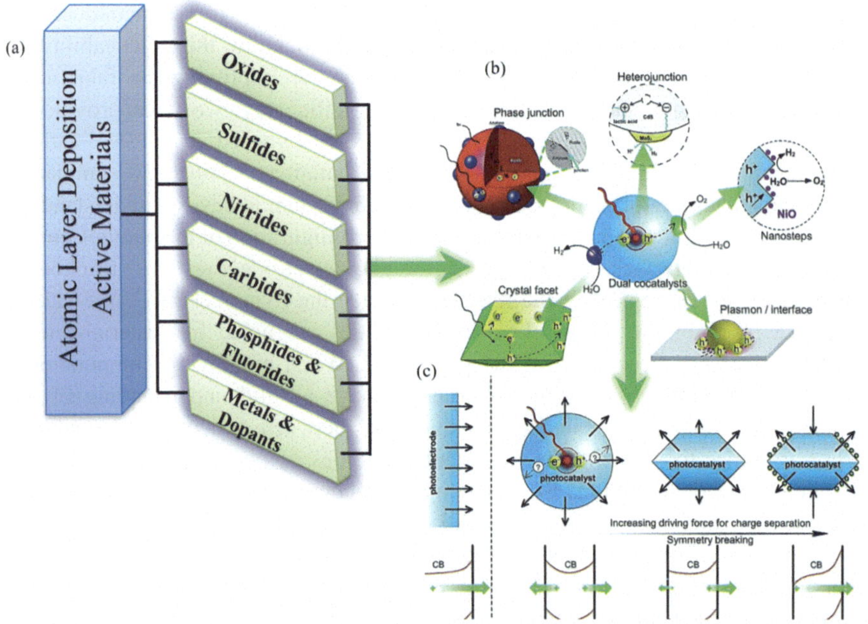

Fig. 3.8 (a) A broad spectrum of ALD active materials for PEC applications. (b) The potential to create photocatalysts with tailorable interfaces. (c) Potential pathways for the driving force behind charge separation at interfaces with a wide variety of built-in electric fields. [(b–c) Reprinted with permission from Ref. [199]]

TiAlN, HaN, and others; (4) Carbides [112, 192]: Mo_2C, WC, Ni_3C, Ti_2C, and others; (4) Metals [193]; Al, Ta, Ni, Ir, Pt, Ru, Cu, W, and others; (5) fluorides [194]: ZnF_2, ZrF_4, HfF_4, MgF_2, MnF_2, and others; (6) phosphides [195]: GaP, InP, CoP, and others; (7) dopants [196, 197]: Ti, In, N, Ga, Sn, Al, Hf, P, B, and others. Table 3.3 provides an overview of a few current results produced using a limited number of materials [198].

With ALD, it is possible to precisely tune the characteristics of materials used in water-splitting applications by combining them to create oxynitrides, oxysulfides, ternary metal oxides, and other alternative doping processes. Some oxides-related materials have been extensively investigated in PEC water splitting, and all of the aforementioned compounds may be deposited using the ALD method. In addition, we tried to detail the tuning of the created base materials for efficient PEC water splitting by using various ALD materials as direct light absorbers, photoelectrodes, and passivation or protection layers, such as metals, carbides, phosphide, sulfides, and dopants. Figure 3.8b illustrates the wide range of possible interfaces and charge transfer parameters among heterojunction-forming catalysts [199]. As demonstrated in Fig. 3.8c, the deposited particles or layers experience a different magnitude and direction of the inherent electric field depending on which side of the parent material they are on. In industrially relevant catalytic processes, embedding active photocatalysts within porous support materials is advantageous in addition to top-layer passivation. There are a few materials that have not been publicly presented but have been studied in the PEC environment without ALD. MnO_2, SnO_2, WO_3, $SrTiO_3$, ZnO, $BiVO_4$, and TiO_2 are examples of n-type oxides that are commonly used as anodes in PEC applications, while p-type oxides like Bi_2O_3, $CaFe_2O_4$, $LaFeO_3$, $CuBi_2O_4$, p-type Fe_2O_2, Cu_2O, and CuO are typically used as photocathodes. It is also common practice to use ALD to create PEC devices out of other materials, including as CdSe, CdS, GaAs, GaP, and InP. Having both n-type and p-type in a material increases the possibility that it will have a wider range of properties. The bandgap and expected band-edge positions of the two types of semiconductors must be perfectly aligned beforehand for the manufactured photoelectrode to be more workable. The goal of developing new ideas for photoelectrode design is to increase the manufactured electrode's efficiency; to do this, one must understand the impact of various nanostructures, factors influencing device production, and catalyst surface reactions [140]. Materials deposited without the use of ALD for photoelectrode fabrication may vary widely in their functionality. The PEC efficiency is generally limited by surface recombination, slow kinetics, and photocorrosion, despite the chemical composition of the intended photoelectrodes. On the flip side, ALD provides the chance to avoid using noble metal nanoparticles while still overcoming these challenges and extending the device's efficiency. Devices can be built with structural engineering, band-gap engineering (p-n-heterojunctions), and films for regulating electrolyte/electrode interfaces thanks to the atomic precision of the ALD process.

ALD, on the other hand, offers the possibility of overcoming such obstacles and keeping the device's effectiveness for a longer period of time without the use of any noble metal nanoparticles. Atomic precision in the ALD process for device

Table 3.3 PEC water splitting using ALD-grown metal oxides

Material	Reactants & substrate	Growth parameter/thickness	PEC features	Ref.
A-Fe_2O_3	$Fe_2(O^tBu)_6$ H_2O & $TiSi_2$ nanosheet	5 s 50 ms, 180 °C & 20 nm(400 cycles)	1.6/2.7 mA cm^{-2} photocurrentat 1.23/1.53 V	[202]
A-Fe_2O_3	$Fe(Cp)_2$ O_3 & ITO	2 × 30 s 2 × 620 s, 200 °C	Red. of applied bias by 200 mV, tripling of photocurrent	[203]
A-Fe_2O_3 Mg-doped (p-type)	$Fe_2(O^tBu)6$ H2O $Mg(EtCp)_2$ & F doped SnO_2 (FTO)	5 s 50 ms 15 ms, 180 °C, & 20 nm (400 cycles)	Reduction of applied bias by 200 mV	[204]
α-Fe_2O_3Ti-doped	$Fe(Cp)_2$ H_2O O_3 $Ti(O^iPr)4$ & FTO	20 s 10 × (15 ms + 6 s) 0.2 s, 200 °C & 20.8 nm (300 cycles)	Ti doping of very thin films doubles the photoresponse, APCE of 50%	[205]
CoPi@*a*-Fe_2O_3	$Fe(Cp)_2$ H_2O + O_3 & FTO	20 s 10 × (15 ms + 1 s) & 60 nm (1000 cycles)	Shift in photocurrent onset potential of 0.23 V	[206]
MnO_x@Fe_2O_3	$Fe_2(O^tBu)_6$ H_2O $(Mn(EtCp)_2)$ & Ti foil	5 s 50 ms 0.3 s, 180/250 °C, Fe_2O_3/MnO_x & 20/2 nm Fe_2O_3/MnO_x (400/20 cycles)	Significant photocurrent suppression due to Fermi-level pinning by MnO_x	[207]
α-Fe_2O_3/ Fe_3O4	$(Me_2Cp)_2Fe$ O_3/H_2O_2 for α-Fe_2O_3/Fe_3O_4 & Si, quartz, FTO, porous silica	10 × 1 s 15/0.5 s, 120 °C & 40.7/37.6 nm a-Fe_2O_3/Fe_3O_4 (600 cycles, quartz)	H_2O ox. onset potential of 1.1 V, current density at 300 mV of 0.4 (H_2O_2) and O.5 (O_3) mA	[208]
TiO_2ALD@n-Si	TiCl4 H_2O & Si	300 °C & 25–40 nm	1 mA cm^{-2} (planar), 0.25 mA cm^{-2} (nanostructures)	[209]
TiO_2 ALD@TiO_2	TiCl4 H_2O & FTO	300 °C & 60–400 cycles	1.1 mA cm^{-2} (1.5 V vs RHE)	[210]
TiO2@$TiSi_2$	Ti(OiPr)4 H_2O & Ti foil	0.1 s 0.01 s, 275 °C & 27 nm	−0.6 mA cm^{-2} (0 V vs Ag/AgCl), Voc = −0.8 V	[204]

Pt/TiO_2/Al: ZnO/Cu_2O	TiO_2 [* Ti(i-PrO)4, * H_2O] ZnO (* DEZ, * H_2O) Al_2O_3 (* TMA, * H_2O) & FTO	4 nm ZnO, 11 nm TiO2	−7.6 mA cm^{-2} (0 V vs RHE)	[211]
TiO2 nanotube	$TiCl_4$ H_2O//GaQ_3	50–200 °C	0.45 mA cm^{-2} (0.4 V vs SCE)	[212]
WO3/Mn	$(tBuN)_2(Me_2N)_2W$ H_2O [(H_2O)(terpy)$Mn(O)_2Mn(H_2O)$(terpy)] $(NO_3)_3$ (terpy = 2,2′:6′,2″-terpyridine) & ITO	2 s 50 s, 350 °C & 1600 cycles		[213]
Al2O3@SiOx	TMA H_2O & Si	230 °C & 1.1–4.6 nm	26.2 mA cm^{-2} (vs RHE) Anodic shift of 170 mV	[214]
Al2O3@WO3	TMA O_2 & FTO	100 °C & 10–300 cycles	Photocurrent enhancement 3 times with 30 cycles	[215]
TiO2@ZnO nanowire	TMA H_2O & ITO	250 °C & 100 cycles	0.7 mA cm^{-2} (1.23 V vs RHE)	[216]

fabrication allows for structure engineering (multidimensional nanostructures), band-gap engineering via the formation of p-n-heterojunctions, and the development of films for tuning interfaces between electrolyte and electrode. Although Wen et al. [139] and other prominent researchers in the ALD area [200, 201] have provided illustrative descriptions of ALD's usefulness in energy storage devices, a short overview is provided in the following section.

3.4.3 ALD of Photoactive Electrodes in PEC Applications

Developing innovative nanostructured photoactive electrodes with surface functionalization that aids material stability and protection against disintegration is one of the most intriguing approaches to improving PEC device performance. Next-generation PEC devices necessitate material surface thickness control, the capacity to apply coatings over 3D intricate nanostructures, and core-shell structures production. In light of these needs, ALD has emerged as a formidable technology that meets and exceeds the aforementioned requirements and those for other deposition techniques by a wide margin. Many reviews have indicated the benefits of ALD for solar cells and photocatalysis [217, 218]. As a result, the goal of this section is to gain an understanding of recent advances in the field of PEC water-splitting, with a focus on designs that employ ALD methodologies and their potential to achieve desired efficiencies.

The photocathode and photoanode are the most crucial components of a PEC system's fully integrated photodevice. The photoanode and photocathode both serve distinct functions in PEC devices. Below, we explore the impact of various ALD coatings on the photocathode, as well as methods for improving the PEC device's efficiency.

3.4.3.1 Active Photocathode Materials

It was recently reported that the hydrogen production efficiency of a photocathode having a p-n junction sandwiched between Ga_2O_3 and Cu_2O is 80%. This high-efficiency results from 20 nm conformal ALD TiO_2 layer on the photocathode and the formation of a quantum field in relation to absorbing over 600 nm visible light, which together allow for operation for 100 hours. Light absorption and p-n junction protection contribute to the extraordinary stability seen in reported photocathodes based on Cu_2O [128]. Another team of researchers has proposed gallium nitride (GaN) for usage in the application of PEC water-splitting because of its advantageous band locations and energy bandgap exclusive of external bias. Though, photocorrosion makes stability issues with GaN a major worry in real-world applications. From this vantage point, a layer of ALD Al_2O_3 on the GaN surface improved the durability of the material against corrosion and increased hydrogen generation [219]. To further understand the influence of cocatalysts (MoS_2) on H_2

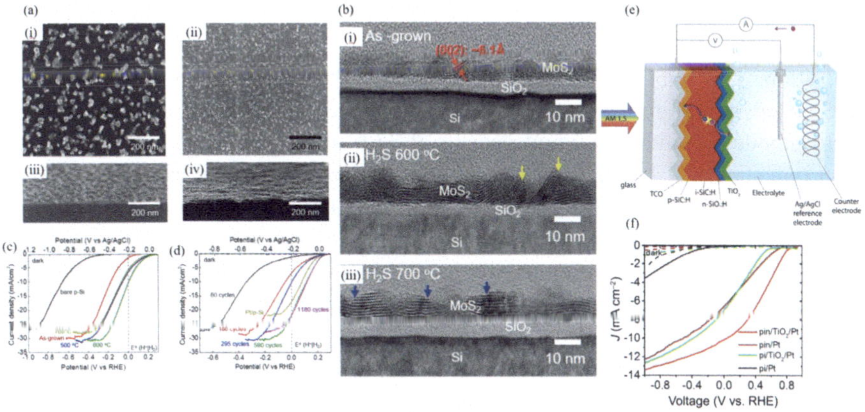

Fig. 3.9 (**a**) SEM images of ALD MoS_2 film growth on p-Si wafers for (i) 180 cycles, (ii) 590 cycles, (iii) 15°-tilted views at 180 cycles, and (iv) 15°-tilted views at 590 cycles. MoS_2, which initially forms as nanoparticles during early ALD cycles, undergoes a transition into a continuous thin film during later cycles. (**b**) TEM images of (i) ALD MoS_2 film growth on p-Si wafers at (ii) 600 °C and (iii) 700 °C. Edge locations of the (002) basal plane are shown by yellow and blue arrows. (**c**) The current density-potential (j-V) curves showing PEC efficiency of ALD-deposited MoS_2 film and sulfurized ALD-deposited MoS_2 on p-type Si wafers at different temperatures in a 0.5 M H_2SO_4 and 1-sun illumination PEC HER (295 cycles). (**d**) j-V curves measured by a PEC HER as the number of ALD cycles increased from 60 to 1180. [(**a–d**) Reprinted with permission from Ref. [181]]. (**e**) The three-electrode arrangement [p = (p)a-SiC:H, i = (i)a-SiC:H, n = (n)nc-SiO_x:H] used in PEC experiments. Electrolyte: 0.2 M aqueous potassium hydrogen phthalate (KPH) (pH 4). (**f**) j-V curves measured by a PEC devices for (n)nc-SiO_x:H and/or TiO_2. [(**e–f**) Reprinted with permission from Ref. [228]]

evolution, Oh et al. [181] investigated ALD MoS_2 growth on a Silicon photocathode (Fig. 3.9a, d). High-performance HER was shown by crystalline MoS_2 with a thickness of 14 nm in a 0.5 M H_2SO_4 electrolyte of 1 sun illumination for 24 hours. In comparison to Si photocathodes decorated by Pt, the 630 mV overpotential decrease obtained with the optimized cocatalyst, MoS_2 on Si is impressive. The crystallinity, stoichiometry, and edge site density of the utilized cocatalyst significantly impact the performance as a result of restricted electrical and optical losses. The research also discussed the potential for reaching state-of-the-art performance by generating cocatalysts using ALD on Si nanomaterials. For this reason, Cheng et al. [220] investigated the possibility of a monolithic photocathode that featured an ALD-coated titania protective layer. The goal of this research was to mitigate the detrimental effect that the monolithic electrode surface has on the cathode–electrolyte interface. Titania's bandgap is useful for water reduction since it eases electron transport, and the material was selected to prevent corrosion. The design of such a device displayed lower losses owing to the light that was absorbed and the surface reflectivity. PEC performance may be enhanced by processes including synergy via cocatalyst functionalization, as shown in a study of Si photocathodes coated with cocatalysts [221], which discriminated between non-layered and layered transition-metal chalcogenides. An active photocathode composed of MoS_2/Al_2O_3/n + p-Si was created by Fan et al. [172], in which the Al_2O_3 layer serves to both protect the Si surface and passivate it during MoS_2 deposition. The stability of the Si-based photocathode containing a H_2 evolution MoS_2 catalyst was found to be above average after 120 hours. Additionally, the photocathode exhibited a current density of 35.6 mA cm^{-2} when subjected to an onset potential of 0.4 VRHE. No catalytic effects are seen when the alumina layer between MoS_2 and n + p-Si is removed, indicating that the lack of the alumina layer is preventing charge carrier transfer. Another group utilized a comparable MoS2 cocatalyst integrated on Si pyramids in their work; however, they enhanced the catalytic efficacy by adding phosphorus atoms. An ALD-deposited titania layer, in the shape of $MoS_{1.75}P_{0.25}$/10-nm TiO2/Si pyramids, was applied to the photocathode, resulting in an improved photocurrent retention of around 84% after 2 hours [222]. The onset potential for the $MoS_{1.75}P_{0.25}$/10-nm TiO2/Si pyramids in the MoS_2/ALD layer/Si photocathode system was 0.29 V.

To increase the photocurrent and reach a photovoltage of more than 0.8 V, a photocathode device was created using a layer of n-doped nanocrystalline SiO_2 on top of a layer of p-doped a-SiC:H. This research proved to internally boost the photocathode electric field by restricting recombination of Shockley–Read–Hall using a TiO_2 protective covering. This coating also assisted in better aligning bands at significant interfaces, which resulted in an increase in the amount of charge carriers that were incorporated into the electrolyte. Figure 3.9b, c shows a simplified schematic representation and j-V curves measured by a PEC HER system, clearly illustrating the sandwiched structural structure and the generation of H_2 and O_2. To bring down the price of producing photovoltaic energy, Liu et al. [223] created a heterojunction photocathode of p-Si/n-CdS coated in a thin layer of TiO_2 using the ALD technique. By acting as a passivation layer, TiO_2 prevents surface carriers from recombining, prevents hole transfer for effective separation of charge, and shields against electrolyte degradation. Open-circuit voltages in the range of 49–425 mV were observed in the p-Si/n-CdS/n-TiO2/Pt photocathode, mostly as a result of the valence band locations of ALD passivation layer and TiO_2 overlayer. Many methods have been proposed for improving the stability and lowering the losses of the Si photocathodes often used in PEC devices. InAs nanowires were integrated with p-Si using a catalyst-free growing approach to generate heterojunctions with superior light absorption. Since InAs is a great material for increasing electron mobility, the nanowires are able to reduce protons at the cathode–electrolyte interface. Given that a Pt cocatalyst was used, it was necessary to protect the as-grown InAs by coating the InAs nanowires with TiO_2 passivation layer in order to prevent the excessive buildup of photoelectrons. The onset potential was shifted by 0.91 V vs. RHE compared to that of p-Si, which increased the stability of the Pt/TiO_2/InAs NWs/p-Si photocathode from 10 hours to 20 hours [224].

From what has been said above, it may deduce that PEC devices place a premium on catalyst regulation, charge carrier separation/transfer, and visible-light harvesting. In this setting, increasing photocurrent density might be aided by using a bifacially constructed Si photocathode that takes part within electrochemical process and produces hydrogen. The most significant result of this work is the dramatic increase in photocurrent density, which was achieved by creating a two-sided photocathode that can absorb sunlight thanks to a semi-transparent Pt layer. Under dual-sun illumination, the constructed device produces hydrogen at a rate that is about double that of a PEC system with single-faced (~56.88%). The system also has an Al_2O_3 passivation layer and a Si_3N_4 antireflective coating, both of which were deposited via ALD. When designing bifacial p-Si photocathodes, it may be possible to use synergy to enhance PEC-device systems that generate hydrogen [225].

To achieve the goal of creating a thermoelectrically active hybrid device, a p-Si photocathode containing bilayers HfO_x/SiO_x was developed. HfO_x improves the system's stability and corrosion resistance by acting as an ALD coating. According to photocathode activity, electron tunneling through HfO_x is independent of coating thickness, and charge is transported to the electrolyte quickly and efficiently from the Si thanks to improved kinetics. PEC is quick and unimpeded by the HfO_x layer because its energy level has no impact on Si [226]. The creation of multilayer cathode materials that facilitate favorable and continuous electron transport to the electrolyte is another strategy for improving the PEC device's performance. Following the passivation process using a TiO_2 ALD layer, a multilayer CdS/copper indium gallium selenide (CIGS) photocathode was fabricated. This photocathode exhibited significantly improved stability across a broad spectrum of pH levels. Another combination of p-Si photocathodes, specifically p-Si/AZO/TiO_2/CoP_2, exhibited a photocurrent density of 16.7 mA cm^{-2} and a notable output photovoltage of 0.54 V at 0 V vs. RHE. However, the efficiency of this combination was enhanced through the use of a non-noble-metal cocatalyst, specifically Ni-Mo [226]. A connection formed by AZO (aluminum-doped zinc oxide) and p-Si improved stability and performance, while TiO2 helped prevent corrosion. CoP2 nanocrystals placed on the photocathode surface sped up the hydrogen evolution reaction [227]. diverse combinations of Pt, p-Si, and prospective metal oxides result in diverse photocathode materials, as shown in Table 3.4, which also displays their efficiencies, onset potentials, and current densities [145]. Phosphorus-doped MoS_2 with interlayer thicknesses (TiO_2) between 10 and 30 nm has shown the greatest performance among p-Si photocathode systems with cocatalyst/electrocatalyst.

3.4.3.2 Active Photoanode Materials

$BiVO_4$ is a popular choice as a photoanode material in PEC systems. Zhong et al. [179] successfully coated a $BiVO_4$ film on a single-layer crystal In-doped tin oxide (InTO) substrate.

When the manufactured film was doped with an ALD-deposited layer of NiO and subsequently impregnated with CoOx, the efficiency of the half-cell hydrogen conversion surpassed 1.5%. The production of p-n junction using CuO and $BiVO_4$ is another method of developing a photoanode. In this method, a spin-coated $BiVO_4$ film and CuO nanocones are formed by means of a reaction that takes place in a water bath. The resulting photoanode created an unconventional heterojunction that increased PEC-cell efficiency. The optimization of the synergy between the injection

Table 3.4 Features of photocathode active materials

Active photocathode material	ALD-deposited material & thickness	Efficiency/ Stability	Current density (mA/cm^2)	Potential V vs. RHE	Functionality	Ref.
n^+p-Si/2nmAl_2O_3/MoS_2	Al_2O_3 & 2 nm	120 hour	35.6	0.4 V onset	Protective and passivation layer	[172]
$^{MoS}1.75^{P}0.25'$10nmTiO_2/Si	TiO_2 & 10 nm	84%retentionafter 2 h	−23.8at 0 V	0.29 V onset	Passivation layer	[222]
p-Si/n-CdS/n-TiO_2/Pt	TiO_2 & 10 nm	STH2.07%	10	49–425 mV open	Passivation and low valence band position	[223]
Pt/TiO_2/InAsNWs/p-Si	TiO_2 & 50 nm	20 hours	8.6at 0 V	0.91 V onset	Passivation layer	[224]
p-Si/HfO_x/SiO_x	HfOx & 5-nm	–	42 at 0.75 V		Protection layer	[226]
Ni-Mo/TiO_2/CdS/CIGS	TiO_2 & 20 nm	100 minutes	15–25 at 0 V	0.5 V onset	Passivation layer	[229]
p-Si/AZO/TiO_2/CoP_2	ZO/TiO_2 & 45 nm	–	−16.7 at 0 V	0.54 V onset	Passivation and protection layer	[227]
p-doped/intrinsic a-SiC:H	TiO_2 & 10 nm	–	10 at 0 V	0.8 V onset	Improve band alignment	[228]
Pt/C60/TiO_2/Sb_2Se_3	TiO_2 & 30 nm	10 hours	17		Protective layer	[230]
n^+np^+Si/Pt/TiO_2	TiO_2 & 15 nm	11.5%CE	–	–	Stability enhancement	[160]
CIGS/CdS/Al_2O_3/TiO_2/Pt	TiO_2 & 18 nm	8 h[4.5%]/STH 1.01%	–	–	Prevents photocathode and facilitates electron transfer	[231]
TiO_2HITcell	TiO_2 & 20 nm	>120 h/STH–10%	–	–	Passivation layer	[232]

and carrier separation underlies the observed improvement [233], which results from internal processes at photoelectrode interfaces. ALD overlayers play a unique part in each scenario. Since NiO may function as a protective layer and can serve as cocatalysts for the O_2 evolution process, the first scenario exemplifies the impact of ALD NiO and cocatalyst CoO_x for accelerated hydrogen and oxygen evolution. However, there has not been much effort put into evaluating the thickness of the protective layer, and it has been argued that depositions made beyond 300 cycles are counterproductive. However, the $BiVO_4$/CuO p-n junction was successfully passivated after the optimal 9 cycles of TiO_2. It has been theorized that a higher thickness of TiO_2 discourages hole transport to the electrolyte, leading to recombination at the CuO/TiO_2 interface. Furthermore, the valence band of TiO_2 is more positively charged than that of CuO, which means that thicker TiO_2 may not permit hole tunneling, but thin amorphous TiO_2 sheets may more easily tunnel holes from CuO. Therefore, thinner protective ALD layers is preferred over thicker films in light of the advantageous hole tunneling mechanism in PEC systems; however, it is crucial to acknowledge that optimization will be necessary, taking into account the base materials' morphological orientation and thickness.

In addition to improving stability, ALD layers are sometimes used to establish coating thickness, as detailed in the following research. To measure the quantity of H_2 evolved during the integration of perovskite solar cell and a PEC, a CdS photoanode was developed. The mesoporous TiO_2 layer and CdS in a tandem device were fabricated using a novel method to create a full photoanode. In addition, the device was combined with a $C_{s0.05}$ $(MA_{0.17}FA_{0.83})_{0.95}Pb(I_{0.83}Br_{0.17})_3$ triple-cation perovskite solar cell to combine the photoanode's high transparency with the perovskite's high absorption of light at longer wavelengths, resulting in improved PEC-conversion performance. When the tandem device was exposed to AM 1.5 G light, a photocurrent of an unbiased 7.8 mA cm^{-2} was produced [234]. This research employs ZnO as sacrificial films to better understand the changes in morphology that occur during the process of ion-exchange, and to optimize the deposition of CdS coated on the mesoporous TiO_2. The maximum saturated photocurrent was achieved in CdS-based electrodes with a sacrificial coating of ALD ZnO after 10 cycles. For the purpose of preventing nonuniformity and pore blockage in the coated samples, the highest saturated photocurrent in CdS-based electrodes was attained by performing 10 cycles of ALD ZnO immediately followed by ion exchange.

It has been shown that the PEC reactivities of photoanodes have been increased by combining them with active oxygen evolution catalysts. This is analogous to the phenomenon that occurred when photocathodes were mixed with active H_2 evolution catalysts. A black-Si/TiO_2/$Co(OH)_2$ heterostructured photoanode was produced by Yu et al., in which an ALD-deposited layer of TiO_2 played the role of a shield for the Si and $Co(OH)_2$ that were responsible for triggering the OER. The photocurrent that was generated was measured to be 32.3 mA cm^{-2}, and it was generated by using an external potential that was 1.48 V relative to RHE. The ALD-deposited TiO_2 coating on the b-Si/TiO_2/$Co(OH)_2$ photoanode was measured to have a thickness of 8 nm using cross-section of high-resolution imaging. The observed improvement in efficiency may be attributed mostly to the excellent defect covering on the charge-carrier separation and b-Si via the ALD layer that increased the lifespan to 4 h [235].

There are not much literature on the use of ALD for the direct fabrication of photoanodes due to the lengthy optimization time required for film production and analysis. PEC water-splitting applications have found success with the use of monoclinic bismuth vanadate photoanodes. Lamm et al. [236] used Vanadium (V) triisopropoxy oxide (VTIP; a vanadium source) with the $BiPh_3$ catalyst, followed by the deposition of ALD $BiVO_4$. As a result of the decomposition and the creation of numerous phases, the host material crystallized. Finally, $BiVO_4$ was formed by ALD, with film thickness restricted to 60 nm to serve as a barrier; this increased PEC activity and optoelectronic characteristics. Tantalum oxynitride (Ta_3N_5), another common photoanode material, was immediately produced by ALD and then converted to tantalum nitride through ammonolysis (750 °C/ 30 min) to measure PEC activity. In addition, Hajibabaei et al. [237] were fore-runners in developing the first Tantalum-doped TiO_2 film growth on conducting transparent base materials that could survive ammonolysis. It has been discovered that ALD-deposited Ta_3N_5 may be used as a photoanode; however, further in-depth study is needed to fully comprehend this system. This same team also performed Ta_3N_5 deposition using a custom-built method, and they discovered that Ta_3N_5 with a length of about 50 nm is optimal for hole diffusion, which is the performance-limiting factor for Ta_3N_5-based PEC devices [238]. In conclusion, studies suggest that 50–60 nm is an optimal thickness for developing direct photoanodes for PEC process, albeit this may vary with the material, growth pattern, and its band gap. Using ALD, Hannula et al. created a $TiO_2/TiSi_2/Si$ heterojunction that transfers charge between the photoanode materials through a Z-scheme mechanism. This Z-scheme relies on the $TiSi_2$ contact between Si and TiO_2 as its recombination center. According to the research, the stability of the photoanode is enhanced by the TiO_2 ALD layer, and the TiO_2, Si, and $TiSi_2$ bands are in a state of harmonic alignment for such charge transfer. The created technology served dual purposes as a protective layer and a photoactive medium. The $TiSi_2$ interlayers grown from between 3 and 10 nm TiO_2 films displayed significant band bending and substantial band offsets. $TiSi_2$ islands enabled a pinch off action that would be impossible with a thicker interlayer (greater than 30 nm) [239]. A similar kind of strategic electrical system might be used as a buffer layer in between the tiny band gap semiconductors and the electrolyte systems. Similar strategies of band bending and passivation are found in the Ta_3N_5-Cu_2O/Al-doped ZnO-TiO_2 (ALD)/CoOOH PEC system. In this system, a type II heterojunction occurs between Cu_2O and Ta_3N_5. Ta_3N_5's valence band holes tunnel to Cu_2O through Al-doped ZnO, and TiO_2's strong upward band bending enables charge extraction to the CoOOH cocatalyst for the water oxidation process, yielding 4.61 mA cm^{-2} at 1.23 V vs. RHE [240]. In a photoanode for water oxidation, interlayers play a critical function in facilitating the movement of charge carriers amongst the various semiconductor systems involved.

Another problem that needs quick attention for the stability of material and lengthy reactivity time is the photoelectrodes' corrosion. Silicon and other semicon-ductor photoanodes were protected from corrosion thanks to layers produced by Ros et al. [241] in their research, NiO was employed as a corrosion protector because of its unique combination of transparent, conductive, chemically stable, and catalytic behavior at the electrolyte/electrode interface. Thus, the produced ALD-deposited NiO protected the photoanode from significant deterioration, maintaining a high efficiency of 85% for more than 1000 hours at a 10 mA cm^{-2} photocurrent density. As a photoanode for PEC application was constructed between hierarchical CdS NR and WO_3 nanosheets (also known as "WO_3-NS/CdS-NR"). Wang et al. [242] demonstrated that the different components of the hierarchical system oxidize sulfite in a synergistic manner. Moreover, at 0.8 V vs. RHE, the produced photoanode generated a photocurrent density of approximately 5.4 mA cm^{-2}. In order to enhance the photostability of the heterojunction photoanode, an ALD layer of very thin TiO_2 consisting of 20 cycles was deposited on top of the WO_3-NS/CdS-NR array. Through the use of an ALD-deposited TiO_2 layer as the protective layer and a unified Al-ZnO/CdS/TiO_2 photoanode as the active component, Wang et al. [243] accomplished the creation of a photostable device that exhibited enhanced photostability. An approximate 11.7 mA cm^{-2} photocurrent was obtained at 1.23 V vs. RHE after 60 TiO_2-ALD cycles, with a conversion efficiency of 5.9% at 0.48 V. The active electrode is shielded from the electrolyte by the ALD-deposited layer of TiO_2, which also passivates the surface of the CdS and facilitates simple reactions. All of these factors contribute to the enhanced stability of the photoanode that is formed. In ALD-integrated device systems, thickness control may be employed to maximize output. Table 3.5 lists the functions and qualities attained by different kinds of photoanode materials combination with single ALD-deposited layers.

Metal oxides including iron oxide, titanium oxide, and zinc oxide are often employed in the PEC procedure because their small bandgaps are ideal for absorbing light. The combination of Fe_2O_3/ZnO/TiO_2 as a photoanode is recognize for its use in water-splitting processes. The structure that was produced is useful because it produces high levels of photocurrent density and considerable oxygen evolution after the Fe_2O_3 surface has been passivated with TiO_2 and ZnO. Unexpectedly, the photocurrent density is improved by the passivation layers by a factor of roughly seven, to approximately 4.25 mA cm^{-2}, because they prevent the recombination of photogenerated electron–hole pairs. Using the system described in Fig. 3.10a, b for the oxygen evolution process [244], one can see the differential influence of electrolyte type of materials surfaces adsorption. Nandanapalli et al. [245] designed a ZnO/CoO core-shell nanomaterial for use in PEC applications. Their research asserts that the self-assembled ZnO nanostructure, which was seed-deposited with an ALD layer of ZnO, outperformed the bare ZnO nanostructure by a significant margin.

Figure 3.10c is a schematic of the photoanode manufacturing process, and it demonstrates how the ALD increases the stability of the nanomaterial [140]. As seen in Fig. 3.10(i), the samples' colors changed noticeably from white to a light greenish blue as the CoO layer grew over the ZnO nanostructures during deposition. The surface morphology and the extent to which CoO covers ZnO nanostructures both alter subtly with increasing deposition time, as shown in SEM images. Therefore, the ZnO nanostructures are totally concealed by the CoO layers produced at a time

Table 3.5 Features of photoanode active materials

Active photoanode material	ALD material & thickness	Efficiency/ stability	Current density (mA/cm^2)	Functionality	Ref.
Perovskite CdS/TiO_2 integrated PV	ZnO & 10ALD cycle	–	7.8	Using ALD with ion-exchange to create CdS Ignores the problem of clogged pores and uneven covering	[234]
b-Si/ 8nmTiO_2/Co $(OH)_2$	TiO_2 & 8 nm	4 hours	32.3 at 1.48 V	Extends the operational lifetime	[235]
Ta_3N_5/FTO	TaO_x & 28 nm	–	2.4 at 1.23 V	Enhanced conductivity	[238]
NiO/Ni $(OH)_2$/ NiOOH	NiO & 50 nm	85% retention after 1000 hours	10	Enhanced stability	[241]
WO_3-NS/ CdS-NR/ TiO_2	Ultrathin TiO_2 & 20 cycles	–	5.4 at 0.8 V	Enhanced photostability	[242]
Al-ZnO/ CdS/TiO_2	TiO_2 & 3 nm	5.9% CE at 0.48 V	11.7 at 1.23 V	Isolating the point of direct touch, passing a charge through the surface, Spotting holes created by light	[243]
Fe_2O_3/ZnO/ TiO_2	ZnO & 30 nm	–	4.25	Passivation and reduce surface recombination	[244]
ZnO/CuO	TiO_2 & 12 nm ZnO & 50–60 nm	IPCE 16%	–	Seed layer	[245]
α-Fe_2O_3/Co (dca)/TiO_2	TiO_2 & 50 cycles	IPCE 16% at 1.6 V	0.35 at 1.23 V	Passivates surface defects and suppress recombination	[246]
CdS@ 3.5 nm TiO_2	TiO_2 & 3.5 nm	9% decay after 9 hours	1.8 at 0 V	Limit the charge recombination and enhance the wettability of the electrodes	[247]
Si/Graphene/ TiO_2/	TiO_2 & 10 nm	–	19 at 1.5 V	Protective layer	[248]

period of 180 s. X-ray diffraction analysis shows that CoO growth has no effect on the crystal structure of ZnO nanostructures since all of the diffraction peaks of the CoO-encapsulated nanostructures are identical to those of pure ZnO nanostructures. Except for CoO structures developed in 30 s, their intensities decreased with increasing CoO layer thickness (Fig. 3.10c(ii)). The XRD peak intensity was substantially higher in the instance of ZnO nanostructures growing in 30 s of CoO. Passivation of surface imperfections and physisorbed molecules such as -OH, H_2O, likely account for this. Because of this, ZnO crystals became more

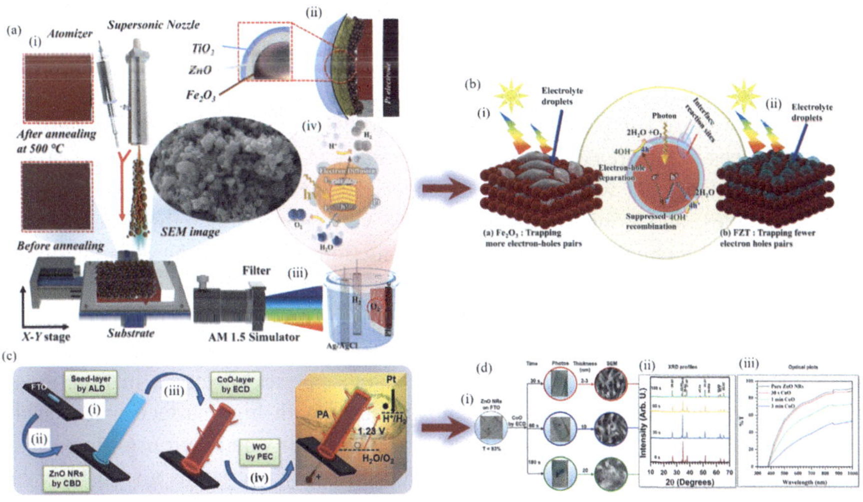

Fig. 3.10 (**a**) SEM images of the sample's surface morphology. (i) Cold-spraying setup schematic showing Fe_2O_3 films before and after annealing. (ii) ALD-deposited ZnO/TiO_2 layers on Fe2O3. (iii) PEC setup with developed photoanodes. (iv) The enlarged image depicts electron hole creation caused by solar radiation. (**b**) Schematic depiction of the PEC water-splitting mechanism showing (i) a Fe_2O_3 surface with elliptical electrolyte droplets and (ii) an FZT surface with spherical droplets. The middle image illustrate O_2-evolution process, which occurs when FZT surface is illuminated. [(**a–b**) Reprinted with permission from Ref. [244]]. (**c**) Schematic illustration of the fabricated CoO-encapsulated ZnO nanostructures and PEC devices: (i) ALD ZnO film growth on an FTO substrate, (ii) vertically aligned CBD grown ZnO nanostructures, (iii) CoO-layered ZnO nanostructures grown by ECD, and (iv) water-oxidation (WO) performance of FTO/ZnO-film/CoO-passivated ZnO NR structures. (d) The effects of CoO growth on the morphology and optical characteristics of ZnO nanostructures. (i) Photographs and SEM images depicting the physical appearance and surface morphology of CoO-produced ZnO nanostructures at various times (30, 60, and 180 s). ZnO nanostructures produced with and without CoO layers show significant variation in (ii) their structural and (iii) their optical characteristics as a function of deposition time. [(**c–d**) Reprinted with permission from Ref. [245]]

pure. However, with the exception of 30 s CoO developed structures, the optical transmittance of ZnO nanostructures considerably dropped as the CoO deposition time increased (Fig. 3.10c(i),). To improve PEC efficiency, Alfakes et al. [142] employed ALD to Hf-doped ZnO as the base material. It was argued [30] that Hf, as a metal dopant, is useful for improving PEC reactivity as it allows for fine-tuning of the overall composition of the material, resulting in improved optical and electrical characteristics.

ALD's contributions to improving PEC system reactivity as a result of its ability to regulate band parameters, electrolyte/material protection, corrosion protection, hole channeling, core-shell structures, heterojunction, band alignments, and electrode/electrolyte interface reaction cannot be overstated. However, it is challenging to solve all the difficulties in a single system, thus depending on the selection of base material, tackling the main becomes priority.

3.5 Summary

The ALD process has been successfully applied on a small scale in the lab, and it holds great promise for paving the way towards significant, ultimately beneficial future advances in the area of green fuel technology. As ALD has progressed, it has been used to deposit a wide variety of materials, including oxides, nitrides, sulphides, carbides, and even elementary metals, verifying the incredible benefits of expanding ALD technology in hydrogen production. However, the compositional and structural designs of ALD are delicate. For first of all, surface alterations rely heavily on the ultra-thin deposition made possible by the self-limiting aspect of the ALD process. Therefore, optimizing the unit interface leads to better energy device performance. Furthermore, ALD may create nanoscale architectures in functional materials, not only film deposition. Nanostructures with high surface areas, such electrodes, were used because of the materials' adaptability and their usefulness in a variety of contexts. Lastly, the ALD-deposited film may operate as a seed layer or reactant to direct the uniform development of additional materials in a film form with the right posttreatment. Electrolytic processes often make use of metal-organic frameworks (MOFs) and transition metal dichalcogenides (TMDs). The need for electrochemical energy devices is extremely high and rising, therefore, even if most of these experiments are still in the lab, mass manufacturing of the materials and structures via ALD has enormous promise. Researchers have looked at new ways to increase the throughput and decrease the cost, and various novel configurations of ALD systems have been developed to fulfil the needs of industrialization.

When looking at one important component, that of increasing PEC-photoelectrode efficiency, many of the early researchers in the field of PEC-device-technology have noted the need to adjust interfaces to enable smooth charge flow across the system, which increases reactivity. The charge-transfer reaction mechanism in a PEC device relies on materials that are electrochemically and photonically active for PEC processes (e.g., Fe_2O_3, TiO_2, ZnO), innovative design strategies that guarantee strong interfacial bonding between the photoelectrode materials, and materials-design approaches like nanostructure self-assembly, 3D printing, heterojunction, and Z-scheme formation. It is very complex and difficult to modify bulk interfacial characteristics. Since ALD is self-limiting and can heal surface defects more efficiently than any other thin-film fabrication technique, it is currently the most useful thin-film technology for controlling atomic-layer processing on the surface of any host PEC-active substrate. Since the ultrathin ALD layer does not alter the characteristics of the constituent components in the system, ALD makes it much simpler to fabricate and tune the parameters of a PEC device. The PEC water-splitting application of ALD has shown the technique's viability for future usage in solar hydrogen generation. Various treatments, such as thermal treatment and atomic bombardment, can change interfacial properties, which aid in fabricating the photoanode and the photocathode. ALD-deposited layers have been widely used as potential light absorbers in PEC applications. The main emphasis was on how ALD changes the characteristics of a material to improve

PEC efficiency or material stability, two of the most important aspects of ALD for manipulating photoelectrodes for PEC water-splitting applications.

ALD materials have much greater peak hydrogen reaction rates than most thermally and chemically modified materials with smaller surface areas because of their stability and large surface area. Maghemite (γF_2O_3) and spinel cobalt oxide ($Co_xFe_{3-x}O_4$) nanoscale (2–5 nm) conformal films produced by ALD on porous m-ZrO_2 substrates demonstrate the capacity to be cycled swiftly and frequently in chemical looping cycles for the generation of hydrogen. Cobalt ferrites are the better option between these two materials for several reasons. First, they may be completely reduced to metallic Co and Fe, which would greatly increase the hydrogen yields. Second, these materials are robust enough to withstand up to seven redox cycles, which suggests well for their long-term use. Finally, their hydrogen generation rates are noticeably higher than powders' because of their totally exposed surfaces to the gas, as opposed to a dense microcrystalline structure whose production is rate-limited by oxygen transport along grain boundaries or through the bulk. Altogether, these findings point the way towards long-lasting materials that can withstand a wide range of heat redox and chemical looping cycles. What's more, they demonstrate that ALD thin films provide a promising platform for exploring basic reaction chemistry without the constraints of transport through the bulk or along grain boundaries. However, the substantial surface area of the supports enables significant oxide loadings, demonstrating that realistic hydrogen generation rates are possible with these materials, despite the films' thinness. ALD techniques already exist for depositing many different materials, demonstrating the platform's adaptability.

References

1. W.M.M. Kessels, H.C.M. Knoops, S.A.F. Dielissen, A.J.M. MacKus, M.C.M. Van De Sanden, Surface reactions during atomic layer deposition of Pt derived from gas phase infrared spectroscopy. Appl. Phys. Lett. **95**, 93–96 (2009). https://doi.org/10.1063/1.3176946
2. A.J.M. MacKus, M.A. Verheijen, N. Leick, A.A. Bol, W.M.M. Kessels, Influence of oxygen exposure on the nucleation of platinum atomic layer deposition: Consequences for film growth, nanopatterning, and nanoparticle synthesis. Chem. Mater. **25**, 1905–1911 (2013). https://doi.org/10.1021/cm400562u
3. J.W. Clancey, A.S. Cavanagh, R.S. Kukreja, A. Kongkanand, S.M. George, Atomic layer deposition of ultrathin platinum films on tungsten atomic layer deposition adhesion layers: Application to high surface area substrates. J. Vac. Sci. Technol. A Vac. Surf. Film. **33**, 01A130 (2015). https://doi.org/10.1116/1.4901459
4. J.A. Enterkin, W. Setthapun, J.W. Elam, S.T. Christensen, F.A. Rabuffetti, L.D. Marks, P.C. Stair, K.R. Poeppelmeier, C.L. Marshall, Propane oxidation over Pt/SrTiO3 nanocuboids. ACS Catal. **1**, 629–635 (2011). https://doi.org/10.1021/cs200092c
5. J. Lu, B. Liu, J.P. Greeley, Z. Feng, J.A. Libera, Y. Lei, M.J. Bedzyk, P.C. Stair, J.W. Elam, Porous alumina protective coatings on palladium nanoparticles by self-poisoned atomic layer deposition. Chem. Mater. **24**, 2047–2055 (2012). https://doi.org/10.1021/cm300203s

6. S.T. Christensen, H. Feng, J.L. Libera, N. Guo, J.T. Miller, P.C. Stair, J.W. Elam, Supported Ru-Pt bimetallic nanoparticle catalysts prepared by atomic layer deposition. Nano Lett. **10**, 3047–3051 (2010). https://doi.org/10.1021/nl101567m
7. J. Lu, K. Bin Low, Y. Lei, J.A. Libera, A. Nicholls, P.C. Stair, J.W. Elam, Toward atomically-precise synthesis of supported bimetallic nanoparticles using atomic layer deposition. Nat. Commun. **5**, 1–9 (2014). https://doi.org/10.1038/ncomms4264
8. J. Lu, B. Fu, M.C. Kung, G. Xiao, J.W. Elam, H.H. Kung, P.C. Stair, Coking- and sintering-resistant palladium catalysts achieved through atomic layer deposition. Science **355**(80), 1205–1208 (2012). https://doi.org/10.1126/science.1212906
9. Y. Lei, B. Liu, J. Lu, R.J. Lobo-Lapidus, T. Wu, H. Feng, X. Xia, A.U. Mane, J.A. Libera, J.P. Greeley, J.T. Miller, J.W. Elam, Synthesis of Pt-Pd core-shell nanostructures by atomic layer deposition: Application in propane oxidative dehydrogenation to propylene. Chem. Mater. **24**, 3525–3533 (2012). https://doi.org/10.1021/cm300080w
10. B.J. O'Neill, D.H.K. Jackson, A.J. Crisci, C.A. Farberow, F. Shi, A.C. Alba-Rubio, J. Lu, P.J. Dietrich, X. Gu, C.L. Marshall, P.C. Stair, J.W. Elam, J.T. Miller, F.H. Ribeiro, P.M. Voyles, J. Greeley, M. Mavrikakis, S.L. Scott, T.F. Kuech, J.A. Dumesic, Stabilization of copper catalysts for liquid-phase reactions by atomic layer deposition. Angew. Chem. Int. Ed. **52**, 13808–13812 (2013). https://doi.org/10.1002/anie.201308245
11. A.C. Alba-Rubio, B.J. O'Neill, F. Shi, C. Akatay, C. Canlas, T. Li, R. Winans, J.W. Elam, E.A. Stach, P.M. Voyles, J.A. Dumesic, Pore structure and bifunctional catalyst activity of overlayers applied by atomic layer deposition on copper nanoparticles. ACS Catal. **4**, 1554–1557 (2014). https://doi.org/10.1021/cs500330p
12. B.J. O'Neill, C. Sener, D.H.K. Jackson, T.F. Kuech, J.A. Dumesic, Control of thickness and chemical properties of atomic layer deposition overcoats for stabilizing Cu/γ-Al_2O_3 catalysts. ChemSusChem **7**, 3247–3251 (2014). https://doi.org/10.1002/cssc.201402832
13. T.D. Gould, A. Izar, A.W. Weimer, J.L. Falconer, J.W. Medlin, Stabilizing Ni catalysts by molecular layer deposition for harsh, dry reforming conditions. ACS Catal. **4**, 2714–2717 (2014). https://doi.org/10.1021/cs500809w
14. J. Lee, D.H.K. Jackson, T. Li, R.E. Winans, J.A. Dumesic, T.F. Kuech, G.W. Huber, Enhanced stability of cobalt catalysts by atomic layer deposition for aqueous-phase reactions. Energy Environ. Sci. **7**, 1657–1660 (2014). https://doi.org/10.1039/c4ee00379a
15. J. Chavez-Galan, R. Almanza, Solar filters based on iron oxides used as efficient windows for energy savings. Sol. Energy **81**, 13–19 (2007). https://doi.org/10.1016/j.solener.2006.06.009
16. C.D. Park, D. Magana, A.E. Stiegman, High-quality Fe and γ-Fe_2O_3 magnetic thin films from an epoxide-catalyzed sol-gel process. Chem. Mater. **19**, 677–683 (2007). https://doi.org/10.1021/cm0617079
17. J.A. Glasscock, P.R.F. Barnes, I.C. Plumb, A. Bendavid, P.J. Martin, Structural, optical and electrical properties of undoped polycrystalline hematite thin films produced using filtered arc deposition. Thin Solid Films **516**, 1716–1724 (2008). https://doi.org/10.1016/j.tsf.2007.05.020
18. B. Mauvernay, L. Presmanes, S. Capdeville, V.G. de Resende, E. De Grave, C. Bonningue, P. Tailhades, Elaboration and characterization of Fe_{1-x} thin films sputter deposited from magnetite target, Thin Solid Films **515** (2007), 6532–6536. https://doi.org/10.1016/j.tsf.2006.11.131
19. V. Meille, Review on methods to deposit catalysts on structured surfaces. Appl. Catal. A Gen. **315**, 1–17 (2006). https://doi.org/10.1016/j.apcata.2006.08.031
20. S. Park, S. Lim, H. Choi, Chemical vapor deposition of iron and iron oxide thin films from Fe (II) dihydride complexes. Chem. Mater. **18**, 5150–5152 (2006). https://doi.org/10.1021/cm0601990
21. E. Kester, B. Gillot, Cation distribution, thermodynamic and kinetics considerations in nanoscaled copper ferrite spinels. New experimental approach by XPS and new results both in the bulk and on the grain boundary. J. Phys. Chem. Solids **59**, 1259–1269 (1998). https://doi.org/10.1016/S0022-3697(98)00061-4

22. B. Fahlman, Recent advances in chemical vapor deposition. Curr. Org. Chem. **10**, 1021–1033 (2006). https://doi.org/10.2174/138527206777435481
23. L.F. Hakim, S.M. George, A.W. Weimer, Conformal nanocoating of zirconia nanoparticles by atomic layer deposition in a fluidized bed reactor. Nanotechnology **16**, S375 (2005). https://doi.org/10.1088/0957-4484/16/7/010
24. Z.H. Barber, The control of thin film deposition and recent developments in oxide film growth. J. Mater. Chem. **16**, 334–344 (2006). https://doi.org/10.1039/b506228d
25. L.F. Hakim, J. Blackson, S.M. George, A.W. Weimer, Nanocoating individual silica nanoparticles by atomic layer deposition in a fluidized bed reactor. Chem. Vap. Depos. **11**, 420–425 (2005). https://doi.org/10.1002/cvde.200506392
26. P.O. Oviroh, R. Akbarzadeh, D. Pan, R.A.M. Coetzee, T.C. Jen, New development of atomic layer deposition: processes, methods and applications. Sci. Technol. Adv. Mater. **20**, 465–496 (2019). https://doi.org/10.1080/14686996.2019.1599694
27. S.M. George, A.W. Ott, J.W. Klaus, Surface chemistry for atomic layer growth. J. Phys. Chem. **100**, 13121–13131 (1996). https://doi.org/10.1021/jp9536763
28. L. Niinistö, J. Päiväsaari, J. Niinistö, M. Putkonen, M. Nieminen, Advanced electronic and optoelectronic materials by atomic layer deposition: An overview with special emphasis on recent progress in processing of high-k dielectrics and other oxide materials. Phys. Status Solidi Appl. Res. **201**, 1443–1452 (2004). https://doi.org/10.1002/pssa.200406798
29. D.M. King, J.A. Spencer, X. Liang, L.F. Hakim, A.W. Weimer, Atomic layer deposition on particles using a fluidized bed reactor with in situ mass spectrometry. Surf. Coat. Technol. **201**, 9163–9171 (2007). https://doi.org/10.1016/j.surfcoat.2007.05.002
30. J.H. Han, G. Gao, Y. Widjaja, E. Garfunkel, C.B. Musgrave, A quantum chemical study of ZrO_2 atomic layer deposition growth reactions on the SiO_2 surface. Surf. Sci. **550**, 199–212 (2004). https://doi.org/10.1016/j.susc.2003.12.030
31. D. Reeves, Energy requirements in rye production. Handb. Energy Util. Agric. (1980). http://www.cabdirect.org/abstracts/19810719500.html
32. T. Nakumura, Hydrogen production from water utilizing. Sol. Energy **19**, 467–475 (1977)
33. M. Lie, H. Fjellvåg, A. Kjekshus, Growth of Fe_2O_3 thin films by atomic layer deposition. Thin Solid Films **488**, 74–81 (2005). https://doi.org/10.1016/j.tsf.2005.04.063
34. M. De Ridder, P.C. Van De Ven, R.G. Van Welzenis, H.H. Brongersma, S. Helfensteyn, C. Creemers, P. Van Der Voort, M. Baltes, M. Mathieu, E.F. Vansant, Growth of iron oxide on yttria-stabilized zirconia by atomic layer deposition. J. Phys. Chem. B **106**, 13146–13153 (2002). https://doi.org/10.1021/jp0211640
35. L.F. Hakim, J.L. Portman, M.D. Casper, A.W. Weimer, Aggregation behavior of nanoparticles in fluidized beds. Powder Technol. **160**, 149–160 (2005). https://doi.org/10.1016/j.powtec.2005.08.019
36. D. Fino, N. Russo, G. Saracco, V. Specchia, Catalytic removal of NO_x and diesel soot over nanostructured spinel-type oxides. J. Catal. **242**, 38–47 (2006). https://doi.org/10.1016/j.jcat.2006.05.023
37. J.B. Silva, C.F. Diniz, R.M. Lago, N.D.S. Mohallem, Catalytic properties of nanocomposites based on cobalt ferrites dispersed in sol-gel silica. J. Non-Cryst. Solids **348**, 201–204 (2004). https://doi.org/10.1016/j.jnoncrysol.2004.08.169
38. Q. Yang, H. Choi, S.R. Al-Abed, D.D. Dionysiou, Iron-cobalt mixed oxide nanocatalysts: Heterogeneous peroxymonosulfate activation, cobalt leaching, and ferromagnetic properties for environmental applications. Appl. Catal. B Environ. **88**, 462–469 (2009). https://doi.org/10.1016/j.apcatb.2008.10.013
39. M. Azhar Uddin, H. Tsuda, S. Wu, E. Sasaoka, Catalytic decomposition of biomass tars with iron oxide catalysts. Fuel **87**, 451–459 (2008). https://doi.org/10.1016/j.fuel.2007.06.021
40. I. Baldychev, J.M. Vohs, R.J. Gorte, The effect of thermodynamic properties of zirconia-supported Fe_3O_4 on water-gas shift activity. Appl. Catal. A Gen. **356**, 225–230 (2009). https://doi.org/10.1016/j.apcata.2009.01.012

41. J.Y. Lee, D.W. Lee, K.Y. Lee, Y. Wang, Cr-free Fe-based metal oxide catalysts for high temperature water gas shift reaction of fuel processor using LPG. Catal. Today **146**, 260–264 (2009). https://doi.org/10.1016/j.cattod.2009.01.041
42. T.J. Yoon, W. Lee, Y.S. Oh, J.K. Lee, Magnetic nanoparticles as a catalyst vehicle for simple and easy recycling. New J. Chem. **27**, 227–229 (2003). https://doi.org/10.1039/b209391j
43. K.B. Klepper, O. Nilsen, H. Fjellvåg, Growth of thin films of Co_3O_4 by atomic layer deposition. Thin Solid Films **515**, 7772–7781 (2007). https://doi.org/10.1016/j.tsf.2007.03.182
44. K.B. Klepper, O. Nilsen, H. Fjellvåg, Deposition of thin films of organic-inorganic hybrid materials based on aromatic carboxylic acids by atomic layer deposition. Dalt. Trans. **39**, 11628–11635 (2010). https://doi.org/10.1039/c0dt00817f
45. M. Rooth, A. Johansson, K. Kukli, J. Aarik, M. Boman, A. Hårsta, Atomic layer deposition of iron oxide thin films and nanotubes using ferrocene and oxygen as precursors. Chem. Vap. Depos. **14**, 67–70 (2008). https://doi.org/10.1002/cvde.200706649
46. B.S. Lim, A. Rahtu, R.G. Gordon, Atomic layer deposition of transition metals. Nat. Mater. **2**, 749–754 (2003). https://doi.org/10.1038/nmat1000
47. J.R. Scheffe, A. Francés, D.M. King, X. Liang, B.A. Branch, A.S. Cavanagh, S.M. George, A.W. Weimer, Atomic layer deposition of iron(III) oxide on zirconia nanoparticles in a fluidized bed reactor using ferrocene and oxygen. Thin Solid Films **517**, 1874–1879 (2009). https://doi.org/10.1016/j.tsf.2008.09.086
48. J. Bachmann, J. Jing, M. Knez, S. Barth, H. Shen, S. Mathur, U. Gösele, K. Nielsch, Ordered iron oxide nanotube arrays of controlled geometry and tunable magnetism by atomic layer deposition. J. Am. Chem. Soc. **129**, 9554–9555 (2007). https://doi.org/10.1021/ja072465w
49. O. Nilsen, M. Lie, S. Foss, H. Fjellvåg, A. Kjekshus, Effect of magnetic field on the growth of α-Fe_2O_3 thin films by atomic layer deposition. Appl. Surf. Sci. **227**, 40–47 (2004). https://doi.org/10.1016/j.apsusc.2003.10.045
50. M. Lie, K. Barnholt Klepper, O. Nilsen, H. Fjellvåg, A. Kjekshus, Growth of iron cobalt oxides by atomic layer deposition. J. Chem. Soc. Dalt. Trans., 253–259 (2007). https://doi.org/10.1039/b711718n
51. J.R. Scheffe, M.D. Allendorf, E.N. Coker, B.W. Jacobs, A.H. McDaniel, A.W. Weimer, Hydrogen production via chemical looping redox cycles using atomic layer deposition-synthesized iron oxide and cobalt ferrites. Chem. Mater. **23**, 2030–2038 (2011). https://doi.org/10.1021/cm103622e
52. J.R. Scheffe, J. Li, A.W. Weimer, A spinel ferrite/hercynite water-splitting redox cycle. Int. J. Hydrog. Energy **35**, 3333–3340 (2010). https://doi.org/10.1016/j.ijhydene.2010.01.140
53. C.L. Muhich, B.D. Ehrhart, V.A. Witte, S.L. Miller, E.N. Coker, C.B. Musgrave, A.W. Weimer, Predicting the solar thermochemical water splitting ability and reaction mechanism of metal oxides: A case study of the hercynite family of water splitting cycles. Energy Environ. Sci. **8**, 3687–3699 (2015). https://doi.org/10.1039/c5ee01979f
54. C.L. Muhich, B.W. Evanko, K.C. Weston, P. Lichty, X. Liang, J. Martinek, C.B. Musgrave, A.W. Weimer, Efficient generation of H2 by splitting water with an isothermal redox cycle. Science **341**(80), 540–542 (2013). https://doi.org/10.1126/science.1239454
55. Z. Chen, R. Wu, Y. Liu, Y. Ha, Y. Guo, D. Sun, M. Liu, F. Fang, Ultrafine Co nanoparticles encapsulated in carbon-nanotubes-grafted graphene sheets as advanced electrocatalysts for the hydrogen evolution reaction. Adv. Mater. **30**, 1–10 (2018). https://doi.org/10.1002/adma.201802011
56. S. Chandrasekaran, N. Kaeffer, L. Cagnon, D. Aldakov, J. Fize, G. Nonglaton, F. Baleras, P. Mailley, V. Artero, A robust ALD-protected silicon-based hybrid photoelectrode for hydrogen evolution under aqueous conditions. Chem. Sci. **10**, 4469–4475 (2019). https://doi.org/10.1039/c8sc05006f
57. I.S. Kim, M.J. Pellin, A.B.F. Martinson, Acid-compatible halide perovskite photocathodes utilizing atomic layer deposited TiO 2 for solar-driven hydrogen evolution. ACS Energy Lett. **4**, 293–298 (2019). https://doi.org/10.1021/acsenergylett.8b01661

58. N. Cheng, S. Stambula, D. Wang, M.N. Banis, J. Liu, A. Riese, B. Xiao, R. Li, T.K. Sham, L.M. Liu, G.A. Botton, X. Sun, Platinum single-atom and cluster catalysis of the hydrogen evolution reaction. Nat. Commun. **7**, 1–9 (2016). https://doi.org/10.1038/ncomms13638
59. Z. Zhang, L. Deng, Z. Zhao, Y. Zhao, J. Yang, J. Jiang, G. Huang, Y. Mei, Nickel nanograins anchored on a carbon framework for an efficient hydrogen evolution electrocatalyst and a flexible electrode. J. Mater. Chem. A **8**, 3499–3508 (2020). https://doi.org/10.1039/c9ta13632k
60. X. Zhang, B. Shao, Z. Sun, Z. Gao, Y. Qin, C. Zhang, F. Cui, X. Yang, Platinum nanoparticle-deposited Ti3C2Tx MXene for hydrogen evolution reaction. Ind. Eng. Chem. Res. **59**, 1822–1828 (2020). https://doi.org/10.1021/acs.iecr.9b05046
61. K. Cao, L. Shi, M. Gong, J. Cai, X. Liu, S. Chu, Y. Lang, B. Shan, R. Chen, Nanofence stabilized platinum nanoparticles catalyst via facet-selective atomic layer deposition. Small **13**, 1–7 (2017). https://doi.org/10.1002/smll.201700648
62. J. Cai, J. Zhang, K. Cao, M. Gong, Y. Lang, X. Liu, S. Chu, B. Shan, R. Chen, Selective passivation of Pt nanoparticles with enhanced sintering resistance and activity toward CO oxidation via atomic layer deposition. ACS Appl. Nano Mater. **1**, 522–530 (2018). https://doi.org/10.1021/acsanm.7b00026
63. K. Cao, Q. Zhu, B. Shan, R. Chen, Controlled synthesis of Pd/Pt core shell nanoparticles using area-selective atomic layer deposition. Sci. Rep. **5**, 1–7 (2014). https://doi.org/10.1038/srep08470
64. R. Chen, H. Kim, P.C. McIntyre, D.W. Porter, S.F. Bent, Achieving area-selective atomic layer deposition on patterned substrates by selective surface modification. Appl. Phys. Lett. **86**, 1–3 (2005). https://doi.org/10.1063/1.1922076
65. X. Liu, Q. Zhu, Y. Lang, K. Cao, S. Chu, B. Shan, R. Chen, Oxide-nanotrap-anchored platinum nanoparticles with high activity and sintering resistance by area-selective atomic layer deposition. Angew. Chem. Int. Ed. **56**, 1648–1652 (2017). https://doi.org/10.1002/anie.201611559
66. M. Song, H. Tan, X. Li, A.I.Y. Tok, P. Liang, D. Chao, H.J. Fan, Atomic-layer-deposited amorphous MoS_2 for durable and flexible Li–O_2 batteries. Small Methods **4**, 1–10 (2020). https://doi.org/10.1002/smtd.201900274
67. D.H. Kwon, Z. Jin, S. Shin, W.S. Lee, Y.S. Min, A comprehensive study on atomic layer deposition of molybdenum sulfide for electrochemical hydrogen evolution. Nanoscale **8**, 7180–7188 (2016). https://doi.org/10.1039/c5nr09065b
68. S. Balasubramanyam, M. Shirazi, M.A. Bloodgood, L. Wu, M.A. Verheijen, V. Vandalon, W.M.M. Kessels, J.P. Hofmann, A.A. Bol, Edge-site nanoengineering of WS2 by low-temperature plasma-enhanced atomic layer deposition for electrocatalytic hydrogen evolution. Chem. Mater. **31**, 5104–5115 (2019). https://doi.org/10.1021/acs.chemmater.9b01008
69. M.X. Wu, Y.W. Yang, Metal–organic framework (MOF)-based drug/cargo delivery and cancer therapy. Adv. Mater. **29**, 1–20 (2017). https://doi.org/10.1002/adma.201606134
70. M.S. Khan, M. Shahid, Synthesis of Metal-Organic Frameworks (MOFs): Routes to various MOF topologies, morphologies, and composites. Electrochem. Appl. Met. Fram. Adv. Future Potential, 17–35 (2022). https://doi.org/10.1016/B978-0-323-90784-2.00007-1
71. O. Shekhah, J. Liu, R.A. Fischer, C. Wöll, MOF thin films: Existing and future applications. Chem. Soc. Rev. **40**, 1081–1106 (2011). https://doi.org/10.1039/c0cs00147c
72. V. Stavila, A.A. Talin, M.D. Allendorf, MOF-based electronic and opto-electronic devices. Chem. Soc. Rev. **43**, 5994–6010 (2014). https://doi.org/10.1039/c4cs00096j
73. W. Chaikittisilp, K. Ariga, Y. Yamauchi, A new family of carbon materials: Synthesis of MOF-derived nanoporous carbons and their promising applications. J. Mater. Chem. A **1**, 14–19 (2013). https://doi.org/10.1039/c2ta00278g
74. D. Sheberla, J.C. Bachman, J.S. Elias, C.J. Sun, Y. Shao-Horn, M. Dincă, Conductive MOF electrodes for stable supercapacitors with high areal capacitance. Nat. Mater. **16**, 220–224 (2017). https://doi.org/10.1038/nmat4766

75. F. Zou, X. Hu, Z. Li, L. Qie, C. Hu, R. Zeng, Y. Jiang, Y. Huang, MOF-derived porous $ZnO/ZnFe_2O_4/C$ octahedra with hollow interiors for high-rate lithium-ion batteries. Adv. Mater. **26**, 6622–6628 (2014). https://doi.org/10.1002/adma.201402322
76. Y. Liu, J.F. Eubank, A.J. Cairns, J. Eckert, V.C. Kravtsov, R. Luebke, M. Eddaoudi, Assembly of Metal-Organic Frameworks (MOFs) based on indium-trimer building blocks: A porous MOF with soc topology and high hydrogen storage. Angew. Chem. Int. Ed. **46**, 3278–3283 (2007). https://doi.org/10.1002/anie.200604306
77. S.J. Yang, T. Kim, J.H. Im, Y.S. Kim, K. Lee, H. Jung, C.R. Park, MOF-derived hierarchically porous carbon with exceptional porosity and hydrogen storage capacity. Chem. Mater. **24**, 464–470 (2012). https://doi.org/10.1021/cm202554j
78. X. Zhang, A. Chen, M. Zhong, Z. Zhang, X. Zhang, Z. Zhou, X.H. Bu, *Metal–Organic Frameworks (MOFs) and MOF-Derived Materials for Energy Storage and Conversion* (Springer Singapore, 2019). https://doi.org/10.1007/s41918-018-0024-x
79. Z. Zhao, S. Liu, J. Zhu, J. Xu, L. Li, Z. Huang, C. Zhang, T. Liu, Hierarchical nanostructures of nitrogen-doped porous carbon polyhedrons confined in carbon nanosheets for high-performance supercapacitors. ACS Appl. Mater. Interfaces **10**, 19871–19880 (2018). https://doi.org/10.1021/acsami.8b03431
80. H.L. Jiang, Y. Tatsu, Z.H. Lu, Q. Xu, Non-, micro-, and mesoporous metal-organic framework isomers: Reversible transformation, fluorescence sensing, and large molecule separation. J. Am. Chem. Soc. **132**, 5586–5587 (2010). https://doi.org/10.1021/ja101541s
81. X. Ge, Z. Li, L. Yin, Metal-organic frameworks derived porous core/shellCoP@C polyhedrons anchored on 3D reduced graphene oxide networks as anode for sodium-ion battery. Nano Energy **32**, 117–124 (2017). https://doi.org/10.1016/j.nanoen.2016.11.055
82. Y. Yoo, H.K. Jeong, Rapid fabrication of metal organic framework thin films using microwave-induced thermal deposition. Chem. Commun., 2441–2443 (2008). https://doi.org/10.1039/b800061a
83. A. Zhang, X.Y. Li, S. Zhang, Z. Yu, X. Gao, X. Wei, Z. Wu, W.D. Wu, X.D. Chen, Spray-drying-assisted reassembly of uniform and large micro-sized MIL-101 microparticles with controllable morphologies for benzene adsorption. J. Colloid Interface Sci. **506**, 1–9 (2017). https://doi.org/10.1016/j.jcis.2017.07.022
84. M.C. So, S. Jin, H.J. Son, G.P. Wiederrecht, O.K. Farha, J.T. Hupp, Layer-by-layer fabrication of oriented porous thin films based on porphyrin-containing metal-organic frameworks. J. Am. Chem. Soc. **135**, 15698–15701 (2013). https://doi.org/10.1021/ja4078705
85. G. Jia, W. Zhang, G. Fan, Z. Li, D. Fu, W. Hao, C. Yuan, Z. Zou, Three-dimensional hierarchical architectures derived from surface-mounted metal–organic framework membranes for enhanced electrocatalysis. Angew. Chem. Int. Ed. **56**, 13781–13785 (2017). https://doi.org/10.1002/anie.201708385
86. W. Zhang, X. Yao, S. Zhou, X. Li, L. Li, Z. Yu, L. Gu, ZIF-8/ZIF-67-derived Co-Nx-embedded 1D porous carbon nanofibers with graphitic carbon-encased Co nanoparticles as an efficient bifunctional electrocatalyst. Small **14**, 1–8 (2018). https://doi.org/10.1002/smll.201800423
87. H. Liu, H. Wang, T. Chu, M. Yu, Y. Yang, An electrodeposited lanthanide MOF thin film as a luminescent sensor for carbonate detection in aqueous solution. J. Mater. Chem. C **2**, 8683–8690 (2014). https://doi.org/10.1039/c4tc01551g
88. J. Ruan, F. Mo, Z. Chen, M. Liu, S. Zheng, R. Wu, F. Fang, Y. Song, D. Sun, Rational construction of nitrogen-doped hierarchical dual-carbon for advanced potassium-ion hybrid capacitors. Adv. Energy Mater. **10**, 1–10 (2020). https://doi.org/10.1002/aenm.201904045
89. Z. Zhao, Z. Zhang, Y. Zhao, J. Liu, C. Liu, Z. Wang, G. Zheng, G. Huang, Y. Mei, Atomic layer deposition inducing integration of Co, N codoped carbon sphere on 3D foam with hierarchically porous structures for flexible hydrogen producing device. Adv. Funct. Mater. **29**, 1–9 (2019). https://doi.org/10.1002/adfm.201906365

90. Z. Zhao, Y. Kong, Z. Zhang, G. Huang, Y. Mei, Atomic layer-deposited nanostructures and their applications in energy storage and sensing. J. Mater. Res. **35**, 701–719 (2020). https://doi.org/10.1557/jmr.2019.329
91. Z. Zhao, Y. Kong, G. Huang, C. Liu, C. You, Z. Xiao, H. Zhu, J. Tan, B. Xu, J. Cui, X. Liu, Y. Mei, Area-selective and precise assembly of metal organic framework particles by atomic layer deposition induction and its application for ultra-sensitive dopamine sensor. Nano Today **42**, 101347 (2022). https://doi.org/10.1016/j.nantod.2021.101347
92. Z. Zhao, Y. Kong, X. Lin, C. Liu, J. Liu, Y. He, L. Yang, G. Huang, Y. Mei, Oxide nanomembrane induced assembly of a functional smart fiber composite with nanoporosity for an ultra-sensitive flexible glucose sensor. J. Mater. Chem. A **8**, 26119–26129 (2020). https://doi.org/10.1039/d0ta09211h
93. Z. Zhao, Y. Kong, C. Liu, J. Liu, Z. Wang, G. Zhong, G. Huang, Y. Mei, Atomic layer deposition-induced integration of N-doped carbon particles on carbon foam for flexible supercapacitor. J. Mater. **6**, 209–215 (2020). https://doi.org/10.1016/j.jmat.2020.01.011
94. Z. Zhao, Y. Kong, C. Liu, G. Huang, Z. Xiao, H. Zhu, Z. Bao, Y. Mei, Atomic layer deposition-assisted fabrication of 3D Co-doped carbon framework for sensitive enzyme-free lactic acid sensor. Chem. Eng. J. **417**, 129285 (2021). https://doi.org/10.1016/j.cej.2021.129285
95. Y. Kong, Z. Zhao, Y. Wang, S. Yang, G. Huang, Y. Wang, C. Liu, C. You, J. Tan, C. Wang, B. Xu, J. Cui, X. Liu, Y. Mei, Integration of a metal-organic framework film with a tubular whispering-gallery-mode microcavity for effective CO_2 sensing. ACS Appl. Mater. Interfaces **13**, 58104–58113 (2021). https://doi.org/10.1021/acsami.1c16322
96. Z. Zhao, G. Huang, Y. Kong, J. Cui, A.A. Solovev, X. Li, Y. Mei, *Atomic Layer Deposition for Electrochemical Energy: From Design to Industrialization* (Springer Nature, Singapore, 2022). https://doi.org/10.1007/s41918-022-00146-6
97. J. Zhao, D.T. Lee, R.W. Yaga, M.G. Hall, H.F. Barton, I.R. Woodward, C.J. Oldham, H.J. Walls, G.W. Peterson, G.N. Parsons, Ultra-fast degradation of chemical warfare agents using MOF–nanofiber kebabs. Angew. Chem. Int. Ed. **55**, 13224–13228 (2016). https://doi.org/10.1002/anie.201606656
98. I. Stassen, M. Styles, G. Grenci, H. Van Gorp, W. Vanderlinden, S. De Feyter, P. Falcaro, D. De Vos, P. Vereecken, R. Ameloot, Chemical vapour deposition of zeolitic imidazolate framework thin films. Nat. Mater. **15**, 304–310 (2016). https://doi.org/10.1038/nmat4509
99. M. Krishtab, I. Stassen, T. Stassin, A.J. Cruz, O.O. Okudur, S. Armini, C. Wilson, S. De Gendt, R. Ameloot, Vapor-deposited zeolitic imidazolate frameworks as gap-filling ultra-low-k dielectrics. Nat. Commun. **10**, 1–9 (2019). https://doi.org/10.1038/s41467-019-11703-x
100. T. Stassin, I. Stassen, J. Marreiros, A.J. Cruz, R. Verbeke, M. Tu, H. Reinsch, M. Dickmann, W. Egger, I.F.J. Vankelecom, D.E. De Vos, R. Ameloot, Solvent-free powder synthesis and MOF-CVD thin films of the large-pore metal-organic framework MAF-6. Chem. Mater. **32**, 1784–1793 (2020). https://doi.org/10.1021/acs.chemmater.9b03807
101. H. Wang, Z. Lu, S. Xu, D. Kong, J.J. Cha, G. Zheng, P.C. Hsu, K. Yan, D. Bradshaw, F.B. Prinz, Y. Cui, Electrochemical tuning of vertically aligned MoS_2 nanofilms and its application in improving hydrogen evolution reaction. Proc. Natl. Acad. Sci. USA **110**, 19701–19706 (2013). https://doi.org/10.1073/pnas.1316792110
102. W. Ren, H. Zhang, C. Cheng, Ultrafine Pt nanoparticles decorated MoS_2 nanosheets with significantly improved hydrogen evolution activity. Electrochim. Acta **241**, 316–322 (2017). https://doi.org/10.1016/j.electacta.2017.04.145
103. L. Zhang, Z.J. Zhao, M. Norouzi Banis, L. Li, Y. Zhao, Z. Song, Z. Wang, T.K. Sham, R. Li, M. Zheng, J. Gong, X. Sun, Selective atomic layer deposition of RuO: X catalysts on shape-controlled Pd nanocrystals with significantly enhanced hydrogen evolution activity. J. Mater. Chem. A **6**, 24397–24406 (2018). https://doi.org/10.1039/c8ta08931k
104. P. Nayak, Q. Jiang, N. Kurra, X. Wang, U. Buttner, H.N. Alshareef, Monolithic laser scribed graphene scaffolds with atomic layer deposited platinum for the hydrogen evolution reaction. J. Mater. Chem. A **5**, 20422–20427 (2017). https://doi.org/10.1039/c7ta06236b

105. H. Chen, J.T. Chen, L. Shao, L. Wang, X.Z. Fu, J.L. Luo, Minimum and well-dispersed platinum nanoparticles on 3D porous nickel for highly efficient electrocatalytic hydrogen evolution reaction enabled by atomic layer deposition. Appl. Surf. Sci. **494**, 1091–1099 (2019). https://doi.org/10.1016/j.apsusc.2019.07.251
106. S. Shin, Z. Jin, D.H. Kwon, R. Bose, Y.S. Min, High turnover frequency of hydrogen evolution reaction on amorphous MoS_2 thin film directly grown by atomic layer deposition. Langmuir **31**, 1196–1202 (2015). https://doi.org/10.1021/la504162u
107. H. Zhang, D.J. Hagen, X. Li, A. Graff, F. Heyroth, B. Fuhrmann, I. Kostanovskiy, S.L. Schweizer, F. Caddeo, A.W. Maijenburg, S. Parkin, R.B. Wehrspohn, Atomic layer deposition of cobalt phosphide for efficient water splitting. Angew. Chem. Int. Ed. **59**, 17172–17176 (2020). https://doi.org/10.1002/anie.202002280
108. D. Kim, J.G. Song, H. Yang, H. Lee, J. Park, H. Kim, Textile-based high-performance hydrogen evolution of loweratureatomic layer deposition of cobalt sulfide (Nanoscale (2019) 11 (844–850) doi: 10.1039/C8NR08969H). Nanoscale **11**, 7002 (2019). https://doi.org/10.1039/c9nr90074h
109. K.P. Katuri, N.M.S. Bettahalli, X. Wang, G. Matar, S. Chisca, S.P. Nunes, P.E. Saikaly, A microfiltration polymer-based hollow-fiber cathode as a promising advanced material for simultaneous recovery of energy and water. Adv. Mater. **28**, 9504–9511 (2016). https://doi.org/10.1002/adma.201603074
110. I.J. Hsu, Y.C. Kimmel, X. Jiang, B.G. Willis, J.G. Chen, Atomic layer deposition synthesis of platinum–tungsten carbide core–shell catalysts for the hydrogen evolution reaction. Chem. Commun. **48**, 1063–1065 (2012). https://doi.org/10.1039/c1cc15812k
111. T.A. Ho, C. Bae, S. Lee, M. Kim, J.M. Montero-Moreno, J.H. Park, H. Shin, Edge-On MoS_2 thin films by atomic layer deposition for understanding the interplay between the active area and hydrogen evolution reaction. Chem. Mater. **29**, 7604–7614 (2017). https://doi.org/10.1021/acs.chemmater.7b03212
112. W. Xiong, Q. Guo, Z. Guo, H. Li, R. Zhao, Q. Chen, Z. Liu, X. Wang, Atomic layer deposition of nickel carbide for supercapacitors and electrocatalytic hydrogen evolution. J. Mater. Chem. A **6**, 4297–4304 (2018). https://doi.org/10.1039/c7ta10202j
113. W. Xiong, Z. Guo, H. Li, R. Zhao, X. Wang, Rational bottom-up engineering of electrocatalysts by atomic layer deposition: a case study of $Fe_xCo_{1-x}S_y$-based catalysts for electrochemical hydrogen evolution, ACS Energy Lett. **2** (2017), 2778–2785. https://doi.org/10.1021/acsenergylett.7b01056
114. C. Zhen, R. Chen, L. Wang, G. Liu, H.M. Cheng, Tantalum (oxy)nitride based photoanodes for solar-driven water oxidation. J. Mater. Chem. A **4**, 2783–2800 (2016). https://doi.org/10.1039/c5ta07057k
115. S.K. Sahoo, Renewable and sustainable energy reviews solar photovoltaic energy progress in India: A review. Renew. Sust. Energ. Rev. **59**, 927–939 (2016). https://doi.org/10.1016/j.rser.2016.01.049
116. H. He, A. Liao, W. Guo, W. Luo, Y. Zhou, Z. Zou, State-of-the-art progress in the use of ternary metal oxides as photoelectrode materials for water splitting and organic synthesis. Nano Today **28**, 100763 (2019). https://doi.org/10.1016/j.nantod.2019.100763
117. P. Peerakiatkhajohn, J.-H. Yun, S. Wang, L. Wang, Review of recent progress in unassisted photoelectrochemical water splitting: from material modification to configuration design. J. Photon. Energy **7**, 012006 (2016). https://doi.org/10.1117/1.jpe.7.012006
118. S.H. Yoon, D.W. Park, K.S. Kim, Preparation of WO_3, $BiVO_4$ and reduced graphene oxide composite thin films and their photoelectrochemical performance. Korean J. Chem. Eng. **34**, 3220–3225 (2017). https://doi.org/10.1007/s11814-017-0237-4
119. J. Feng, H. Huang, S. Yan, W. Luo, T. Yu, Z. Li, Z. Zou, Non-oxide semiconductors for artificial photosynthesis: Progress on photoelectrochemical water splitting and carbon dioxide reduction. Nano Today **30**, 100830 (2020). https://doi.org/10.1016/j.nantod.2019.100830

120. H. Gerischer, Charge transfer processes at semiconductor-electrolyte interfaces in connection with problems of catalysis. Surf. Sci. **18**, 97–122 (1969). https://doi.org/10.1016/0039-6028(69)90269-6
121. Z. Zhang, X. Wang, J. Long, Q. Gu, Z. Ding, X. Fu, Nitrogen-doped titanium dioxide visible light photocatalyst: Spectroscopic identification of photoactive centers. J. Catal. **276**, 201–214 (2010). https://doi.org/10.1016/j.jcat.2010.07.033
122. O. Khaselev, J.A. Turner, A monolithic photovoltaic-photoelectrochemical device for hydrogen production via water splitting. Science **280**(80), 425–427 (1998). https://doi.org/10.1126/science.280.5362.425
123. A. Fujishima, K. Honda, Electrochemical photolysis of water at a semiconductor electrode. Nature **238**, 37–38 (1972). https://doi.org/10.1038/238037a0
124. M. Leskelä, M. Mattinen, M. Ritala, Review article: Atomic layer deposition of optoelectronic materials. J. Vac. Sci. Technol. B **37**, 030801 (2019). https://doi.org/10.1116/1.5083692
125. J.B. Lee, S. Choi, J. Kim, Y.S. Nam, Plasmonically-assisted nanoarchitectures for solar water splitting: Obstacles and breakthroughs. Nano Today **16**, 61–81 (2017). https://doi.org/10.1016/j.nantod.2017.08.008
126. M. Zhou, X.W. Lou, Y. Xie, Two-dimensional nanosheets for photoelectrochemical water splitting: Possibilities and opportunities. Nano Today **8**, 598–618 (2013). https://doi.org/10.1016/j.nantod.2013.12.002
127. J. Brillet, J.H. Yum, M. Cornuz, T. Hisatomi, R. Solarska, J. Augustynski, M. Graetzel, K. Sivula, Highly efficient water splitting by a dual-absorber tandem cell. Nat. Photonics **6**, 824–828 (2012). https://doi.org/10.1038/nphoton.2012.265
128. L. Pan, J.H. Kim, M.T. Mayer, M.K. Son, A. Ummadisingu, J.S. Lee, A. Hagfeldt, J. Luo, M. Grätzel, Boosting the performance of Cu_2O photocathodes for unassisted solar water splitting devices. Nat. Catal. **1**, 412–420 (2018). https://doi.org/10.1038/s41929-018-0077-6
129. J. Kim, W. Yang, Y. Oh, H. Lee, S. Lee, H. Shin, J. Kim, J. Moon, Self-oriented Sb_2Se_3 nanoneedle photocathodes for water splitting obtained by a simple spin-coating method. J. Mater. Chem. A **5**, 2180–2187 (2017). https://doi.org/10.1039/c6ta09602f
130. Y. Pi, Z. Li, D. Xu, J. Liu, Y. Li, F. Zhang, G. Zhang, W. Peng, X. Fan, 1T-phase MoS_2 nanosheets on TiO_2 nanorod arrays: 3D photoanode with extraordinary catalytic performance. ACS Sustain. Chem. Eng. **5**, 5175–5182 (2017). https://doi.org/10.1021/acssuschemeng.7b00518
131. J.Y. Kim, D.H. Youn, K. Kang, J.S. Lee, Highly conformal deposition of an ultrathin FeOOH layer on a hematite nanostructure for efficient solar water splitting. Angew. Chem. Int. Ed. **55**, 10854–10858 (2016). https://doi.org/10.1002/anie.201605924
132. L. Yao, W. Wang, L. Wang, Y. Liang, J. Fu, H. Shi, Chemical bath deposition synthesis of TiO_2/Cu_2O core/shell nanowire arrays with enhanced photoelectrochemical water splitting for H_2 evolution and photostability. Int. J. Hydrog. Energy **43**, 15907–15917 (2018). https://doi.org/10.1016/j.ijhydene.2018.06.127
133. W. Jian, X. Cheng, Y. Huang, Y. You, R. Zhou, T. Sun, J. Xu, Arrays of ZnO/MoS_2 nanocables and MoS_2 nanotubes with phase engineering for bifunctional photoelectrochemical and electrochemical water splitting. Chem. Eng. J. **328**, 474–483 (2017). https://doi.org/10.1016/j.cej.2017.07.056
134. L. Cai, J. Zhao, H. Li, J. Park, I.S. Cho, H.S. Han, X. Zheng, One-step hydrothermal deposition of Ni:FeOOH onto photoanodes for enhanced water oxidation. ACS Energy Lett. **1**, 624–632 (2016). https://doi.org/10.1021/acsenergylett.6b00303
135. H. Gong, N. Freudenberg, M. Nie, R. Van De Krol, K. Ellmer, BiVO4 photoanodes for water splitting with high injection efficiency, deposited by reactive magnetron co-sputtering. AIP Adv. **6**, 45108 (2016). https://doi.org/10.1063/1.4947121
136. C. Fàbrega, S. Murcia-López, D. Monllor-Satoca, J.D. Prades, M.D. Hernández-Alonso, G. Penelas, J.R. Morante, T. Andreu, Efficient WO3 photoanodes fabricated by pulsed laser deposition for photoelectrochemical water splitting with high faradaic efficiency. Appl. Catal. B Environ. **189**, 133–140 (2016). https://doi.org/10.1016/j.apcatb.2016.02.047

137. D. Hu, J. Xiang, Q. Zhou, S. Su, Z. Zhang, X. Wang, M. Jin, L. Nian, R. Nözel, G. Zhou, Z. Zhang, J. Liu, One-step chemical vapor deposition of MoS_2 nanosheets on SiNWs as photocathodes for efficient and stable solar-driven hydrogen production. Nanoscale **10**, 3518–3525 (2018). https://doi.org/10.1039/c7nr09235k
138. R.J. Britto, J.D. Benck, J.L. Young, C. Hahn, T.G. Deutsch, T.F. Jaramillo, Molybdenum disulfide as a protection layer and catalyst for gallium indium phosphide solar water splitting photocathodes. J. Phys. Chem. Lett. **7**, 2044–2049 (2016). https://doi.org/10.1021/acs.jpclett.6b00563
139. L. Wen, M. Zhou, C. Wang, Y. Mi, Y. Lei, Nanoengineering energy conversion and storage devices via atomic layer deposition. Adv. Energy Mater. **6** (2016). https://doi.org/10.1002/aenm.201600468
140. M.G. Lee, J.S. Park, H.W. Jang, Solution-processed metal oxide thin film nanostructures for water splitting photoelectrodes: A review. J. Korean Ceram. Soc. **55**, 185–202 (2018). https://doi.org/10.4191/kcers.2018.55.3.08
141. C. Glynn, C. O'Dwyer, Solution processable metal oxide thin film deposition and material growth for electronic and photonic devices. Adv. Mater. Interfaces **4** (2017). https://doi.org/10.1002/admi.201600610
142. B. Alfakes, C. Garlisi, J. Villegas, A. Al-Hagri, S. Tamalampudi, N.S. Rajput, J.Y. Lu, E. Lewin, J. Sá, I. Almansouri, G. Palmisano, M. Chiesa, Enhanced photoelectrochemical performance of atomic layer deposited Hf-doped ZnO. Surf. Coat. Technol. **385**, 125352 (2020). https://doi.org/10.1016/j.surfcoat.2020.125352
143. M. Zafar, J.Y. Yun, D.H. Kim, Performance of inverted organic photovoltaic cells with nitrogen doped TiO2 films by atomic layer deposition. Korean J. Chem. Eng. **35**, 567–573 (2018). https://doi.org/10.1007/s11814-017-0285-9
144. S.M. George, Atomic layer deposition: An overview. Chem. Rev. **110**, 111–131 (2010). https://doi.org/10.1021/cr900056b
145. A.T. Sivagurunathan, S. Adhikari, D.H. Kim, Strategies and implications of atomic layer deposition in photoelectrochemical water splitting: Recent advances and prospects. Nano Energy **83**, 105802 (2021). https://doi.org/10.1016/j.nanoen.2021.105802
146. F. Nan, T. Cai, S. Ju, L. Fang, Enhanced photoelectrochemical water splitting of $BiVO_4$ photonic crystal photoanode by decorating with MoS_2 nanosheets. Appl. Phys. Lett. **112**, 173902 (2018). https://doi.org/10.1063/1.5025737
147. P.S. Bassi, R.P. Antony, P.P. Boix, Y. Fang, J. Barber, L.H. Wong, Crystalline Fe_2O_3/Fe_2TiO_5 heterojunction nanorods with efficient charge separation and hole injection as photoanode for solar water oxidation. Nano Energy **22**, 310–318 (2016). https://doi.org/10.1016/j.nanoen.2016.02.013
148. F. Yu, F. Li, T. Yao, J. Du, Y. Liang, Y. Wang, H. Han, L. Sun, Fabrication and kinetic study of a ferrihydrite-modified BiVO4 photoanode. ACS Catal. **7**, 1868–1874 (2017). https://doi.org/10.1021/acscatal.6b03483
149. H. Tong, Y. Jiang, Q. Zhang, W. Jiang, K. Wang, X. Luo, Z. Lin, L. Xia, Boosting photoelectrochemical water oxidation with cobalt phosphide nanosheets on porous BiVO4. ACS Sustain. Chem. Eng. **7**, 769–778 (2019). https://doi.org/10.1021/acssuschemeng.8b04405
150. Y. Gao, Y. Li, G. Yang, S. Li, N. Xiao, B. Xu, S. Liu, P. Qiu, S. Hao, L. Ge, Fe_2TiO_5 as an efficient co-catalyst to improve the photoelectrochemical water splitting performance of $BiVO_4$. ACS Appl. Mater. Interfaces **10**, 39713–39722 (2018). https://doi.org/10.1021/acsami.8b14141
151. M.M. Momeni, Y. Ghayeb, M. Shafiei, Preparation and characterization of $CrFeWTiO_2$ photoanodes and their photoelectrochemical activities for water splitting. Dalt. Trans. **46**, 12527–12536 (2017). https://doi.org/10.1039/C7DT01596H
152. J. Gu, Y. Yan, J.L. Young, K.X. Steirer, N.R. Neale, J.A. Turner, Water reduction by a p-$GaInP_2$ photoelectrode stabilized by an amorphous TiO_2 coating and a molecular cobalt catalyst. Nat. Mater. **15**, 456–460 (2016). https://doi.org/10.1038/nmat4511

153. P.S. Shinde, P.R. Fontenot, J.P. Donahue, J.L. Waters, P. Kung, L.E. McNamara, N.I. Hammer, A. Gupta, S. Pan, Synthesis of MoS_2 from $[Mo_3S_7(S_2CNEt_2)_3]I$ for enhancing photoelectrochemical performance and stability of Cu_2O photocathode toward efficient solar water splitting. J. Mater. Chem. A **6**, 9569–9582 (2018). https://doi.org/10.1039/C8TA01771A
154. P. Sharma, P. Doiphode, O. Bhorade, A. Yengantiwar, Bismuth vanadate thin films for efficient photoelectrochemical water splitting. Emerg. Mater. **3**, 187 (2020). https://doi.org/10.1007/s42247-020-00093-2
155. M. Basu, Z.-W. Zhang, C.-J. Chen, P.-T. Chen, K.-C. Yang, C.-G. Ma, C.C. Lin, S.-F. Hu, R.-S. Liu, Heterostructure of Si and $CoSe_2$: A promising photocathode based on a non-noble metal catalyst for photoelectrochemical hydrogen evolution. Angew. Chem. Int. Ed. **54**, 6211–6218 (2015). https://doi.org/10.1002/anie.201502573
156. X. Wan, J. Su, L. Guo, Enhanced photoelectrochemical water oxidation on $BiVO_4$ with mesoporous cobalt nitride sheets as oxygen-evolution cocatalysts. Eur. J. Inorg. Chem. **2018**, 2557–2563 (2018). https://doi.org/10.1002/ejic.201800392
157. S.M. Thalluri, B. Wei, K. Welter, R. Thomas, V. Smirnov, L. Qiao, Z. Wang, F. Finger, L. Liu, Inverted pyramid textured p-silicon covered with Co_2P as an efficient and stable solar hydrogen evolution photocathode. ACS Energy Lett. **4**, 1755–1762 (2019). https://doi.org/10.1021/acsenergylett.9b00964
158. Y. Bin Park, J.H. Kim, Y.J. Jang, J.H. Lee, M.H. Lee, B.J. Lee, D.H. Youn, J.S. Lee, Exfoliated NiFe layered double hydroxide cocatalyst for enhanced photoelectrochemical water oxidation with hematite photoanode. ChemCatChem **11**, 443–448 (2019). https://doi.org/10.1002/cctc.201801490
159. J. Li, M. Griep, Y. Choi, D. Chu, Photoelectrochemical overall water splitting with textured $CuBi_2O_4$ as a photocathode. Chem. Commun. **54**, 3331–3334 (2018). https://doi.org/10.1039/C7CC09041B
160. Z. Yin, R. Fan, G. Huang, M. Shen, 11.5% efficiency of TiO_2 protected and Pt catalyzed n^+np^+-Si photocathodes for photoelectrochemical water splitting: Manipulating the Pt distribution and Pt/Si contact. Chem. Commun. **54**, 543–546 (2018). https://doi.org/10.1039/c7cc08409a
161. Y. Yang, M. Wang, P. Zhang, W. Wang, H. Han, L. Sun, Evident enhancement of photoelectrochemical hydrogen production by electroless deposition of M-B (M=Ni, Co) catalysts on silicon nanowire arrays. ACS Appl. Mater. Interfaces **8**, 30143–30151 (2016). https://doi.org/10.1021/acsami.6b09600
162. C.G. Morales-Guio, L. Liardet, M.T. Mayer, S.D. Tilley, M. Grätzel, X. Hu, Photoelectrochemical hydrogen production in alkaline solutions using Cu_2O coated with earth-abundant hydrogen evolution catalysts. Angew. Chem. Int. Ed. **54**, 664–667 (2015). https://doi.org/10.1002/anie.201410569
163. X.-Q. Bao, D.Y. Petrovykh, P. Alpuim, D.G. Stroppa, N. Guldris, H. Fonseca, M. Costa, J. Gaspar, C. Jin, L. Liu, Amorphous oxygen-rich molybdenum oxysulfide decorated p-type silicon microwire arrays for efficient photoelectrochemical water reduction. Nano Energy **16**, 130–142 (2015). https://doi.org/10.1016/j.nanoen.2015.06.014
164. Y. Chen, P.D. Tran, P. Boix, Y. Ren, S.Y. Chiam, Z. Li, K. Fu, L.H. Wong, J. Barber, Silicon decorated with amorphous cobalt molybdenum sulfide catalyst as an efficient photocathode for solar hydrogen generation. ACS Nano **9**, 3829–3836 (2015). https://doi.org/10.1021/nn506819m
165. H. Zhang, Q. Ding, D. He, H. Liu, W. Liu, Z. Li, B. Yang, X. Zhang, L. Lei, S. Jin, A p-Si/NiCoSex core/shell nanopillar array photocathode for enhanced photoelectrochemical hydrogen production. Energy Environ. Sci. **9**, 3113–3119 (2016). https://doi.org/10.1039/C6EE02215D
166. S.A. Lee, T.H. Lee, C. Kim, M.G. Lee, M.-J. Choi, H. Park, S. Choi, J. Oh, H.W. Jang, Tailored NiOx/Ni cocatalysts on silicon for highly efficient water splitting photoanodes via pulsed electrodeposition. ACS Catal. **8**, 7261–7269 (2018). https://doi.org/10.1021/acscatal.8b01999

167. J.H. Kim, Y. Jo, J.H. Kim, J.W. Jang, H.J. Kang, Y.H. Lee, D.S. Kim, Y. Jun, J.S. Lee, Wireless solar water splitting device with robust cobalt-catalyzed, dual-doped $BiVO_4$ photoanode and perovskite solar cell in tandem: A dual absorber artificial leaf. ACS Nano **9**, 11820–11829 (2015). https://doi.org/10.1021/acsnano.5b03859
168. L. Liardet, J.E. Katz, J. Luo, M. Grätzel, X. Hu, An ultrathin cobalt–iron oxide catalyst for water oxidation on nanostructured hematite photoanodes. J. Mater. Chem. A **7**, 6012–6020 (2019). https://doi.org/10.1039/C8TA12295D
169. M. Chhetri, S. Dey, C.N.R. Rao, Photoelectrochemical oxygen evolution reaction activity of amorphous Co–La double hydroxide-BiVO4 fabricated by pulse plating electrodeposition. ACS Energy Lett. **2**, 1062–1069 (2017). https://doi.org/10.1021/acsenergylett.7b00247
170. M.-K. Son, H. Seo, M. Watanabe, M. Shiratani, T. Ishihara, Characteristics of crystalline sputtered $LaFeO_3$ thin films as photoelectrochemical water splitting photocathodes. Nanoscale **12**, 9653–9660 (2020). https://doi.org/10.1039/D0NR01762K
171. G. Huang, J. Mao, R. Fan, Z. Yin, X. Wu, J. Jie, Z. Kang, M. Shen, Integrated $MoSe_2$ with n^+p-Si photocathodes for solar water splitting with high efficiency and stability. Appl. Phys. Lett. **112**, 13902 (2018). https://doi.org/10.1063/1.5012110
172. R. Fan, J. Mao, Z. Yin, J. Jie, W. Dong, L. Fang, F. Zheng, M. Shen, Efficient and stable silicon photocathodes coated with vertically standing nano-MoS_2 films for solar hydrogen production. ACS Appl. Mater. Interfaces **9**, 6123–6129 (2017). https://doi.org/10.1021/acsami.6b15854
173. F. Chen, Q. Zhu, Y. Wang, W. Cui, X. Su, Y. Li, Efficient photoelectrochemical hydrogen evolution on silicon photocathodes interfaced with nanostructured NiP_2 cocatalyst films. ACS Appl. Mater. Interfaces **8**, 31025–31031 (2016). https://doi.org/10.1021/acsami.6b11197
174. Q. Zhou, S. Su, D. Hu, L. Lin, Z. Yan, X. Gao, Z. Zhang, J.-M. Liu, Ultrathin MoS_2-coated Ag@Si nanosphere arrays as an efficient and stable photocathode for solar-driven hydrogen production. Nanotechnology **29**, 105402 (2018). https://doi.org/10.1088/1361-6528/aaa48c
175. D.M. Andoshe, G. Jin, C.-S. Lee, C. Kim, K.C. Kwon, S. Choi, W. Sohn, C.W. Moon, S.H. Lee, J.M. Suh, S. Kang, J. Park, H. Heo, J.K. Kim, S. Han, M.-H. Jo, H.W. Jang, Directly assembled 3D molybdenum disulfide on silicon wafer for efficient photoelectrochemical water reduction. Adv. Sustain. Syst. **2**, 1700142 (2018). https://doi.org/10.1002/adsu.201700142
176. L. He, W. Zhou, D. Cai, S.S. Mao, K. Sun, S. Shen, Pulsed laser-deposited n-Si/NiOx photoanodes for stable and efficient photoelectrochemical water splitting. Catal. Sci. Technol. **7**, 2632–2638 (2017). https://doi.org/10.1039/C7CY00114B
177. H. Li, P. Wen, A. Hoxie, C. Dun, S. Adhikari, Q. Li, C. Lu, D.S. Itanze, L. Jiang, D. Carroll, A. Lachgar, Y. Qiu, S.M. Geyer, Interface engineering of colloidal CdSe quantum dot thin films as acid-stable photocathodes for solar-driven hydrogen evolution. ACS Appl. Mater. Interfaces **10**, 17129–17139 (2018). https://doi.org/10.1021/acsami.7b19229
178. J. Yang, J.K. Cooper, F.M. Toma, K.A. Walczak, M. Favaro, J.W. Beeman, L.H. Hess, C. Wang, C. Zhu, S. Gul, J. Yano, C. Kisielowski, A. Schwartzberg, I.D. Sharp, A multifunctional biphasic water splitting catalyst tailored for integration with high-performance semiconductor photoanodes. Nat. Mater. **16**, 335–341 (2017). https://doi.org/10.1038/nmat4794
179. M. Zhong, T. Hisatomi, T. Minegishi, H. Nishiyama, M. Katayama, T. Yamada, K. Domen, Bulky crystalline BiVO4 thin films for efficient solar water splitting. J. Mater. Chem. A **4**, 9858–9864 (2016). https://doi.org/10.1039/c6ta03072f
180. J. Zeng, X. Xu, V. Parameshwaran, J. Baker, S. Bent, H.S.P. Wong, B. Clemens, Photoelectrochemical water oxidation by GaAs nanowire arrays protected with atomic layer deposited NiO_x electrocatalysts. J. Electron. Mater. **47**, 932 (2017). https://doi.org/10.1007/s11664-017-5824-y
181. S. Oh, J.B. Kim, J.T. Song, J. Oh, S.H. Kim, Atomic layer deposited molybdenum disulfide on Si photocathodes for highly efficient photoelectrochemical water reduction reaction. J. Mater. Chem. A **5**, 3304–3310 (2017). https://doi.org/10.1039/C6TA10707A

182. M. Alqahtani, S. Sathasivam, A. Alhassan, F. Cui, S. BenJaber, C. Blackman, B. Zhang, Y. Qin, I.P. Parkin, S. Nakamura, H. Liu, J. Wu, InGaN/GaN multiple quantum well photoanode modified with cobalt oxide for water oxidation. ACS Appl. Energy Mater. **1**, 6417–6424 (2018). https://doi.org/10.1021/acsaem.8b01387
183. S.K. Choi, W.-S. Chae, B. Song, C.-H. Cho, J. Choi, D.S. Han, W. Choi, H. Park, Photoelectrochemical hydrogen production on silicon microwire arrays overlaid with ultrathin titanium nitride. J. Mater. Chem. A **4**, 14008–14016 (2016). https://doi.org/10.1039/C6TA05200B
184. C. Du, J. Wang, X. Liu, J. Yang, K. Cao, Y. Wen, R. Chen, B. Shan, Ultrathin CoOx-modified hematite with low onset potential for solar water oxidation. Phys. Chem. Chem. Phys. **19**, 14178–14184 (2017). https://doi.org/10.1039/C7CP01588G
185. D. Kang, T.W. Kim, S.R. Kubota, A.C. Cardiel, H.G. Cha, K.S. Choi, Electrochemical synthesis of photoelectrodes and catalysts for use in solar water splitting. Chem. Rev. **115**, 12839–12887 (2015). https://doi.org/10.1021/acs.chemrev.5b00498
186. S.K. Saraswat, D.D. Rodene, R.B. Gupta, Recent advancements in semiconductor materials for photoelectrochemical water splitting for hydrogen production using visible light. Renew. Sust. Energ. Rev. **89**, 228–248 (2018). https://doi.org/10.1016/j.rser.2018.03.063
187. D.V. Esposito, J.B. Baxter, J. John, N.S. Lewis, T.P. Moffat, T. Ogitsu, G.D. O'Neil, T.A. Pham, A.A. Talin, J.M. Velazquez, B.C. Wood, Methods of photoelectrode characterization with high spatial and temporal resolution. Energy Environ. Sci. **8**, 2863–2885 (2015). https://doi.org/10.1039/c5ee00835b
188. A.S. Asundi, J.A. Raiford, S.F. Bent, Opportunities for atomic layer deposition in emerging energy technologies. ACS Energy Lett. **4**, 908–925 (2019). https://doi.org/10.1021/acsenergylett.9b00249
189. N.P. Dasgupta, X. Meng, J.W. Elam, A.B.F. Martinson, Atomic layer deposition of metal sulfide materials. Acc. Chem. Res. **48**, 341–348 (2015). https://doi.org/10.1021/ar500360d
190. H. Kim, Atomic layer deposition of metal and nitride thin films: Current research efforts and applications for semiconductor device processing. J. Vac. Sci. Technol. B Microelectron. Nanom. Struct. **21**, 2231 (2003). https://doi.org/10.1116/1.1622676
191. S. Ghosh, S.M. Jeong, S.R. Polaki, A review on metal nitrides/oxynitrides as an emerging supercapacitor electrode beyond oxide. Korean J. Chem. Eng. **35**, 1389–1408 (2018). https://doi.org/10.1007/s11814-018-0089-6
192. K.J. Blakeney, C.H. Winter, Thermal atomic layer deposition of tungsten carbide films from WCl_6 and $AlMe_3$. J. Vac. Sci. Technol. A Vacuum Surf. Film. **36**, 01A104 (2018). https://doi.org/10.1116/1.5002667
193. D.J. Hagen, M.E. Pemble, M. Karppinen, Atomic layer deposition of metals: Precursors and film growth. Appl. Phys. Rev. **6** (2019). https://doi.org/10.1063/1.5087759
194. Y. Lee, H. Sun, M.J. Young, S.M. George, Atomic layer deposition of metal fluorides using HF-pyridine as the fluorine precursor. Chem. Mater. **28**, 2022–2032 (2016). https://doi.org/10.1021/acs.chemmater.5b04360
195. J. Rongé, T. Dobbelaere, L. Henderick, M.M. Minjauw, S.P. Sree, J. Dendooven, J.A. Martens, C. Detavernier, Bifunctional earth-abundant phosphate/phosphide catalysts prepared: Via atomic layer deposition for electrocatalytic water splitting. Nanoscale Adv. **1**, 4166–4172 (2019). https://doi.org/10.1039/c9na00391f
196. S. Beljakowa, P. Pichler, B. Kalkofen, R. Hübner, Diffusion of phosphorus and boron from atomic layer deposition oxides into silicon. Phys. Status Solidi Appl. Mater. Sci. **216**, 1–7 (2019). https://doi.org/10.1002/pssa.201900306
197. Z. Gao, P. Banerjee, Review article: Atomic layer deposition of doped ZnO films. J. Vac. Sci. Technol. A Vacuum Surf. Film. **37** (2019). https://doi.org/10.1116/1.5112777
198. T. Singh, T. Lehnen, T. Leuning, S. Mathur, Atomic layer deposition grown MO_x thin films for solar water splitting: Prospects and challenges. J. Vac. Sci. Technol. A Vacuum Surf Film **33**, 0–9 (2015). https://doi.org/10.1116/1.4904729

199. M. Khan, Atomic layer deposited photocatalysts: Comprehensive review on viable fabrication routes and reactor design approaches for photo – Mediated redox reactions. J. Mater. Chem. A **6**, 121 (2015). https://doi.org/10.1039/C9TA04780H.Volume
200. N. Guijarro, M.S. Prévot, K. Sivula, Surface modification of semiconductor photoelectrodes. Phys. Chem. Chem. Phys. **17**, 15655–15674 (2015). https://doi.org/10.1039/c5cp01992c
201. Y. Zhao, K. Zheng, X. Sun, Addressing interfacial issues in liquid-based and solid-state batteries by atomic and molecular layer deposition. Joule **2**, 2583–2604 (2018). https://doi.org/10.1016/j.joule.2018.11.012
202. Y. Lin, S. Zhou, S.W. Sheehan, D. Wang, Nanonet-based hematite heteronanostructures for efficient solar water splitting. J. Am. Chem. Soc. **133**, 2398–2401 (2011). https://doi.org/10.1021/ja110741z
203. S.C. Riha, M.J. Devries Vermeer, M.J. Pellin, J.T. Hupp, A.B.F. Martinson, Hematite-based photo-oxidation of water using transparent distributed current collectors. ACS Appl. Mater. Interfaces **5**, 360–367 (2013). https://doi.org/10.1021/am302356k
204. Y. Lin, Y. Xu, M.T. Mayer, Z.I. Simpson, G. McMahon, S. Zhou, D. Wang, Growth of p-type hematite by atomic layer deposition and its utilization for improved solar water splitting. J. Am. Chem. Soc. **134**, 5508–5511 (2012). https://doi.org/10.1021/ja300319g
205. O. Zandi, B.M. Klahr, T.W. Hamann, Highly photoactive Ti-doped α-Fe_2O_3 thin film electrodes: Resurrection of the dead layer. Energy Environ. Sci. **6**, 634–642 (2013). https://doi.org/10.1039/c2ee23620f
206. B. Klahr, S. Gimenez, F. Fabregat-Santiago, J. Bisquert, T.W. Hamann, Photoelectrochemical and impedance spectroscopic investigation of water oxidation with "co-Pi"-coated hematite electrodes. J. Am. Chem. Soc. **134**, 16693–16700 (2012). https://doi.org/10.1021/ja306427f
207. X. Yang, C. Du, R. Liu, J. Xie, D. Wang, Balancing photovoltage generation and charge-transfer enhancement for catalyst-decorated photoelectrochemical water splitting: A case study of the hematite/MnOx combination. J. Catal. **304**, 86–91 (2013). https://doi.org/10.1016/j.jcat.2013.04.014
208. S.C. Riha, J.M. Racowski, M.P. Lanci, J.A. Klug, A.S. Hock, A.B.F. Martinson, Phase discrimination through oxidant selection in low-temperature atomic layer deposition of crystalline iron oxides. Langmuir **29**, 3439–3445 (2013). https://doi.org/10.1021/la305027k
209. Y.J. Hwang, A. Boukai, P. Yang, High density n-S/n-TiO_2 core/shell nanowire arrays with enhanced photoactivit. Nano Lett. **9**, 410–415 (2009). https://doi.org/10.1021/nl8032763
210. Y.J. Hwang, C. Hahn, B. Liu, P. Yang, Photoelectrochemical properties of TiO_2 nanowire arrays: A study of the dependence on length and atomic layer deposition coating. ACS Nano **6**, 5060–5069 (2012). https://doi.org/10.1021/nn300679d
211. A. Paracchino, V. Laporte, K. Sivula, M. Grätzel, E. Thimsen, Highly active oxide photocathode for photoelectrochemical water reduction. Nat. Mater. **10**, 456–461 (2011). https://doi.org/10.1038/nmat3017
212. C.C. Wang, C.C. Kei, T.P. Perng, Fabrication of TiO_2 nanotubes by atomic layer deposition and their photocatalytic and photoelectrochemical applications. Nanotechnology **22**, 365702 (2011). https://doi.org/10.1088/0957-4484/22/36/365702
213. R. Liu, Y. Lin, L.-Y. Chou, S.W. Sheehan, W. He, F. Zhang, H.J.M. Hou, D. Wang, Water splitting by tungsten oxide prepared by atomic layer deposition and decorated with an oxygen-evolving catalyst. Angew. Chem. **123**, 519–522 (2011). https://doi.org/10.1002/ange.201004801
214. M.J. Choi, J.Y. Jung, M.J. Park, J.W. Song, J.H. Lee, J.H. Bang, Long-term durable silicon photocathode protected by a thin Al_2O_3/SiOx layer for photoelectrochemical hydrogen evolution. J. Mater. Chem. A **2**, 2928–2933 (2014). https://doi.org/10.1039/c3ta14443g
215. W. Kim, T. Tachikawa, D. Monllor-Satoca, H. Kim, T. Majima, W. Choi, Promoting water photooxidation on transparent WO_3 thin films by an alumina overlayer. Energy Environ. Sci. **6**, 3732 (2013). https://doi.org/10.1039/b000000x

216. M. Liu, C.Y. Nam, C.T. Black, J. Kamcev, L. Zhang, Enhancing water splitting activity and chemical stability of zinc oxide nanowire photoanodes with ultrathin titania shells. J. Phys. Chem. C **117**, 13396–13402 (2013). https://doi.org/10.1021/jp404032p
217. D.H. Kim, Atomic layer deposition for dye-sensitized solar cells (2013). https://doi.org/10.13140/RG.2.1.3096.3286
218. M.A. Hossain, K.T. Khoo, X. Cui, G.K. Poduval, T. Zhang, X. Li, W.M. Li, B. Hoex, Atomic layer deposition enabling higher efficiency solar cells: A review. Nano Mater. Sci. **2**, 204–226 (2020). https://doi.org/10.1016/j.nanoms.2019.10.001
219. H. Kim, H. Bae, S.W. Bang, S. Kim, S.H. Lee, S.-W. Ryu, J.-S. Ha, Enhanced photoelectrochemical stability of GaN photoelectrodes by Al_2O_3 surface passivation layer. Opt. Express **27**, A206 (2019). https://doi.org/10.1364/oe.27.00a206
220. W.H. Cheng, M.H. Richter, M.M. May, J. Ohlmann, D. Lackner, F. Dimroth, T. Hannappel, H.A. Atwater, H.J. Lewerenz, Monolithic photoelectrochemical device for direct water splitting with 19% efficiency. ACS Energy Lett. **3**, 1795–1800 (2018). https://doi.org/10.1021/acsenergylett.8b00920
221. J. Joe, H. Yang, C. Bae, H. Shin, Metal chalcogenides on silicon photocathodes for efficient water splitting: A mini overview. Catalysts **9** (2019). https://doi.org/10.3390/catal9020149
222. C.J. Chen, V. Veeramani, Y.H. Wu, A. Jena, L.C. Yin, H. Chang, S.F. Hu, R.S. Liu, Phosphorous-doped molybdenum disulfide anchored on silicon as an efficient catalyst for photoelectrochemical hydrogen generation. Appl. Catal. B Environ. **263**, 118259 (2020). https://doi.org/10.1016/j.apcatb.2019.118259
223. S. Liu, Z. Luo, L. Li, H. Li, M. Chen, T. Wang, J. Gong, Multifunctional TiO_2 overlayer for p-Si/n-CdS heterojunction photocathode with improved efficiency and stability. Nano Energy **53**, 125–129 (2018). https://doi.org/10.1016/j.nanoen.2018.08.024
224. S. Choi, J. Hwang, T.H. Lee, H.H. Kim, S.P. Hong, C. Kim, M.J. Choi, H.K. Park, S.S.M. Bhat, J.M. Suh, J. Lee, K.S. Choi, S.H. Hong, J.C. Shin, H.W. Jang, Photoelectrochemical hydrogen production at neutral pH phosphate buffer solution using TiO_2 passivated InAs Nanowire/p-Si heterostructure photocathode. Chem. Eng. J. **392**, 123688 (2020). https://doi.org/10.1016/j.cej.2019.123688
225. H.C. Fu, P. Varadhan, M.L. Tsai, W. Li, Q. Ding, C.H. Lin, M. Bonifazi, A. Fratalocchi, S. Jin, J.H. He, Improved performance and stability of photoelectrochemical water-splitting Si system using a bifacial design to decouple light harvesting and electrocatalysis. Nano Energy **70**, 104478 (2020). https://doi.org/10.1016/j.nanoen.2020.104478
226. J.Y. Jung, D. Woong Kim, D.H. Kim, T. Joo Park, R.B. Wehrspohn, J.H. Lee, Seebeck-voltage-triggered self-biased photoelectrochemical water splitting using HfO_x/SiO_x bi-layer protected Si photocathodes. Sci. Rep. **9**, 1–8 (2019). https://doi.org/10.1038/s41598-019-45672-4
227. H. Li, P. Wen, D.S. Itanze, M.W. Kim, S. Adhikari, C. Lu, L. Jiang, Y. Qiu, S.M. Geyer, Phosphorus-rich colloidal cobalt diphosphide (CoP2) nanocrystals for electrochemical and photoelectrochemical hydrogen evolution. Adv. Mater. **31**, e1900813 (2019). https://doi.org/10.1002/adma.201900813
228. P. Perez-Rodriguez, D. Cardenas-Morcoso, I.A. Digdaya, A.M. Raventos, P. Procel, O. Isabella, S. Gimenez, M. Zeman, W.A. Smith, A.H.M. Smets, Improving the back surface field on an amorphous silicon carbide thin-film photocathode for solar water splitting. ChemSusChem **11**, 1797–1804 (2018). https://doi.org/10.1002/cssc.201800782
229. M. Baek, M. Zafar, S. Kim, D.H. Kim, C.W. Jeon, J. Lee, K. Yong, Enhancing durability and photoelectrochemical performance of the earth abundant Ni–Mo/TiO_2/CdS/CIGS photocathode under various pH conditions. ChemSusChem **11**, 3679–3688 (2018). https://doi.org/10.1002/cssc.201801211
230. J. Tan, W. Yang, Y. Oh, H. Lee, J. Park, R. Boppella, J. Kim, J. Moon, Fullerene as a photoelectron transfer promoter enabling stable TiO_2-protected Sb_2 Se_3 photocathodes for photo-electrochemical water splitting. Adv. Energy Mater. **9**, 1–12 (2019). https://doi.org/10.1002/aenm.201900179

231. M. Chen, Y. Liu, C. Li, A. Li, X. Chang, W. Liu, Y. Sun, T. Wang, J. Gong, Spatial control of cocatalysts and elimination of interfacial defects towards efficient and robust CIGS photocathodes for solar water splitting. Energy Environ. Sci. **11**, 2025–2034 (2018). https://doi.org/10.1039/c7ee03650g
232. C.S. Tan, K.W. Kemp, M.R. Braun, A.C. Meng, W. Tan, C.E.D. Chidsey, W. Ma, F. Moghadam, P.C. McIntyre, >10% solar-to-hydrogen efficiency unassisted water splitting on ALD-protected silicon heterojunction solar cells. Sustain. Energy Fuel **3**, 1490–1500 (2019). https://doi.org/10.1039/c9se00110g
233. L. Meng, W. Tian, F. Wu, F. Cao, L. Li, TiO_2 ALD decorated CuO/$BiVO_4$ p-n heterojunction for improved photoelectrochemical water splitting. J. Mater. Sci. Technol. **35**, 1740–1746 (2019). https://doi.org/10.1016/j.jmst.2019.03.008
234. S.K. Karuturi, H. Shen, T. Duong, P.R. Narangari, R. Yew, J. Wong-Leung, K. Catchpole, H.H. Tan, C. Jagadish, Perovskite photovoltaic integrated CdS/TiO_2 photoanode for unbiased photoelectrochemical hydrogen generation. ACS Appl. Mater. Interfaces **10**, 23766–23773 (2018). https://doi.org/10.1021/acsami.8b04855
235. Y. Yu, Z. Zhang, X. Yin, A. Kvit, Q. Liao, Z. Kang, X. Yan, Y. Zhang, X. Wang, Enhanced photoelectrochemical efficiency and stability using a conformal TiO_2 film on a black silicon photoanode. Nat. Energy **2** (2017). https://doi.org/10.1038/nenergy.2017.45
236. B. Lamm, A. Sarkar, M. Stefik, Surface functionalized atomic layer deposition of bismuth vanadate for single-phase scheelite. J. Mater. Chem. A **5**, 6060–6069 (2017). https://doi.org/10.1039/c6ta09485f
237. H. Hajibabaei, O. Zandi, T.W. Hamann, Tantalum nitride films integrated with transparent conductive oxide substrates: Via atomic layer deposition for photoelectrochemical water splitting. Chem. Sci. **7**, 6760–6767 (2016). https://doi.org/10.1039/c6sc02116f
238. H. Hajibabaei, D.J. Little, A. Pandey, D. Wang, Z. Mi, T.W. Hamann, Direct deposition of crystalline $Ta_3 N_5$ thin films on FTO for PEC water splitting. ACS Appl. Mater. Interfaces **11**, 15457–15466 (2019). https://doi.org/10.1021/acsami.8b21194
239. M. Hannula, H. Ali-Löytty, K. Lahtonen, J. Saari, A. Tukiainen, M. Valden, Highly efficient charge separation in model Z-scheme TiO_2/$TiSi_2$/Si photoanode by micropatterned titanium silicide interlayer. Acta Mater. **174**, 237–245 (2019). https://doi.org/10.1016/j.actamat.2019.05.032
240. Y. Hou, J. Zhang, Q. Yang, C. Wang, H. Tang, S. Yan, Z. Zou, ALD-grown oxide protective layers on Ta_3N_5-Cu_2O n-p nanoarray heterojunction for improved photoelectrochemical water splitting. Appl. Phys. Lett. **117** (2020). https://doi.org/10.1063/5.0027906
241. C. Ros, T. Andreu, J. David, J. Arbiol, J.R. Morante, Degradation and regeneration mechanisms of NiO protective layers deposited by ALD on photoanodes, J. Mater. Chem. A **7**, 21892–21902 (2019). https://doi.org/10.1039/c9ta08638b
242. Z. Wang, G. Yang, C.K. Tan, T.D. Nguyen, A.I.Y. Tok, Amorphous TiO_2 coated hierarchical WO_3 nanosheet/CdS nanorod arrays for improved photoelectrochemical performance. Appl. Surf. Sci. **490**, 411–419 (2019). https://doi.org/10.1016/j.apsusc.2019.06.090
243. R. Wang, L. Wang, Y. Zhou, Z. Zou, Al-ZnO/CdS photoanode modified with a triple functions conformal TiO_2 film for enhanced photoelectrochemical efficiency and stability. Appl. Catal. B Environ. **255**, 2–10 (2019). https://doi.org/10.1016/j.apcatb.2019.05.040
244. T.G. Kim, B. Joshi, C.W. Park, E. Samuel, M.W. Kim, M.T. Swihart, S.S. Yoon, Supersonically sprayed iron oxide nanoparticles with atomic-layer-deposited ZnO/TiO_2 layers for solar water splitting. J. Alloys Compd. **798**, 35–44 (2019). https://doi.org/10.1016/j.jallcom.2019.05.255
245. K.R. Nandanapalli, D. Mudusu, J.S. Yu, S. Lee, Stable and sustainable photoanodes using zinc oxide and cobalt oxide chemically gradient nanostructures for water-splitting applications. J. Colloid Interface Sci. **558**, 9–20 (2019). https://doi.org/10.1016/j.jcis.2019.09.086

246. X. Chen, Y. Fu, T. Kong, Y. Shang, F. Niu, Z. Diao, S. Shen, Protected hematite nanorod arrays with molecular complex co-catalyst for efficient and stable photoelectrochemical water oxidation. Eur. J. Inorg. Chem. **2019**, 2078–2085 (2019). https://doi.org/10.1002/ejic.201801200
247. L. Liu, H. Hou, L. Wang, R. Xu, Y. Lei, S. Shen, D. Yang, W. Yang, A transparent CdS@TiO_2 nanotextile photoanode with boosted photoelectrocatalytic efficiency and stability. Nanoscale **9**, 15650–15657 (2017). https://doi.org/10.1039/c7nr05658c
248. C. Li, Y. Xiao, L. Zhang, Y. Li, J.J. Delaunay, H. Zhu, Efficient photoelectrochemical water oxidation enabled by an amorphous metal oxide-catalyzed graphene/silicon heterojunction photoanode. Sustain. Energy Fuel **2**, 663–672 (2018). https://doi.org/10.1039/c7se00504k

Chapter 4
Hydrogen Storage

4.1 Introduction

Global warming is an obvious warning sign for critical ecological degradation. Throughout the years, humanity has used natural resources such as fossil fuels, oil, coal, and gas, without regard for their environmental consequences. An alternative solution to mitigate this issue is to engage hydrogen as a source of energy. The interest in using hydrogen as a fuel for vehicles originated in the year 1800, but it gained significant momentum during the 1970s due to the oil crisis and further advancements in technology during the 1980s [1].

Nowadays, the global demand for renewable energy as a substitute for fossil fuels has seen a significant increase over the last several decades. Hydrogen exhibits great potential as an energy carrier in forthcoming energy systems. Hydrogen is considered a very promising energy carrier due to its great energy density per unit mass, widespread availability, and little environmental effect. Renewable resources, such as wind energy, biological and photoelectrochemical processes, can be utilized to generate hydrogen [2]. Hydrogen offers an energy density per unit mass that is approximately three times greater than that of petroleum and nearly seven times greater than that of coal of the same mass [3]. Nevertheless, the storage of hydrogen poses a major challenge in the advancement of hydrogen-powered technology, such as fuel cells, and vehicles specially for the utilization in systems with proton-exchange membrane (PEM) fuel cells. To be able to effectively use this energy, a reliable and secure storage system is necessary, that remains the primary barrier for hydrogen to supplant fossil fuels as a prominent energy source. Moreover, finding a safe and technically economical hydrogen storage technique is a fundamental barrier to move towards a hydrogen-based energy economy. Hydrogen storage is a crucial field of study that receives significant global attention and focus, and there are many possibilities to exploit and potential paths to pursue in the future.

P. O. Oviroh et al., *Emerging Atomic Layer Deposition for Hydrogen Energy*,
Springer Series in Materials Science 348,
https://doi.org/10.1007/978-3-031-67774-8_4

It is important to highlight that although liquid- or gaseous-form hydrogen is currently being considered for prototype automobiles and bigger commercial vehicles, solid-based hydrogen storage has the potential to be superior in terms of its storage capacity (both in weight and volume), cost effectiveness, and safety. However, the most challenging aspect of implementing the hydrogen market for mobile applications is the effective hydrogen storage in solid-state materials. Material-based storage refers to the process of hydrogen being attached to solid materials [4]. The strength of this attachment can range from weak physical adsorption to strong chemical bonding in substances like zeolites, metal hydrides, and metal–organic frameworks (MOFs) [5]. The U.S. Department of Energy (DOE) has defined ambitious goals for this approach, aiming for a gravimetric density of 6.5 wt% and a volumetric capacity of 40 g L^{-1} at 100 bar [6]. The technology should be able to operate within a temperature range of −40 °C to +40 °C and achieve reversible storage/release durations of less than 5 minutes.

The study on materials has been concentrated on the hydrogen storage in the form of physisorbed or chemisorbed hydrogen. While several materials meet most of the criteria set by the DOE for use in vehicles, such as hydrogen storage capacity, other parameters including poor kinetics, non-reversibility, and occasionally thermodynamically unfavorable restrict or hinder their practical application. Extensive and scientific investigation aimed at determining the mechanism, structure, and thermodynamics, as well as their interconnections, is essential for enabling innovation and advancing materials development. Several of these materials have achieved one or more of the objectives; however, it has been difficult for a particular material system to fulfil all of these demanding criteria. In the past, researchers have used nano-encapsulation and low-concentration chemical additions as independent methods to address these difficulties [7].

Vapor-phase techniques, such as atomic layer deposition (ALD) [8], provide significant advantages compared to other methods for functionalizing H_2 storage materials. This opens up many options in the sector. In order to take use of these prospects, it is necessary to overcome the existing constraints on room temperature and waterless ALD techniques. Doing so will significantly broaden the potential for encapsulating and incorporating additives into organic frameworks and complicated metal hydrides. ALD has distinct features that enable it well-suited for both nano-encapsulation and the incorporation of low-concentration additives through "doping" [9]. ALD technology offers precise control over the thickness at a submonolayer level. It exhibits excellent conformity on materials with high surface area and has a self-limiting nature. In other words, once the precursor in the gaseous form interacts with the surface sites, the surface reactions terminate.

This chapter provides an overview of the current state of different materials and techniques employed for hydrogen storage. Furthermore, it highlights different techniques for storing hydrogen, such as compressed storage, cryogenic-liquid storage, adsorptive storage by high-surface-area adsorbents such as 2D materials, chemical storage in metal hydrides and complex hydrides, Zeolites, MOFs, COFs, etc. In addition, this chapter outlines a range of hydrogen storage techniques, together with their advantages and disadvantages, as well as the latest advancements in hydrogen technology. Furthermore, this chapter will demonstrate instances where the use of ALD, specifically in vapor-phase functionalization, has enhanced the material characteristics of carbon compounds, metal hydrides, and organic frameworks for the purpose of hydrogen storage.

4.2 Hydrogen Storage Mechanism

Hydrogen storage in solid form medium primarily encompasses a variety of phenomena: physisorption, chemisorption, and intermediary processes like Kubas type and spillover. The primary methods for reversible hydrogen storage are the process of adsorption on a solid surface [10] or the process of absorption of hydrogen inside the material [11]. During adsorption, the hydrogen atom retains its original structure mostly because of Van der Waals forces. However, in the process of hydrogen absorption, the originality is sacrificed, and hydrogen makes chemical bonds with the corresponding elements [12]. Various adsorbents, including metal–organic compounds, porous organic polymers, porous carbon compounds, and zeolites, among others, are suited for this purpose. These materials have a positive affinity for hydrogen adsorption and may adjust their ability to store hydrogen based on factors such as specific surface area, pore size, operating pressure, and temperature. The exothermic nature of the adsorption process can be attributed to the poor interaction between the adsorbent and atmospheric gaseous-form hydrogen. Liquid nitrogen, with a boiling point of -196 °C, is typically employed as a refrigerant to cool exothermic processes. For instance, a material with the ability to adsorb hydrogen at an enthalpy of 4 kJ mol^{-1}, equivalent to 0.56 kWh kg^{-1}, necessitates the addition of 10 kg of liquid nitrogen in order to facilitate more heat absorption. The reason for this is that nitrogen has a heat-absorbing capability of 0.056 kWh kg^{-1} [13].

The hydrogen can be molecularly adsorbed over the solid substrate during the adsorption process. Fast adsorption and desorption kinetics are necessary to select a better adsorbent, which will speed up the refueling process. At low temperature and pressure during adsorption, the hydrogen density can be greatly enhanced due to weak VdW interactions [14]. The physisorption process is reversible with increased adsorption and non-adsorption rates because of low binding energy. For these sorts of investigations, several experimental and theoretical evidences are available [15–17]. Moreover, there are three different types of hydrogen storage mechanisms included in solid-state hydrogen storage: chemical, physical, and intermediate. When hydrogen is stored physically, or by physisorption, it is absorbed on the surface through the use of physical forces (VdW), which are frequently present in porous materials with large specific surface areas.

As previously stated, the adsorption phenomenon also corresponds to chemical bonding. Adsorption of hydrogen involves a modification in the atomic hydrogen's initial composition as a result of chemical bonding. Chemical bonding can be classified into three kinds: metal hydrides, chemical hydrides, and complicated metal hydrides. Chemisorption refers to the process in which hydrogen atoms spontaneously react with chemical hydride groups, such as metal hydrides, complex metal hydrides, and chemical hydrides, under ambient conditions. However, chemical bonds are more robust than VdW bonds generated by physical interactions (physisorption). Consequently, a greater amount of energy is required to remove hydrogen by desorption. In addition, several theoretical and experimental investigations have been conducted utilizing metal hydrides, chemical hydrides (consisting of nonmetals such as oxygen, nitrogen, carbon, boron, hydrogen, etc.), and complicated metal hydrides to examine the weight % of hydrogen storage. Chemisorption, although effective, is hindered by sluggish kinetics and limited reversibility. These limitations might be attributed to a low desorption rate at high temperatures and a strong binding energy among hydrogen and the material surface. Once the vehicle is filled with hydrogen, the hydrogen will be consumed at a high temperature (desorption). Therefore, metal hydrides encounter challenges such as the rate at which they release hydrogen, the ability to be reversed back to their original state, the temperature and pressure required for hydrogen release, the cost involved, and other related factors. The quantity of hydrogen accumulated by the processes of physisorption and chemisorption falls short of the desired threshold set by the U.S. Department of Energy (DOE). Hence, it is imperative to always strive towards attaining the objective of reversibility, optimal storage performance, and cost-effectiveness in solid-state materials. Therefore, based on both methods, it is evident that the absorption and release of hydrogen must be a reversible process as mandated. Nevertheless, pure solid-state materials are unable to achieve sufficient hydrogen absorption. As a result, two additional processes, namely spillover and Kubas interaction, have been suggested to enhance the adsorption process.

4.2.1 Spillover Mechanism

In recent years, hydrogen spillover has become a feasible method for storing hydrogen. This phenomenon is commonly employed to improve hydrogen absorption in carbonaceous materials. The hydrogen spillover process is the dissociative chemisorption of gaseous hydrogen molecules from a metal catalyst particle followed by spontaneous migration of atomic hydrogen, known as the active species, to another surface site of the substrate/catalyst support and finally diffusion of hydrogen atoms onto substrate surfaces [18, 19]. In other words, the spillover mechanism involves (i) chemisorption on the surface. (ii) Dissociation occurring on the metal catalyst. (iii) The transfer of hydrogen from a metal catalyst to the surface of phase change material (PCM). (iv) Surface diffusion and desorption on the substrate. Hydrogen chemisorption can be accomplished using a phase nucleation process. The catalyst must possess a high capacity for hydrogen adsorption, with chemisorption of hydrogen on the catalyst requiring 0.8–1.8 eV of energy. The hydrogen atom must surpass a significant energy barrier of 2.45–3.2 eV to go to the substrate. Figure 4.1 illustrates the hydrogen migration over the substrate. Lueking et al. [20] used the term “hydrogen spillover” to specifically refer to the reaction that

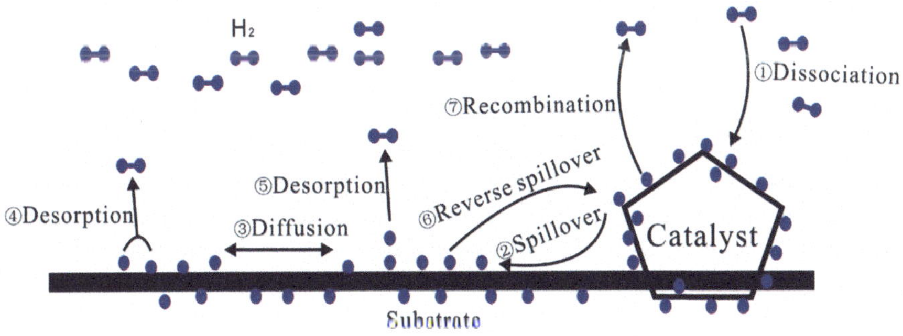

Fig. 4.1 Schematic diagram of the hydrogen diffusion pathways in a spillover hydrogen storage mechanism [22]

occurs among a multi-walled nanotube (MWNT) and a catalyst. Through this procedure, it was discovered that the hydrogen absorption of MWNT gained by up to 40%. Hydrogen adsorption by spillover refers to a procedure where hydrogen molecules are fragmented into hydrogen atoms, which are then adsorbed onto a surface. To optimize the rate of adsorption, the substrate surface can be enhanced by the process of doping or functionalization with various adsorbent species such as catalytic metal clusters (platinum, ruthenium, nickel, etc.) [21]. Spillover is anticipated to occur in some carbon materials that have a carbon ratio of sp^2/sp^3 more than 0.43 and a surface area less than 3500 $m^2\ g^{-1}$. Carbon materials in form of pure or raw are considered unsuitable for effective hydrogen absorption at ambient temperature due to the presence of a large energy barrier. However, the reduction of high-energy spillover barriers to the threshold range can be achieved by the introduction of impurities, defects, and functional groups, such as fluorine atoms, boron atoms, and other elements [22]. Spillover is commonly observed in graphitic materials as well as in functionalized or doped structures [23].

4.2.2 *Kubas Interaction*

As previously mentioned, binding energy is crucial in the storage of hydrogen. Several authors have investigated the optimal binding energy for storing hydrogen at ambient conditions, revealing that it falls within the range of 20–30 kJ mol^{-1} [24]. Chemisorption has challenges related to desorption, whereas physisorption encounters challenges associated with low hydrogen absorption. Therefore, in order to surmount these challenges, researchers have achieved intriguing experimental findings. The limitations of both physisorption and chemisorption can be overcome by the utilization of Kubas interaction [25]. This phenomenon elucidates the interplay between transition metals (TM) (catalysts) and hydrogen molecules. The catalyst in this process forms a chemical link with nanostructures by hybridizing the nanostructure with the d-orbitals of a transition metal. This bonding is referred to as Dewar coordination [26]. Subsequently, the σ and σ* bonds of numerous hydrogens combine with transition metal (TM) d-orbitals, resulting in the formation of TM-hydrogen complexes. The mechanism is referred to as Kubas interaction, and the complexes formed between transition metals and hydrogen are called Kubas-dihydrogen complexes [27]. Transition metals engage in chemical bonding with nanostructures such as carbon nanotubes (CNTs), graphene oxide (GO), single-walled carbon nanotubes (SWCNTs), or multi-walled carbon nanotubes (MWCNTs) through hybridization. As more hydrogen atoms bond together, the weight fraction of hydrogen increases. Currently, nanostructures coated with transition metals are extensively utilized as Kubas-type hydrogen storage materials due to their notable capacity and ability to be reversed [28]. Using density functional theory (DFT) calculations [9], it is anticipated that these materials have the potential to offer substantial hydrogen storage capacity (9–12 wt %) with reduced adsorption energy (−0.4 to −0.2 eV).

Transition metals' capacity for hydrogen storage is determined by the amount of unoccupied d-orbitals, as stated by the 1s electron rule [29]. The rule can be expressed as presented in Eq. (4.1):

$$N_{\max} = \frac{18 - n_v - n_\pi}{2} \tag{4.1}$$

$N_{\max}$ represents the upper limit of hydrogen molecules that the transition metal may take in, n_v indicates the quantity of valence band electrons, and n_π represents the quantity of π electrons. This rule refers to the computation of the quantity of hydrogen in Kubas-type storage. In general, transition metals tend to cluster together due of their high cohesive energy, which is typically about 4 eV. This clustering reduces their ability to store hydrogen. According to reports, calcium atoms can also create complexes with hydrogen molecules, although these complexes have a lower ability to form clusters because of their low cohesive energy (1.83 eV). Thus, Ca atoms are regarded as a highly favorable choice for hydrogen storage through Kubas interaction [30]. Several empirical investigations have been conducted to validate theoretical forecasts, revealing that Kubas-type contacts exhibit greater strength compared to Van der Waals interactions, although are less potent than chemisorption bonds [29].

4.3 Hydrogen Storage Technologies

A desired hydrogen storage material should be cost-effective, possess a substantial hydrogen storage capacity (5.5 wt% and 40 g L^{-1} hydrogen capacity by 2020, DOE), operate at moderate temperatures, be simple to absorb and desorb hydrogen, and have a long lifespan [31]. Figure 4.2 displays various hydrogen storage systems and compared their operational conditions [32]. Metal hydrides offer significant promise for commercially storing hydrogen among various techniques. MgH_2 is

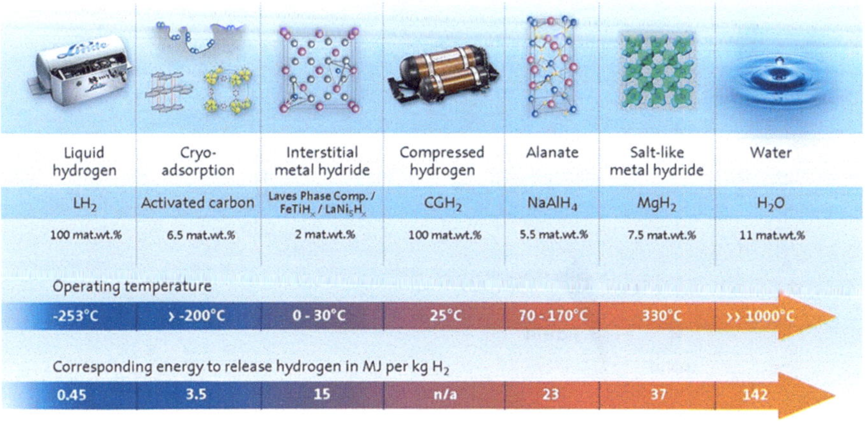

Fig. 4.2 Comparison of various hydrogen storage techniques and operating circumstances [32]

highly regarded as a good choice for hydrogen storage due to its large capacity (7.6 wt%, 110 g L^{-1} H_2), inexpensive, and abundance [33, 34].

4.3.1 Physical Storage

Within the concept of "physical storage" of hydrogen, three subtechnologies are typically classified as follows:

(a) CGH_2 (compressed gaseous hydrogen), operating at pressures of 35–70 MPa and at room temperature.
(b) Liquid hydrogen (LH_2) stored at a pressure range of 0.1–1 MPa and a temperature of around −253 °C.
(c) Cryoadsorption process utilizing high-surface-area materials at a pressure between 0.2 and 0.5 MPa and a temperature around −193 °C.

There are not any significant covalent or ionic interactions or chemical bonds among hydrogen and the substrate in any of these cases. Interstitial metal hydrides are occasionally included within this classification.

Hydrogen molecule H_2 exists in several states based on temperature and pressure, as seen in the phase diagram (Fig. 4.3). Hydrogen exists as a solid with a density of 70.6 kg m^3 at temperature of 262 °C and as a gas with a density of 0.089886 kg m^3 at 0 °C and a pressure of 1 bar. Liquid hydrogen with a density of 70.8 kg^3 m^3 at a temperature of 253 °C is present in the region bounded by the triple point and the critical point. Hydrogen is a gas at 298.15 K and can be characterized using the Van der Waals equation:

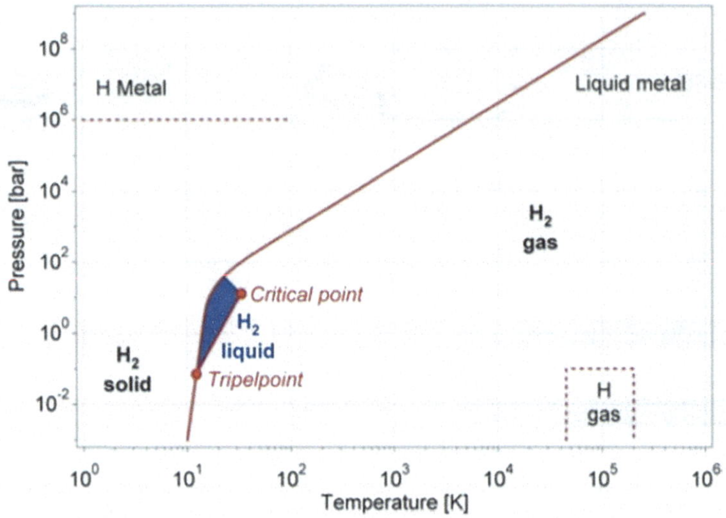

Fig. 4.3 Basic hydrogen phase diagram. Liquid hydrogen is present within the temperature range defined by the solid line, the critical point at 32 K and the triple point at 21.2 K [35]

$$\rho(V) = \frac{nRT}{V - nb} - \alpha.\frac{n^2}{V^2} \quad (4.2)$$

where p represents the pressure of gas, T is the absolute temperature, V denotes volume, n the number of moles, a denotes repulsion constant or dipole interaction ($a = 2.476 \times 10^{-2}$ m^6 Pa mol^{-2}), b is the occupied volume by the hydrogen molecules ($b = 2.661 \times 10^{-5}$ m^3 mol^{-1}), and the R is gas constant ($R = 8.314$ J K^{-1} mol^{-1}) [35].

4.3.1.1 Compressed Gas Storage

High-pressure gas cylinders, which can function at a maximum pressure of 20 MPa, are the most often used type of storage system. The materials used to make compressed gas storage containers include carbon fiber reinforced plastic composite, steel, and aluminum. The gravimetric hydrogen density decreases as pressure increases, as a result of the thicker cylinder walls. Novel lightweight composite cylinders have been engineered with the capacity to resist pressures of up to 80 MPa [36]. Consequently, the hydrogen contained within these cylinders can reach a volumetric density of 36 kg m^3, which is almost half the density of hydrogen in its liquid state at its typical boiling point [4]. This technology is advantageous for fuel needs, as it allows for compact storage without sacrificing its energy efficiency. Increasing the pressure of the gas enhances the energy density per unit volume. Despite the simplicity of the technology, the process suffers lack of efficiency in terms of volume and weight. In 2010, the compressed gas storage technology was employed by over 80% of hydrogen refueling stations worldwide [37].

The wall thickness of a cylinder with two hemispheres on top can be determined using the equation [35]:

$$\frac{d_w}{d_o} = \frac{\Delta \rho}{2.\sigma_v + \Delta \rho} \tag{4.3}$$

where d_w denotes the wall thickness, d_o is the cylinder's outer diameter, Δp is the overpressure, and σ_V is the material's tensile strength. The tensile strengths of different materials vary, with aluminum exhibiting a range of 50 MPa to premium steel surpassing 1100 MPa. Some materials have a density of just 2370 kg m^{-3} and a tensile strength of up to 2410 MPa [38]. The development of new lightweight composite cylinders, which possess the ability to endure pressures of up to 80 MPa, has facilitated the achievement of a volumetric density of 36 kg^3 m^{-3} for hydrogen. This density is about half of the volumetric density observed in its liquid state at the standard boiling point.

An ideal material for a high-pressure cylinder should possess low density, high tensile strength, and be nonreactive with hydrogen, preventing hydrogen diffusion. Typically, pressure cylinders have mostly utilized austenitic stainless steel such as AISI 316, AISI 304, AISI 316L, and AISI 304L over 300 °C to prevent carbon grain-boundary dissociation [39]. Copper or aluminum alloys have also been employed, as they are generally resistant to hydrogen effects at room temperatures. Figure 4.4 displays the volumetric density of hydrogen within the cylinder and the ratio of wall thickness to the outside diameter of the pressure cylinder made of stainless steel with a tensile strength of 460 MPa. The volumetric density rises with pressure and peaks beyond 1000 bar, dependent on the material's tensile strength. The gravimetric density depends as pressure increases, reaching its maximum at zero overpressure. The gain in density of volumetric storage is compromised by the decrease in gravimetric density in pressured gas systems.

Safety of pressurized cylinders is a major problem, specifically in regions with high populations. It is anticipated that next pressure vessels would consist of three distinct layers: an inner polymer liner, a carbon-fiber composite overwrap designed to bear tension, and an outside layer constructed of aramid-material capable of withstanding mechanical and corrosion-induced damage. The industry aims to develop a 110 kg, 70 MPa cylinder with a volumetric storage density of 30 kg m^{-3} and gravimetric storage density of 6%. Conventional mechanical compressors with pistons can be used to compress hydrogen.

A cylindrical storage tank is not the optimal configuration for accommodating the restricted space within a vehicle. With the usage of demonstration buses, this problem is mitigated by equipping the bus with a rack of pressurized cylinders located on the roof. However, in a car, a significant amount of luggage space has to be sacrificed. Compressed gas storage is now the most optimal solution for storing hydrogen onboard, but it will not be capable of fulfilling the goals defined by the DOE. Compressed gas storage is a cost-effective solution for applications that

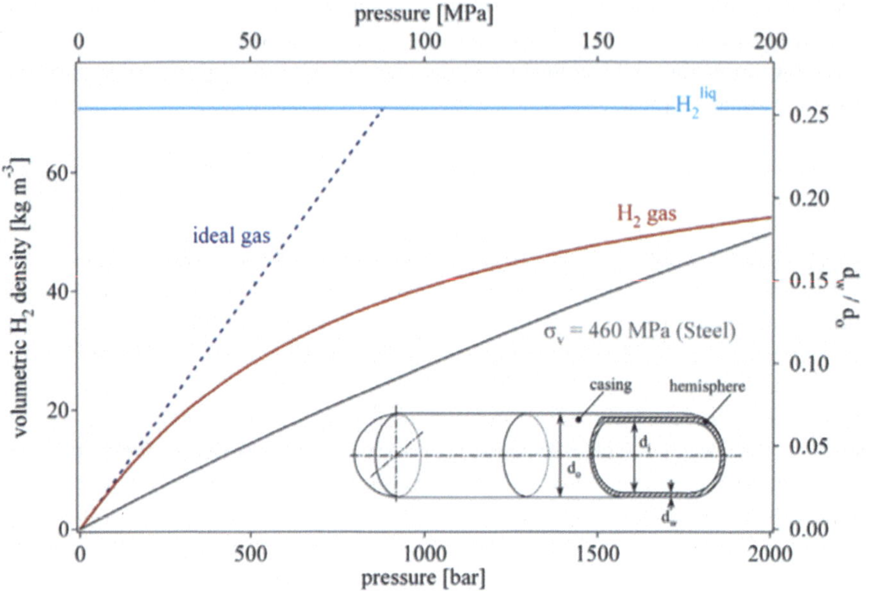

Fig. 4.4 Shows the volumetric density of compressed hydrogen gas, which includes both liquid hydrogen and the ideal gas, versus gas pressure. For steel having a tensile strength of 460 MPa, the wall thickness ratio to outside diameter of the pressure cylinder is displayed on the right. An inset displays a schematic design of the pressure cylinder [38]

require ample space, such as stationary storage at a fueling station and distributed generation. However, it is not suitable for portable applications, due to its excessive size.

4.3.1.2 Liquid Hydrogen

Cryogenic tanks are employed for the purpose of storing liquid hydrogen at conditions of 21.2 K temperature and ambient pressure. Because hydrogen has a low critical temperature of 33 K, it can only be stored in its liquid state in open systems. Liquid hydrogen can be stored at a temperature of −253 °C, whereas compressed gaseous hydrogen can be stored at normal ambient temperature. This is attributed to the absence of a liquid phase beyond the critical temperature. At ambient temperature, the pressure in a closed storage system can reach around 10^4 bar. The boiling point of hydrogen is low in the liquid hydrogen storage technique. Consequently, it is necessary to use specialized containers equipped with insulating systems in order to safeguard the efficiency of hydrogen.

The storage capacity of liquid hydrogen tanks is 0.070 Kg/L, but compressed gas tanks can only hold 0.030 Kg/L. To preserve the subzero temperature, it is necessary to ensure that the storage tanks are well insulated. Liquefaction is the process in which hydrogen atoms or molecules form strong bonds with other elements. Currently, the focus of research is on the advancement of composite tank materials that would lead to tanks that are both lighter and more durable [40, 41]. While the technology shows promise in terms of its efficiency in both gravimetric and volumetric aspects, further investigation is needed to address challenges associated to the absorption and desorption of hydrogen, the significant energy loss caused by high hydrogen liquefaction degrees, hydrogen boil-off, and expense of tanks. The significant energy requirements for liquefying and continuously evaporating hydrogen restrict the potential uses of liquid hydrogen storage systems to situations where the cost of hydrogen is not a significant concern and the hydrogen is used up relatively quickly, such as in aerospace applications. Table 4.1 summarizes the advantages and disadvantages of hydrogen storage methods.

4.3.2 Physisorption

The materials' hydrogen storage technique is often categorized as chemisorption or physisorption based on their interaction with hydrogen. Chemisorption involves the dissociation of hydrogen molecules into hydrogen atoms, which subsequently form chemical interactions with materials. Physisorption is the process of hydrogen molecules being adsorbed onto high specific surface area materials by VdW forces. The estimation of interaction energy between solid material and hydrogen molecules can be achieved by the following equation:

$$E = \frac{\alpha_{H_2} \alpha_{\text{substrate}}}{R^6} \tag{4.4}$$

where R represents the interaction distance and α is the polarizability. Since hydrogen's polarizability is fixed in this situation, using highly polarizable materials is the only possible way to raise the interaction energy. The average contact energy among carbon materials and hydrogen molecules is between 4 and 5 kJ mol^{-1}, indicating a poor connection. Because of this, rising temperatures have the potential to desorb hydrogen, which results in lower storage capacity at higher temperatures.

To enhance storage kinetics, maintaining the molecular structure of hydrogen during the process can be achieved using physisorption. The most widely investigated materials include porous materials like zeolites, carbon compounds (graphene, nanotubes, and fullerenes), covalent organic frameworks (COFs), metal–organic frameworks (MOFs), clathrates, organotransition metal complexes, and microporous metal coordination materials (MMOMs).

The compressed gas approach depends on a high initial pressure that leads to safety concerns, whereas cryogenic storage needs a significant energy input for the initial condensing of hydrogen. Complex hydrides, such as Mg_2NiH_4, are expensive, sensitive to impurities, have a poor reversible gravimetric capacity, and high temperature as 600 K is needed to release hydrogen.

Table 4.1 Pros and cons of hydrogen storage techniques

Storage method	Advantages	Disadvantages	References
Compressed gas	The pressurized hydrogen storage technology enables rapid filling and discharging Well-developed technology, particularly when implemented on a very limited scale	Compression of low to moderate intensity with losses of 15% Storage of hydrogen necessitates either low temperature or high pressure due to its low density High-pressure delivery of hydrogen Unsafety (leakage possibility)	[42, 43]
Liquid hydrogen	Compressibility Require low pressure Inexpensive Reliable High liquid density and high storage capacity	Require cooling systems Boil-off in days Energy losses about 30%	[23, 44–46]
Solid state	High quantity of hydrogen can be kept inside a small container The possibility of achieving a high hydrogen density under mild temperatures and pressures	Volume and weight of the storage material	[42, 47]
Physisorption	Great reversibility, quick refilling time and cycle life, free impurities accumulation	Aggregation Poor interaction with H_2 molecule Needs low temperature or very high pressure	[48]
Chemisorption	Low reactivity, large storage density, short storage time	Poor kinetics of dehydrogenation High temperature needed for desorption Highly reactive to moist air Hard to handle No reversibility Impurities absorption	[49, 50]

4.3.2.1 Zeolites

The term "Zeolite" originates from the Greek words "zeo" which means "to boil" and "lithos" which means "stone." Within zeolites, hydrogen molecules are compelled to enter the voids of the molecular network when subjected to high pressure and temperatures, which possess varying pore structures and compositions [51]. Zeolites hold H_2 inside their cavities when cooled to room temperature, and the gas can be desorbed by raising the temperature of the system. The findings indicate that zeolites with sodalite cages had a hydrogen storage capacity of 9.2 $cm^3\ g^{-1}$ at

10.0 MPa and 573 K [52]. Zeolites are recognized for their excellent heat resistance, affordability, and tunable chemical structure [53].

Figure 4.5 displays the basic structures of zeolites H-OFF, Na-LEV, NaX, NaA, Li-ABW, and Na-MAZ as presented in the Atlas of Zeolite Framework Types. Zeolites NaA and Na-LEV are 3D porous structures. The zeolite Na-LEV, with a suitable channel diameter and high pore size, achieved the highest hydrogen storage capacity of 2.07 wt%. Zeolites Na-MAZ and H-OFF have hydrogen storage capacities of 1.64 wt% and 1.75 wt%, respectively, which were considered intermediate. Zeolite Li-ABW had the minimum hydrogen storage capacity of 1.02 wt% due to its singular channel structure.

4.3.2.2 Metal–Organic Frameworks (MOFs)

Metal–Organic Frameworks are species of coordination polymers made up of secondary building units and organic linkers in their lattices. Hydrogen storage in MOFs occurs via weak VdW forces, which can reduce the pressure needed for hydrogen storage. MOFs offers high surface area, excellent stability, good thermal stability, well-organized customizable cavities of uniform size, huge void volume, controllable pore size, and tunable characteristics [55, 56]. The adsorption of hydrogen is closely correlated with both the gravimetric and volumetric surface area. MOFs' hydrogen storage capacity is modeled using GMDH-PNN, a statistical approach that provides the final mathematical structure of the model, in contrast to artificial neural networks (ANN). The model utilized hydrogen storage as the output variable and adsorption enthalpy, pressure, and surface area as input variables. The model exhibited greater sensitivity to pressure and surface area. Optimizing MOF surface area and pressures improves the storage of hydrogen [57].

The weight percentages of hydrogen absorption are determined using the formula shown in Eq. (4.5):

$$\mathrm{Wt\%} = \frac{(\text{mass of } H_2)}{(\text{mass of MOF} + \text{mass of } H_2)} \times 100 \tag{4.5}$$

Hydrogen spillover adsorption on porous carbon materials (PCM) can overcome the weak interaction of hydrogen adsorption. The spillover mechanism is more effective for hydrogen storage at room temperature and can be improved by utilizing metal-based catalysts. Hydrogen molecules migrate from the catalyst to surface of material to improve hydrogen storage characteristics [58]. Hole doping can decrease the energy barrier. Hydrogen atom transport in the substrate is hindered by the robust C-H bond, while the MOFs' surface presents an energy barrier ranging from 1.05 to 2.16 eV.

To select a material for storing H_2, one needs to understand its adsorption feature, which is influenced by its shape and pore size, as demonstrated in MOFs. MOFs are nano-porous materials that consist of certain building blocks, including polar metal

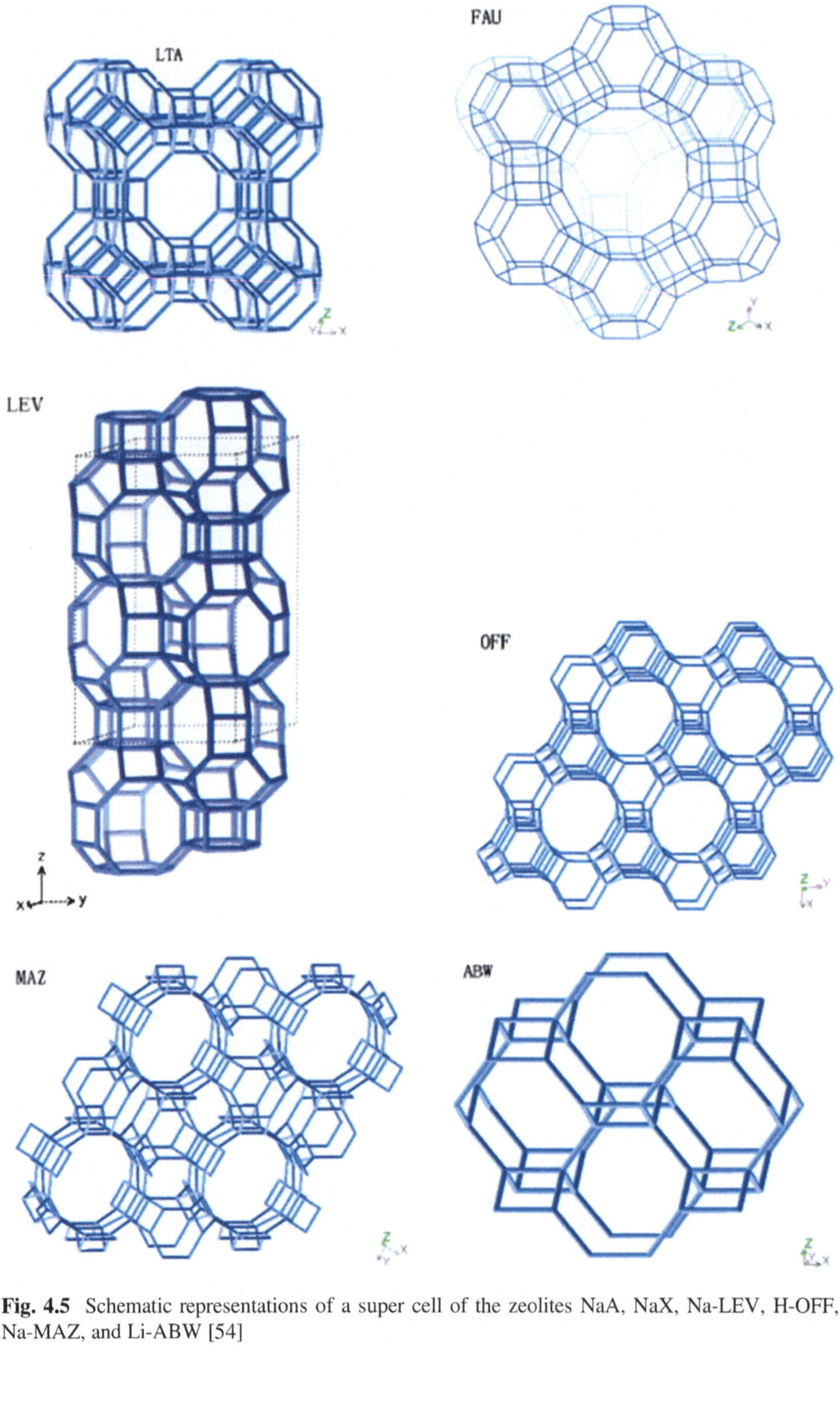

Fig. 4.5 Schematic representations of a super cell of the zeolites NaA, NaX, Na-LEV, H-OFF, Na-MAZ, and Li-ABW [54]

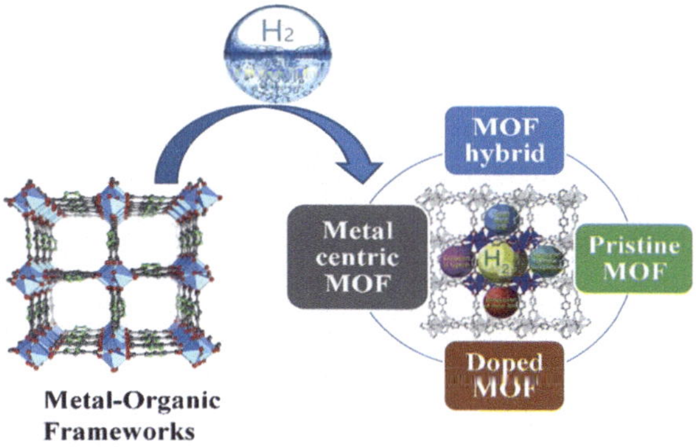

Fig. 4.6 Schematic view of MOF structure and different modification methods [68]

oxide connectors and nonpolar organic linkers. MOFs containing oxides exhibit stability even after the removal of their pores or exposure to heat. There has been a growing interest in hydrogen storage using MOFs since 2003 [59], when the first hydrogen storage system based on MOFs was reported. Adding electropositive metals to MOFs is a promising approach for efficient hydrogen storage.

Li-MOF has the strongest electron dependence due to its aromatic sp^2 carbon structure, which facilitates charge segregation. This results in Li being positively charged (acidic) and offering robust stabilization to molecular hydrogen. Several MOFs have been utilized for effective hydrogen storage, including isoreticular MOFs (IRMOFs) such as IRMOF-1 (1.3 wt%) [60], MOF-177 (7.5 wt%) [61], IRMOF-20 (6.5%) [62], and Li-doped MOF-C-30 (6 wt%) at −30 °C wt% at 50 pressures [63].

MOFs including boron substitution and calcium decoration exhibit increased binding energy due to Kubas interactions occurring within these structures [64]. Differences in sample purity, spillover, pore size, surface area, ligand structure, and catenation have led to varying levels of hydrogen absorption in MOFs [65]. However, they exhibit lower volumetric efficiency compared to COFs (COF-102) because of the larger pore volume they have, leading to empty space within the material.

The BET surface area, porosity, and pore volume all closely correlate with the hydrogen adsorption. To improve their ability to absorb hydrogen, inert metals such as Ru, Pd, and Pt can be doped in MOFs [66, 67]. Hydrogen adsorption via cryo-adsorption, which combines adsorption with cryogenic storage, can result in increased capacity. MOFs' ability to store hydrogen may thus be enhanced by a variety of techniques, including doping with metal ions, using nanoparticles, and creating composites. Metal–organic framework optimization will be straightforward if the effects of these tactics are recognized. Factors such as pressure, temperature, and composition can have an essential effect in hydrogen adsorption on MOFs. On the other hand, moisture instability and structural collapse brought on by solvent removal may hinder hydrogen adsorption. For usage in fuel cells, guest doping can improve the low conductivity of MOFs. Therefore, by switching to fuel cells, this green technology may aid in lowering the carbon footprint in the industry. Figure 4.6

Table 4.2 The advantages and disadvantages of various catalyst additive MOFs

Catalysts	Advantages	Disadvantages
Doping in MOF	Great hydrogen storage capacity	The stability of inert metals is insufficient. Steric effect
Metal-centered MOFs	Improved stability, acceptable binding energy	Poor resistance against corrosion, formation of interlinked pores
MOF hybrids	Controllability, efficient loading capacity, better hydrogen storage capacity improved surface area, exceptional porosity	No active centers, low stability costly linkers
MOF catalysts	Easy tunability, high porosity large surface area, good thermal stability, high hydrogen storage capacity	Poor active metal centers, insolubility

demonstrates the schematic view of MOF structure and different modification methods to improve its hydrogen storge capability.

Table 4.2 suggests that MOF hybrids have greater promise due to their greater selectivity in terms of materials with increased porosity, stability, and surface area.

4.3.2.3 Covalent Organic Frameworks (COFs)

COFs are the organic building blocks connected by firm covalent bonds (Si–C, B–O, C–O, C–C) instead of metal ions, resulting in materials with low crystal density and high porosity [69, 70]. These sorbent materials have been specifically engineered to have pore sizes similar to the molecular diameter of hydrogen. Their lightweight nature is due to their structure consisting of open channels made of aromatic carbon. 3D-COF's primary benefit compared to other light porous organic materials is its crystalline structure, which produces a large surface area. COF-1 and COF-5 have a 2D structure, while COF-102, COF-105, and COF-108 have 3D structures with a storage capacity three times higher than the 2D structures. COF-102 exhibits a gravimetric capacity of 9.95 wt% at 100 bar and 77 K. Replacing the phenylene groups in COF-102 with pyrene, triphenyl, diphenyl, and naphthalene groups results in COF-102-2, COF-102-3, COF-102-4, and COF-102-5, respectively. COF-102-3 has a gravimetric adsorption of 26.7 wt% and 6.5 wt% at 77 K and 300 K, respectively, at 100 bar, which is the highest among the samples [71].

The potential uses of COFs for storing hydrogen were first assessed using a variety of advanced multiscale theoretical methods. Most of these investigations anticipated that 3D COFs would exhibit higher gravimetric H_2 absorption at 77 K than large surface area MOFs. They selected a 3D COF, COF-102, as a prime instance of this group because of its favorable hydrogen storage characteristics. The self-condensation of tetra(4-dihydroxyborylphenyl)methane at 85 °C produces the COF-102 framework, which is structured around the theoretical C_3N_4 net (ctn type, refer to Fig. 4.7). Palladium (Pd) is a hydrogen storage material that is able to absorb hydrogen to generate PdHx ($x = 0.1$ to 0.6) [72]. Pd nanoparticles were

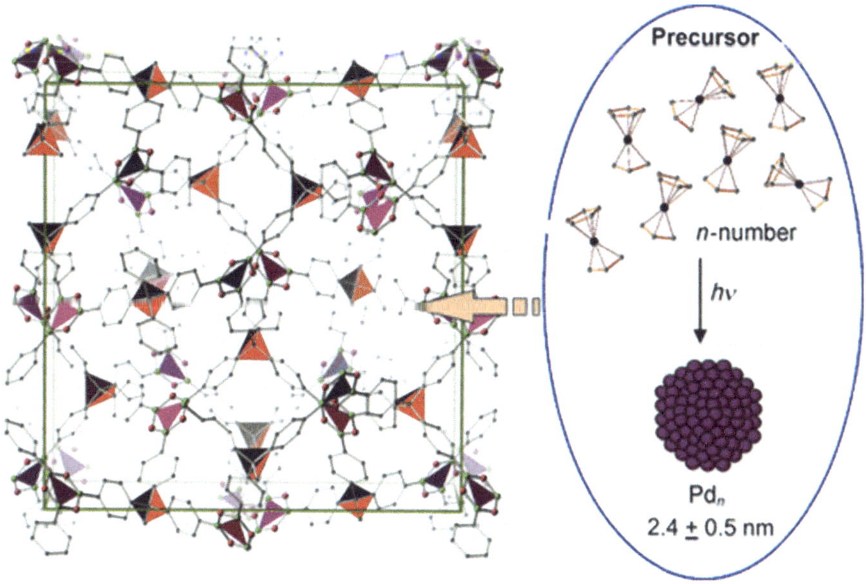

Fig. 4.7 Diagram illustrating the development of the Pd@COF-102 hybrid material. The image shows the crystal structure of COF-102, with oxygen, carbon, and boron atoms shown in green, gray, and red colors, respectively. The reaction occurs within the COF-102 pores to synthesize Pd nanoparticles [73]

applied to the inside of COF-102 microcrystals utilizing a precursor chemical infiltration technique. The hybrid material's hydrogen storage capabilities were assessed at 77 K and ambient temperature.

4.3.2.4 Polymers

Polymers are amorphous, free of metals, lightweight, and have high porosity micro- or nanopores with varied pore diameters and shapes. Furthermore, a multitude of synthetic techniques has remarkable scalability for porous organic polymers. Porous organic polymers, hypercrosslinked polymers, and polymers with inherent microporosity can store and release hydrogen by adsorbing hydrogen onto their highly porous structures. The hypercrosslinked polymers are able to store around 5 wt% of hydrogen at 8 MPa and 77 K, but only store 0.2 wt% of hydrogen at 9 MPa at ambient temperature [74].

N-heterocycle and Ketone polymers trapped and stored hydrogen at normal pressure through forming chemical linkages to produce hydrogenated N-heterocycle polymers and alcohol, respectively. The utilization of water as a hydrogen source in electrochemical hydrogenation has proven to be successful, wherein the polymer acts as a framework for the hydrogenation process. The hydrogenated polymers release hydrogen when exposed to catalysts under moderate

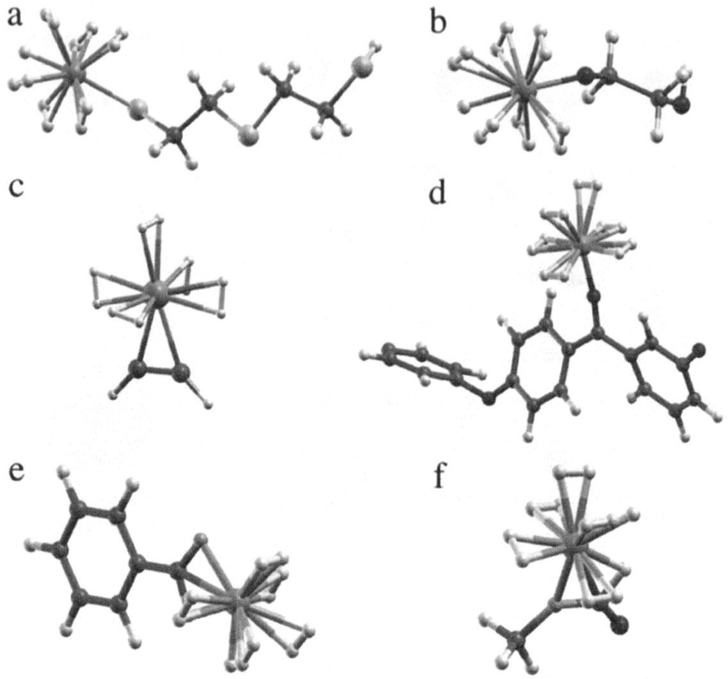

Fig. 4.8 Atomic frameworks of titanium-functional group compound with the highest possible amount of H_2 molecules bound. Pink, white, yellow, red, gray, and green dots represent titanium, sulfur, oxygen, nitrogen, carbon, and hydrogen atoms, respectively [5]

conditions. Organic polymers have the potential to be used in developing innovative, safe, and portable materials for transporting and storing hydrogen.

Integrating metal nanoparticles into polymer hosts has become more important in technical applications, particularly in sophisticated functional materials such as microwave absorbers, hydrogen storage systems, and optical sensors. The use of functionalized organic molecules doped with titanium atoms as high-capacity hydrogen storage materials has been proposed by researchers [5]. They examined six functional groups that may bond with titanium atoms and discovered that each complex can attach up to six hydrogen molecules, as shown in Fig. 4.8. Ti-decorated ethane-1,2-diol is a complex that has hydrogen storage capacity with a maximum gravimetric density of 13 wt% and a realistically useful capacity of 5.5 wt% under ambient conditions. They have showcased several storage materials derived from the modification of established nanomaterials utilizing Ti-functional group complexes.

DOE plans to explore the potential use of carbon doping using organo-metallic complexes containing electronegative nitrogen atoms in a rigid planar form. The findings of Cabasso and Yuan [75] suggest that these materials might facilitate the development of the Department of Energy's hydrogen storage system under specific pressure and temperature conditions.

4.3.2.5 Carbon Nanomaterials

Hydrogen can be absorbed onto the carbon surface through Van der Waals bonding with an energy of 6 kJ mol^{-1}. Various carbon structures like carbon aerogels, carbon nanotubes, carbon foam, and activated carbon have high surface area and low volumetric density. In order to achieve a high packing density, fullerenes demand a large surface area. Figure 4.9 present the view for different kinds of carbon-based materials for hydrogen storage. Various appropriate adsorbents include metal–organic compounds, zeolites, porous organic polymer materials, and porous carbon structures among others. The aforementioned materials exhibit favorable characteristics for hydrogen adsorption and possess the ability to modulate their capacity in accordance with factors such as specific surface area, operating pressure, temperature, and pore volume. The exothermic nature of the adsorption process can be attributed to the very weak binding mechanism between the adsorbent and hydrogen gas.

Hydrogen can be adsorbed onto the solid surface in a molecule form during adsorption. To select a superior adsorbent, rapid adsorption and desorption kinetics are essential to speed up the refueling process. Adsorption can significantly raise hydrogen density at low temperature and pressure due to weak VdW forces

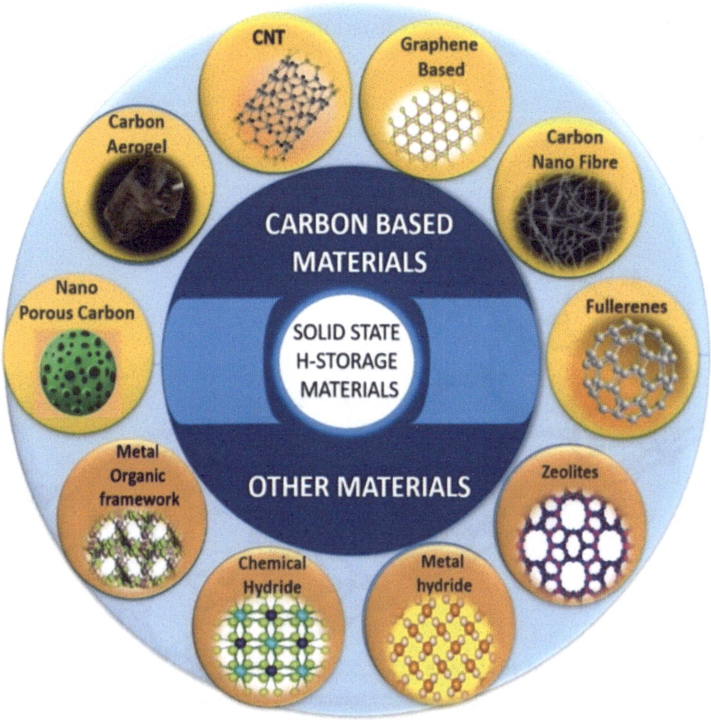

Fig. 4.9 Illustration depicting several carbon-based compounds utilized for hydrogen storage [77]

[14]. Low binding energy enables the physisorption process to be reversible, leading to increased rates of adsorption and non-adsorption. There are several theoretical and experimental evidences accessible for these types of investigations [15, 76].

Carbon Nanotubes

Carbon Nanotubes (CNTs) are microscopic carbon tubes that are two nanometers thick and can store hydrogen in their pores or within their frameworks. CNTs have a single or multiple wall structure, various adsorption sites, high packing density, and are expected to have a capacity of 6 wt% [78]. Figure 4.10 demonstrates H_2 adsorption for CNT configuration. Transition metals or alkali metals have been used to modify CNTs and buckyballs to enhance the binding of H_2 molecules to metal-CNT hybrids. Chen et al. [79] found that K-doped and Li-doped multi-walled nanotubes (MWNTs) have uptake capacities of 14 wt% and 20 wt%, respectively. The chemical stability of K-doped MWNTs is compromised, whereas Li-doped MWNTs exhibit chemical stability but require elevated temperatures ranging from 473 to 673 K to achieve optimum adsorption and desorption of hydrogen. Nanotubes have unpredictability in outcomes, material instability, metal clustering, low synthetic purity, release temperatures, and processing uncertainties issues, which restrict their utilization. The hydrogen storage capacity of several MWNTs and SWNTs at varying pressures and temperatures is listed in Table 4.3.

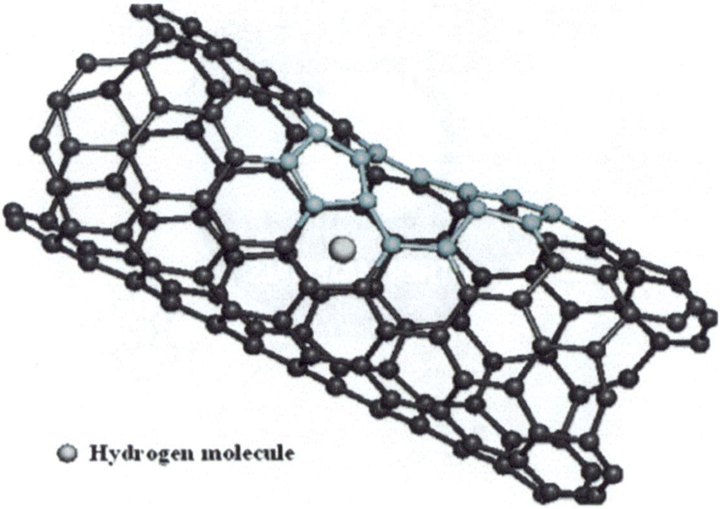

Fig. 4.10 H_2 adsorption for CNT configuration [80]

Table 4.3 Practical estimated hydrogen storage performance of CNTs

Material	Hydrogen storage (wt%)	Condition (temp K, pressure MPa)	References
SWNT (high purity)	8.25	80/7	[81]
SWNT (high purity)	3.5–4.5	298/0.04	[82]
SWNT (low purity)	5–10	406/0.04	[83]
SWNT (50% purity)	4.2	300/10.1	[84]
SWNT (80–90% purity)	7	273/0.04	[82]
Aligned SWCNT (purified)	4	298/11	[85]
SWCNT (Monte Carlo simulation)	11.2	77/10	[86]
Li-doped MWCNT	20	200–400/0.1	[87]
K-doped MWCNT	14	300/0.1	[87]
K-doped MWCNT	1.8	<313/0.1	[88]
Well-aligned MWCNT	3	290/10	[89]

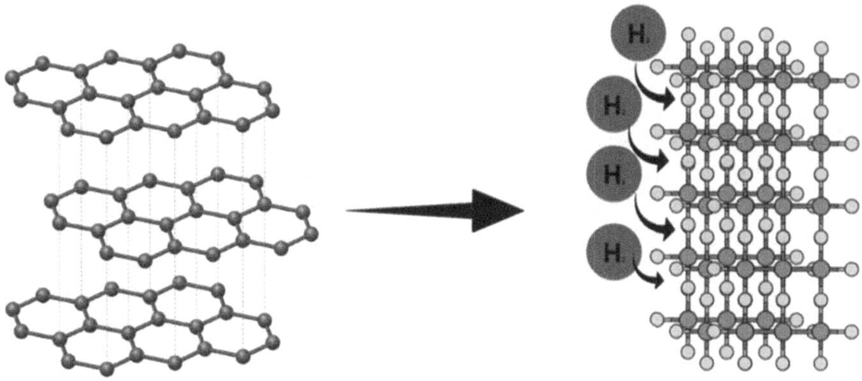

Fig. 4.11 Hydrogen storage mechanism of graphite structure [94]

Graphite

Hydrogen's interaction with graphite can be modified by altering the length among layers, adjusting the sheet curvature, or chemically functionalizing the material. This allows for regulated adsorption and desorption of hydrogen. Hydrogen can be stored between graphite layers and is released when heated to around 450 °C. It surpasses CNTs in efficiency since it is inexpensive, safe, and simple to manufacture. Figure 4.11 presents the hydrogen storing in graphene structure.

Kim et al. [90] discovered that the presence of Pt decoration in porous graphite nanofiber (GNF) results in a reduction in specific surface area (SSA) as the metal content increases, primarily due to the blocking of pores. Given that gas adsorption primarily takes place in micropore volume, the SSA of carbon materials becomes a matter of great importance [91]. Consequently, the presence of Pt decoration led to a reduction in both the SSA and micropore volume, resulting in a decrease in the

Table 4.4 Hydrogen storage capability of some doped graphite

Material	Hydrogen storage (wt%)	Condition (temp, pressure)	References
G (Ni-doped)	0.92	–	[89]
G (Cu-doped)	0.97	–	[89]
G (Co-doped)	1.8	–	[89]
G (Fe-doped)	3.9	–	[89]
G (Al-doped)	3.48	300 K, 100 MPa	[95]
G (Li-doped)	6.5	298 K, 2 MPa	[96]
Ni/porous graphite	4.48	298 K, 10 MPa	[97]

hydrogen storage capacity. According to an additional research, the hydrogen storage capacity was shown to rise with increasing Pt content up to a certain point, and thereafter drop [92]. The findings indicate that the hydrogen storage capacity of graphite is affected by both the metal content and the rate of dispersion. In addition, the overall hydrogen adsorption can be reduced if the hydrogen storage capacity depends only on the pore volume and SSA. The results indicated that the presence of Pt functions as a catalyst, leading to reduction in the hydrogen storage capacity of porous graphite nanofibers (PGNF). The study utilized DFT to explore the use of Al-doped graphite with a larger layer length for storing molecular H_2 at mild temperature [93]. The study reported a hydrogen storage capacity of 3.48 wt% at 300 K and 0.1 GPa conditions. However, N-ST5 was found to have a hydrogen storage capacity of 1.2 wt% at 77 K. Table 4.4 summarizes the hydrogen storage capabilities of several graphite samples.

Fullerenes

Fullerenes consist of pentagonal or hexagonal rings that generate the spherical curvature necessary for the fabrication of the C_{60} molecule. When a metal atom is supported on a carbon fullerene, the high C_{60} electronegativity promotes the transfer of electrons from the metal atom to C_{60}, resulting in the metal atom being in a cationic state. Therefore, the metal ion traps molecular hydrogen through charge polarization (Fig. 4.12). Theoretical studies indicate that the metal atom coated on C_{60} mostly stays in an isolated state [98]. Titanium atoms tend to aggregate on the C_{60} fullerene, but hydrogen bonding is not influenced by this clustering. However, aggregation of titanium atoms decreases the weight fraction of hydrogen storage [99].

DFT calculations demonstrate that lithium-coated fullerenes (Li_2C_{60}) have the lithium atom embedded into their pentagonal sides. Clusters are stable and can accommodate a maximum of 120 hydrogen atoms in molecular configuration, accompanied by a binding energy 0.075 eV/H_2. The minimum energy forms include a lithium atom being embedded on a five-membered ring of a fullerene, which is then bonded to a six-membered ring of another fullerene [100]. Nanostructured materials decorated with alkali metals (K, Ca, Na, and Li) are able to interact with hydrogen

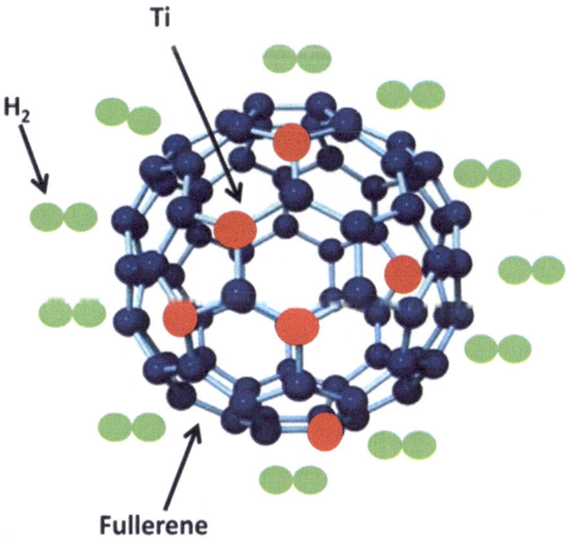

Fig. 4.12 Schematic diagram of Ti-doped fullerene capacity on hydrogen storage [104]

molecules with a binding energy ranging from 0.1–0.2 eV [101–103] and a gravimetric capacity of 8–9 wt% hydrogen.

Activated Carbon

Activated carbon (AC) is a type of carbon that has been processed of natural materials like peat, coal, wood, or waste of agricultural such as oil seeds, nut shells, jute fibers, coconut shell, etc. [105–107]. AC consists of graphite crystallites and amorphous carbon. It typically has a pore diameter of less than 1 nm and a specific surface area of 3000 $m^2\ g^{-1}$. AC can be derived from carbonaceous substances by the process of dry distillation and its pore volume can be enhanced through chemical or thermal activation methods [108]. AC has been identified as a viable option for hydrogen storage due to its cost-effectiveness and widespread availability in the commercial sector. The morphology and shape of activated carbon, such as powder, fiber, and granular forms, can affect its hydrogen storage rate and capacity. Research has demonstrated that the rate at which hydrogen is adsorbed is 2–50 times faster when in fiber form compared to when in granular form [105]. Hydrogen adsorption on these materials is significantly influenced by SSA and pore volume. The hydrogen storage capacity measured experimentally is influenced by the material used, such as doping, and typically varies between 0.5 and 5 wt% [78].

One of the ongoing challenges is the advancement of porous materials that can store sufficient hydrogen to meet the requirements for onboard applications, primarily related to storage temperature and pressure. Maximum adsorption capacity is achieved only under conditions of extremely low cryogenic temperature and high pressure [109]. One of the most researched material classes in this specific area of

Table 4.5 Experimentally determined activated carbon's hydrogen storage capacity

Material	Hydrogen storage (wt%)	Condition (temp, pressure)	References
AC (Pd-doped)	5.5	298 K, 8 MPa	[111]
AC (Pt-doped)	2.3	298 K, 10 MPa	[112]
AC (Ni-doped)	1.8	77 K, 0.1 MPa	[113]
AC (KOH-treated)	6.6	77 K, 4 MPa	[114]
AC	5.7	77 K, 3 MPa	[115]
AC	4.5	77.4 K	[116]
AC	1.6	296 K, 13 MPa	[117]
AC	1.4	77 K, 0.1 MPa	[113]
AC	0.67	303 K, 10 MPa	[115]

Abbreviation: *AC* activated carbon

study is activated carbon. Activated carbons can store hydrogen at ambient temperature, but their current capacity is far less than what is intended for onboard use. While it is evident that improving the textural characteristics (pore size and surface area) might enhance hydrogen storage at mild temperature, modifying textural characteristics by alone would not be sufficient to achieve significant advancements. Doping activated carbons with metal nanoparticles that can cause spillover is one approach that shows promise for strengthening the hydrogen's interaction with the sorbent and enabling room temperature hydrogen storage [110]. However, more basic study is required to fully comprehend spillover as it relates to hydrogen storage procedures.

Table 4.5 displays practical and theoretical hydrogen storage capacity estimates for raw and treated/doped AC from various research. Moreover, it shows that hydrogen storage capabilities vary even under comparable circumstances. Chemical treatment and doping can boost the storage capacity. Nevertheless, none of the findings met the DOE goal.

Various transition metals such as palladium, nickel, vanadium, and cobalt have been investigated [111, 118–120]. Zielinski et al. [121] researched hydrogen storage at ambient temperature using a commercially available AC doped with Ni. The pure AC, with a pore volume of 0.6 $cm^3\ g^{-1}$ and surface area of 1073 $m^2\ g^{-1}$, held approximately 0.15 wt% of hydrogen at 30 bar. The metal-doped AC with 1 wt% Ni can store 0.53 wt% hydrogen at 30 bar. The increase in hydrogen storage was credited to the loosely chemisorbed hydrogen stored as spilt-over species resulting from the breakdown of molecular H_2 on the nickel and migration onto the AC surface. In a recent study conducted by Yang et al. [122], it was shown that the storage of hydrogen in Ni-doped activated carbon increased by a factor of three at room temperature. At a pressure of 300 bar, the Ni-doped activated carbon exhibited a maximum storage capacity of 0.79 wt%. The dimensions of the nickel (Ni) particle had a pivotal role in enhancing the capacity for hydrogen storage under ambient conditions.

4.3.2.6 Glass Microstructures

At first glance, glass microspheres, also known as glass microballoons, seem to be a desired material for storing hydrogen. Glasses often possess lower density and higher tensile strength than steel, making them a potentially suitable material for storing hydrogen. When using small hollow spheres with a diameter less than a millimeter, thin walls can be created that can withstand high pressures [123]. This allows for achieving theoretical storage capacities exceeding 20% or even 30%, depending on factors such as sphere diameter, loading pressure, and wall thickness [124].

Upon closer inspection, the volumetric storage capacities, which are less than 40 kg m^{-3}, are shown to be inadequate. The volumetric density is further reduced by random sphere packings, which have packing densities somewhat above 60% [125], so even this number is optimistic. Furthermore, for the materials to diffuse through the glass walls sufficiently quickly, they must be filled at high temperatures. Since the loading temperature determines the maximum permissible pressure, it is not possible to reach the theoretical storage density at ambient temperature because the pressure drops as it cools. Moreover, high temperatures exceeding 250 °C are necessary to release hydrogen at sufficient rates, depending on the specific material properties and sizes. This leads to additional issues during operation. While it has been suggested that doping glasses could enable improved diffusion from infrared radiation [126]. The overall drawbacks, including limited kinetics and volumetric capacity as well as the need to handle solids, which may limit technological applications of such systems.

First, glass microspheres are rapidly cooled to room temperature after being filled with hydrogen at a high pressure in range of around 350–700 bar at 300 °C. Subsequently, the spheres are transferred to the low-pressure cylinder and subjected to a temperature range of 200–300 °C to allow for a monitored release of hydrogen. In addition to having a low volumetric capacity, the materials also need to be filled at a high pressure [127].

The steel pressure-resistant tank contains glass capillary arrays. The capillaries are achieved by the process of melting one end and subsequently sealing the other end using an alloy (Fig. 4.13a). Figure 4.13b shows an array of fused hexagonal capillaries that were sealed with a thin covering of an appropriate material. Hydrogen gas is pumped into the steel container until the desired storage pressure is attained. Currently, glass capillary arrays are utilized for secure infusing, storing, and controlled releasing hydrogen in mobile applications. Utilizing flexible glass capillaries and cryo-compressed hydrogen storage technology can assist in achieving the (DOE) 2015 targets. Defects in the glass configurations, such as grooves, cracks, or, bubbles reduce the functionality of this technology [128].

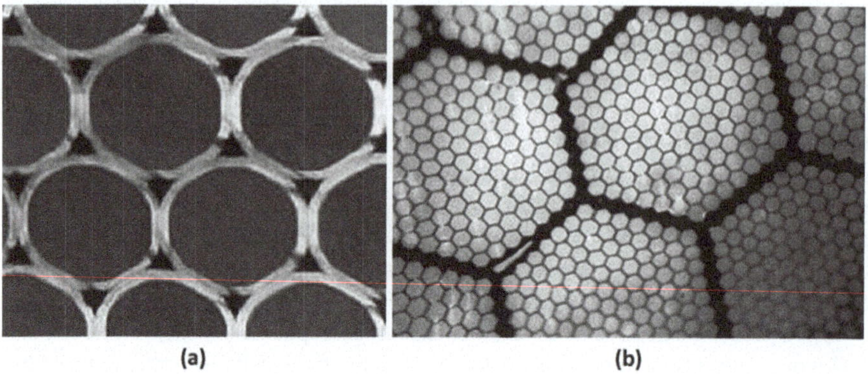

Fig. 4.13 (**a**) Cross-sectional image of the array of tightly packed capillary cylinder. Dark regions denote the inter-tube space. (**b**) Fusion of hexagonal capillaries into a honeycomb array. Block interfaces are represented by dark stripes [128]

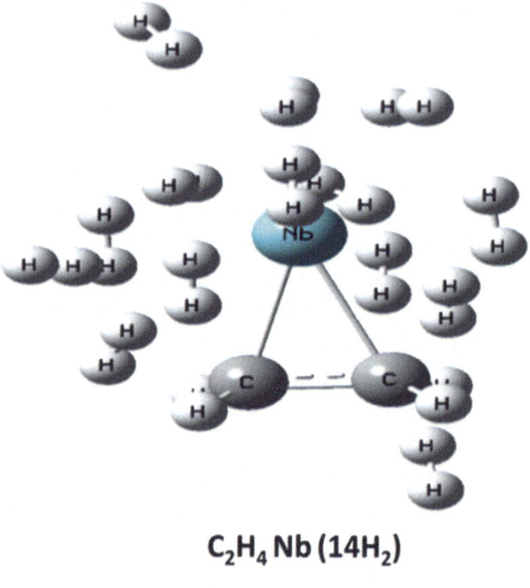

Fig. 4.14 The optimized configuration of C_2H_4Nb $(14H_2)$ by DFT [134]

4.3.2.7 Metal Complexes

Organotransition metal complexes refer to carbon-based materials that incorporate transition metals, enhancing the complex's ability to store hydrogen (Fig. 4.14). Various complexes include Vanadium and Scandium-based propane and ethylene complexes, alkane complexes [129–131], metal-decorated bucky balls [98], Ti atoms and Sc decorated on the $C_{48}B_{12}$ and C_{60} [99, 132], TM-doped organosilica [133], niobium-based ethene [134], Ti-polyacetylene [135], and multidecker. Organometallic compounds are recognized as effective hydrogen storage compounds. These compounds exhibit both high binding energy and rapid hydrogen adsorption and release cycles. Multitiered organometallic complexes are optimal in terms of thermodynamics, kinetics, and binding energy [136]. Niu et al. [137] stated that a metal cation can bind six H_2 molecules by a charge polarization mechanism. P-electron-deficient groups in organometallic systems interact strongly with transition metals through Dewar coordination. Kubas interactions facilitate hydrogen adsorption between transition metals and dihydrogen ligands.

4.3.3 Chemisorption

Chemisorption in solid-state hydrogen storage involves bonding by a chemical mechanism. Hydrogen molecules decompose into hydrogen atoms that occupy the spaces between atoms in materials to create new chemical interactions. Storing hydrogen at near room temperature has the advantage of being an effective method for conserving energy. The process of chemisorption, which involves the bonding of hydrogen to a storage material, is a key property for hydrogen storage [138, 139]. The process of absorbing hydrogen is characterized by an exothermic reaction, meaning it releases heat. On the other hand, the process of desorbing hydrogen is an endothermic process, meaning it requires the input of heat. The reaction can be expressed as [140]

$$\frac{2}{y-x}M\mathrm{H}_x + \mathrm{H}_2 \leftrightarrow \frac{2}{y-x}M\mathrm{H}_y + Q \tag{4.6}$$

where x and y represent the hydrogen numbers of the complex, M represents the metal element, H is the hydrogen atom, and Q denotes the reaction heat.

Chemisorption is defined as the binding of atoms or molecules in the gas and atoms on the material's surface through the sharing and exchange of electrons, also known as covalent and ionic bonds, respectively. The potential-energy plot indicates that there is a distance, commonly referred to as physisorption, at which the molecule's total external force equals zero. The bonding formation among H_2 molecules and the material surface observes if the H_2 molecule has enough energy above the activation energy. The system ultimately reaches a state of equilibrium at the distance where the total energy of the system is minimized. This distance is characterized by a shorter intermolecular distance and exhibits greater stability compared to physisorption.

The density of hydrogen stored in solid-state form is larger than that of hydrogen stored in gaseous and liquid states. According to Wheaton et al. [50], the hydrogen storage density of MgH_2 is measured to be 6.5×10^{22} H atoms/cm^3, whereas the densities for liquid gas and hydrogen are 4.2×10^{22} and 0.99×10^{22} and H atoms/cm^3, respectively [50].

Chemisorption technique has a higher bonding energy of almost 40 kJ/mol compared to hydrogen physisorption. Hydrogen chemisorption requires higher working temperatures, and hydrides could demonstrate either excessive stability or

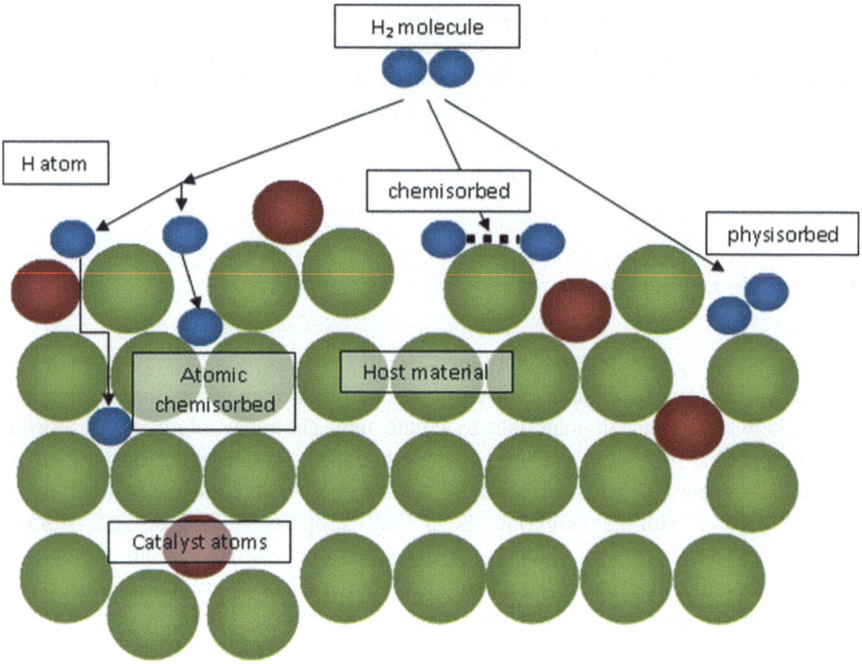

Fig. 4.15 Diagram illustrating physisorption and chemisorption using catalyst materials [141]

instability. The "too stable" hydrides have higher dissociation temperatures but are easier to absorb hydrogen. Hydrides with high instability have a propensity for hydrogen release at ambient temperature or lower. However, the absorption process necessitates elevated hydrogen pressures and temperatures for its occurrence. The adsorption mechanism differences between chemisorption and physisorption sites and effect of catalyst are shown in Fig. 4.15.

Hydrogen adsorption and release on carbon structure are influenced by bonding energy. The lower adsorption energy indicates a poor connection among hydrogen and carbon, enabling hydrogen release at lower temperatures. Thus, hydrogen adsorption at room temperature can only happen at high-pressure conditions. Higher adsorption energy results in a greater hydrogen storage capacity but makes desorption more difficult. Subsequently, the materials must be synthesized in a manner that maximizes hydrogen adsorption while simultaneously minimizing the interaction energy during desorption. The kinetics of sorption materials can also be characterized by activation energy. The activation energy is negligible during adsorption, but a large barrier must be overcome during desorption. The optimal bonding energy at ambient temperature ranges from 10 to 50 kJ mol^{-1}. This bonding energy can be achieved by improved physisorption, spillover, and Kubas binding.

4.3.3.1 Hydrides

Any metal having a redox potential lower than that of the H+/H_2 system at the specific pH level can be considered as a hydrogen storage system. When these metals react with water, they release hydrogen and produce metal hydroxide or oxide. It is more advantageous to utilize metal hydrides in practice as they release both the hydrogen produced by the metal's oxidation and the hydride hydrogen, allowing for high storage capacities to be obtained on a materials basis, such as 8.4 wt% for the reaction of $LiBH_4$ with $4H_2O$. Reference [142] provides a survey of systems that seem appealing in this context.

Hydrolytic systems cannot be regenerated onboard and must be exchanged after releasing hydrogen. These systems must be returned to the facility and subjected to an expensive and energy-intensive industrial procedure for regeneration. Additionally, these hydrolytic reactions are challenging to regulate since one solid is transformed into another solid, potentially causing crust development that restricts reagent access and results in significant control issues. Furthermore, the maximum capacity may not be fully utilized due to this issue.

A potential resolution to these problems has been identified for a particular instance: NaBH4 has notable stability in alkaline water and may be preserved without further measures, save for the preservation of its alkalinity. Hydrogen can be released by bringing the solution into touch with a catalyst, a well-established fact for over 50 years [143, 144]. The Millennium Cell utilized this method for storing hydrogen in cars powered by fuel cells. A 25% weight solution of $NaBH_4$ in 2% NaOH was utilized and passed over a catalyst based on ruthenium to generate hydrogen [145]. Based on Eq. (4.7), the stoichiometric reaction yields a theoretical storage capacity of 7.3 wt%.

$$NaBH_4 + 4H_2O \longrightarrow NaOH + H_3BO_3 + 4H_2 \tag{4.7}$$

The utilization of a 25% NaBF4 solution results in a decrease in the practical storage capacity to about 5.3 wt%. The relation between storage capacity and content is not directly proportional, as water functions as a reagent and releases hydrogen. This storage capacity is one of the highest ever obtained. Additionally, the aqueous $NaBH_4$ system offers the benefit of having a liquid fuel, similar to conventional gasoline or diesel fuel. Unlike other fuels, the result of $NaBH_4$ breakdown is not in a gaseous form but in a slurry of $NaBO_2$, which needs to be extracted from the vehicle and reused. A system, when coupled with a fuel cell, was tested in a prototype Chrysler Voyager vehicle. However, the results were not convincing enough to proceed with a second prototype test, leading to the discontinuation of its development for use in cars. Hydrolytic storage devices could be beneficial for specific uses like remote power supplies, but they do not appear to be a viable option for large-scale transportation applications.

Metal Hydrides

Classical metal hydrides typically contain around 2 wt% H_2 due to their composition as intermetallic TM compounds, which have high atomic weight. Despite the potential for storing multiple hydrogen atoms per metal atom, such as in rhenium complexes like $K_2[ReH_9]$ [146], the hydrogen storage capacity remains limited. Binary hydrides are traditionally classified as metallic hydrides, covalent, and salt-like. In the later complexes, hydrogen usually fills interstitial positions, resulting in various degrees of non-stoichiometry. While this classification may aid in instruction, there are several issues linked to it, as outlined by Grochala and Edwards [147].

In this context, the categories will not be utilized, and the hydrides will solely be categorized as metal hydrides, regardless of whether they exhibit metallic, covalent, or salt-like properties. This classification also applies regardless of whether they are binary, multinary, or complex hydrides containing a hydridometallate (or boron, a hydridoboranate) anion as a different structural component.

Hydrogen atoms in numerous important alloys are located at interstitial positions inside the host lattices. The number of spaces between particles in a lattice is typically more than the quantity of hydrogen absorbed (Fig. 4.16). However, the complete occupation of these gaps has never been observed, despite the potential to significantly boost the storage capacity for hydrogen. Electronic and geometric characteristics are suggested to elucidate this phenomenon [148].

In Table 4.6, the hydrides are denoted as AxByHz. These intermetallic compounds have the advantage of allowing the host materials' characteristics to be modified by replacing partial elements in order to meet the DOE objective, which makes the materials considerably safer to handle and store. The ability to store hydrogen (absorbance properties), charge and discharge pressures and temperature, the ability to form bonds between metal and hydrogen molecules, the size, shape, surface area, and molecular types and structures all influence the interface between the interstitial H_2 and the metal molecules [149].

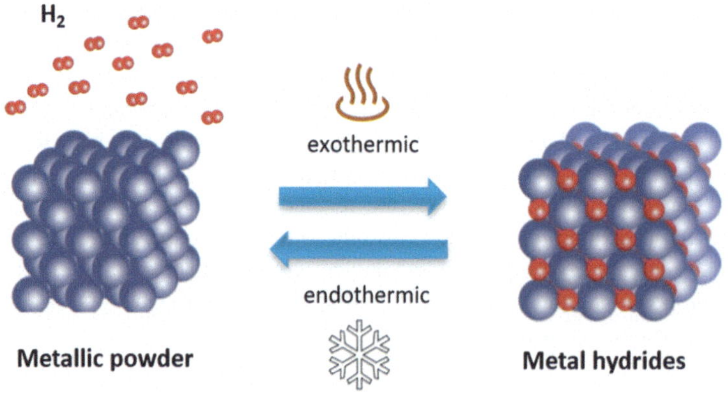

Fig. 4.16 Schematic H_2 adsorption and desorption of metal hydrides [150]

Table 4.6 Summary of pros and cons of intermetallic alloys

Alloy	Advantages	Disadvantages
AB	Reversibility, high weight capacity, light molar mass	Relatively low hydrogen storage capacity, high equilibrium pressure
AB_5	Enhanced hydrogen storage capacity, enhanced stability, immunity to impurities, adjustability	Poor gravimetric hydrogen storage capacity
AB_2	Very simple intermetallic compounds, nonmagnetic metals	Costly due to their status as rare earth metals
A_2B	Excellent magnetic and structural properties, high hydrogen storage capacity	Need high temperature and pressure, and cyclic stability
AB_3	Good electrochemical properties	Lower stability, poor hydrogen storage capacity

Metal hydrides store hydrogen by forming bonds with metal particles. Hydride absorption of hydrogen is an exothermic reaction, resulting in the release of heat. The hydrogen desorption from hydrides is an endothermic process that results in the absorption of heat. Hence, it is crucial to prioritize thermal management. These operations are reversible, enabling us to recycle the metal powder for hydrogen storage forever. Common alloys used for storing hydrogen are AB_5 compounds, including $LaNi_5$, which contain approximately 1.4 wt% of hydrogen and have an equilibrium pressure of around 0.2 MPa at ambient temperature [151]. This material is not ideal for storing hydrogen on cars due to the high cost of lanthanum and the alloy's small storage capacity. The same applies to AB_2 compounds, where the hydrogen storage capacity can reach up to 2 wt% H_2. The materials are more affordable than the $LaNi_5$ alloy (A = Mg, Ti, Zr; B = Mn, Fe, Cr, V), but their hydrogen equilibrium pressures are inadequate for fuel cell use at ambient temperature.

An innovative method for increasing the storage capacity of a mobile system involves combining a high-pressure tank with a metal hydride [152]. This tank is constructed using the principles of a traditional high-pressure gas tank. The combination of the V/Ti/Cr alloy, which has a storage capacity of about 2.2 wt% H_2, allows for a total storage capacity of 7.3 kg hydrogen in a tank volume of 180 L.

Intermetallic hydrides do not have enough storage capacity to meet the requirements of an onboard storage system. However, in specific applications, these materials are utilized in fuel cell systems that have solid-state hydrogen storage. Some interesting examples are a German Navy submarine that uses a fuel cell with a special alloy for hydrogen storage, and fork lift vehicles that are equipped with fuel cells and metal hydride storage systems, which need a strong counterbalance. A high ballast weight is required for submarine diving missions, making the low storage capacity of about 1.5 wt% alloy not a negative. Figure 4.17 and Table 4.7 provide a comparison of hydrogen storage rates and temperatures for various metal hydrides.

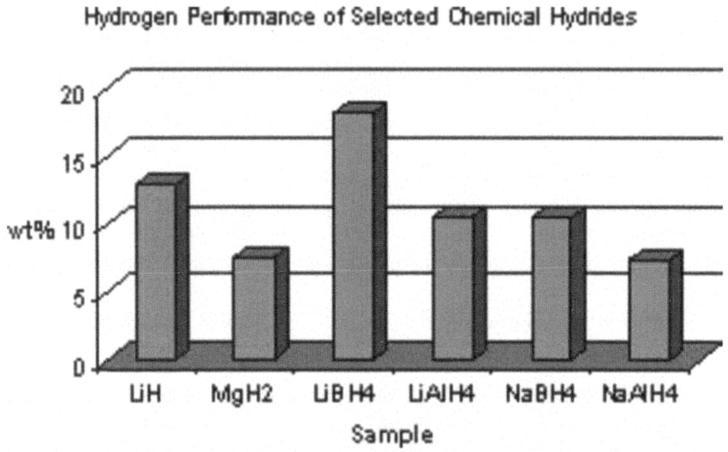

Fig. 4.17 Hydrogen capacity of some selected chemical hydrides [153]

Table 4.7 The potential for storing hydrogen in various metal hydrides

Hydrides	Storage capacities (wt%)	Temperature (K)	References
$CeNi_4Cr$	4.3	293–333	[154]
$CeNi_4Zr$	4	293–333	[155]
TiNi	1.2	400	[156, 157]
$ZrMn_2$	1.77	440	[154]
Mg_2Ni	3.59	255	[154]
TiFe	1.9	185	[158, 159]
$LaNi_5$	1.2	295	[160]
$Ti_{0.47}V_{0.46}Mn$	1.53	303	[161]
$Ti_{0.5}V_{0.5}Mn$	1.9	360	[161]

Borohydrides

A study on the production of hydrogen from sodium borohydride, or $NaBH_4$, was published in 1952 by Schlesinger et al. [162], and due to its high hydrogen level, which amounts to 10.8 wt% [163]. Aside from the unwanted formation of ammonia, borohydrides also have unacceptably poor kinetics. Moreover, amides suffer from capacity and kinetic constraints. In addition, all complex hydrides have the same issue with temperature control during recharge. $NaBH_4$ and $LiBH_4$ both offer a lot of potential as materials for hydrogen storage, according to a feasibility study done by Aiello et al. [164]. In actuality, the business Millennium Cells has already begun producing $NaBH_4$. The waste product, borax ($NaBO_2$), can be hydrolyzed using coke or methane, or it can be regenerated easily by applying hydrogen onto it. $LiBH_4$ has a high gravimetric hydrogen density of 18 wt%, but only releases a little quantity of hydrogen (0.3 wt%) at low temperatures. Furthermore, $LiBH_4$ is a costly chemical

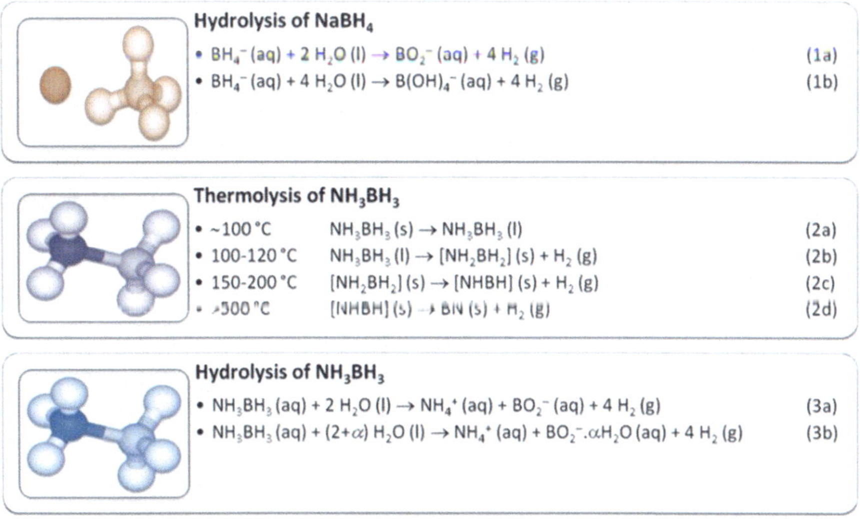

Fig. 4.18 Dehydrogenation of sodium borohydride $NaBH_4$ and of ammonia borane NH_3BH_3 [166]

[165]. Boron-containing materials have a clear advantage in capacity compared to non-boron-containing materials during hydrolysis.

Thermolytic dehydrogenation is not possible due to the material's stability and the high temperatures needed (400–500 °C). Hydrolytic dehydrogenation, facilitated by a metal catalyst, is highly handy since it allows for the release of H_2 at room temperature. The reaction in Eq. (4.1a) in Fig. 4.18 is beneficial as the $NaBH_4$-H_2O pair contains 10.8 wt% hydrogen, with half of the hydrogen derived from water. The borate by-product, BO_2(aq), is thermodynamically stable in its hydrated state, requiring a minimum of 2 more H_2O molecules for the reaction to proceed (Eq. (4.1b) in Fig. 4.18).

Another possibility for storing hydrogen is the boron nitride (BN) nanostructure observed in thin nanoforms and nanoballs [167]. After an 80-hour milling operation, BN can hold up to 2.6 wt% and can desorb at 573 K, a temperature that is unusually high when compared to other nanostructures [168]. Additionally, BN fullerenes have shown a 3 wt% hydrogen storage capability [169].

Ammonium borohydride, NH_4BH_4, is intriguing when considering dynamics and hydrogen storage because of its exceptionally high hydrogen densities of $\rho_m = 24.5$ wt% H_2 and $\rho_V = 157$ g H^2 l^{-1}, as well as the substantial dihydrogen interactions that occur within the compound between NH_4^+ and BH_4^-. NH_4BH_4 is inherently unstable under normal environmental conditions, with a "half-life" of around 6 hours [170]. Figure 4.19 displays the structural variety including structures composed of isolated tetrahedral, five-fold, or octahedral anionic $[M(BH_4)_n]$ complexes, as well as structures composed of one-dimensional chain-like frameworks, two-dimensional layers, and three-dimensional framework structures.

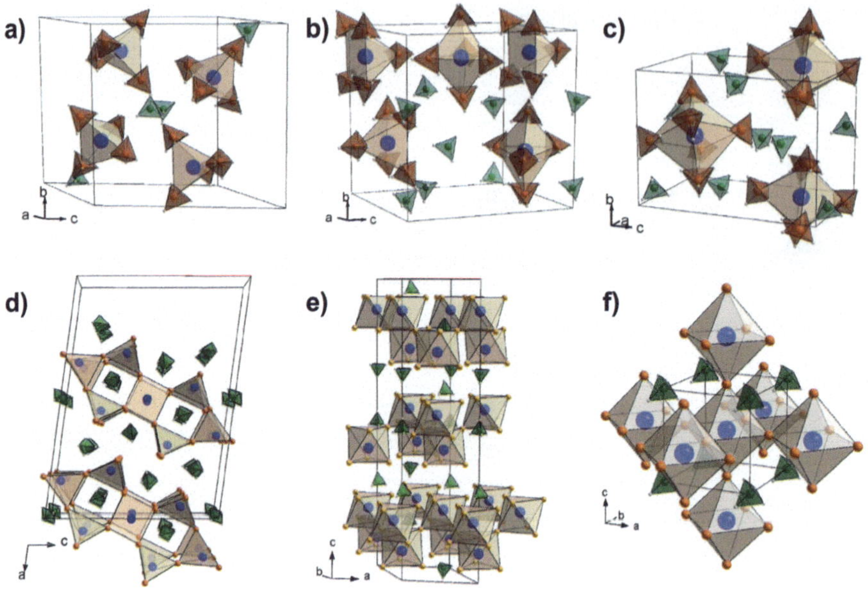

Fig. 4.19 The crystallographic structures of ammonium metal borohydrides provide evidence of the presence of isolated anions of complexes (**a**) $NH_4Y(BH_4)_4$, (**b**) $(NH_4)2Gd(BH_4)_5$, (**c**) $(NH_4)3Gd(BH_4)_6$, (**d**) $NH_4Li(BH_4)_2$, (**e**) $(NH_4)3La_2(BH_4)_9$, and (**f**) $NH_4Ca(BH_4)_3$ [172]

Many reports have demonstrated that the alkali-metal amidoboranes, $NaNH_2BH_3$ and $LiNH_2BH_3$, meet key requirements for hydrogen-storage materials and $NaNH_2BH_3$ and $LiNH_2BH_3$ have large storage capacities of 7.5 wt% and 10.9 wt %, respectively [171]. They may be readily dehydrogenated at around 90 °C without forming the undesirable by-product such as borazine. The materials are eco-friendly compounds that are stable solids, nonexplosive, and nonflammable at ambient conditions. Currently, the only drawback of these materials is their limited reversibility. However, the onboard regeneration process can be facilitated by the heat neutral dehydrogenation of alkali-metal amidoboranes.

Magnesium Hydrides

Magnesium hydride (MgH_2) has a hydrogen storage capacity of 7.7 wt% H_2, low cost, and exhibits strong reversibility during cycle, making it a promising option for onboard hydrogen storage. The thermodynamic properties of the material present a significant challenge for onboard use. Magnesium hydride (MgH_2) is a common metal hydride that exhibits an equilibrium pressure of 0.1 MPa at temperatures reaching up to 300 °C. Twice the value of n, the temperature is too high for PEM fuel cell applications when combined with onboard hydrogen storage, as a high hydrogen pressure is needed at the PEM fuel cell's operating temperature of

Table 4.8 A summary of a few common catalysts in MgH_2

Catalyst	Density	Method	Phenomenon	References
Ge	5 at%	RBM	Mg_2Ge formation	[182]
Fe	5 wt%	BM 20 hours	Unchanged thermodynamics	[183]
Cu	0.4, 1 wt%	SFCD	High kinetic under high T	[184]
Co	10, 20 wt%	RBM 0.5–10 hours	Crystallite and particle sizes	[185]
B	Mg − 2B	BM 59 hours	Significant oxide observed	[186]
V	5 wt%	BM 2 hours	No significant improvement	[187, 188]
Pt	0.5 mol.%	Wet chemi	Spillover mechanism	[189]
Pd	1 wt%	RBM 0–20 hours	Improved due to small size	[190]
Nb	5 at%	BM 20 hours	Unstable NbHx observed	[191]
Mn	5 at%	BM 20 hours	Unchanged thermodynamics	[183]
TiC	2 at%	BM 30 minutes	Improved kinetics	[192]
CuF_2	1–10 mol.%	BM 15 minutes	Cu_2Mg formation	[193]
TiF_3	1–10 mol.%	BM 15 minutes	TiH_2 formation	[193]
VO_2	10 wt%	BM 10 hours	Reduce in crystallite size	[194]
ZnO	6.5–10 wt%	BM 20 hours	No change in enthalpies	[195]
TiO_2	20 wt%	BM 1 hours	γ-MgH_2 formed after 1 h	[196]
SiO_2	5 mol.%	BM 100 hours	0.2 mol% enough catalyst	[197]
Nb_2O_5	1 mol.%	BM 20 hours	Decreased Nb_2O_5	[198]
CuO	5 mol.%	BM 100 hours	0.2 mol% enough catalyst	[199]
Mn_2O_4	5 mol.%	BM 100 hours	0.2 mol% enough catalyst	[200]
In_2O_3	3–4 wt%	BM 20 hours	No change in enthalpies	[195]
Fe_3O_4	5 mol.%	BM 100 hours	No stable phase observed	[201]
Cr_2O_3	0.2–5 mol.%	BM 100 hours	Cr_2O_3 decreased gradually	[202]
CeO_2	10 wt%	RBM 2 hours	20% conversion	[202]
Al_2O_3	1 mol.%	ALD 100 hours	0.2 mol% enough catalyst	[199, 203]
Fe_2O_3	1.5–2.5 wt%	BM 20 hours	No change in enthalpies	[195]

Wet chemi, *SFCD* supercritical fluid chemical deposition wet chemistry, *RBM* random ball-milling, *BM* ball-milling

approximately 90 °C. Despite this, MgH_2 has been extensively studied for over 20 years to enhance the kinetics of hydrogen absorption and desorption, as well as cycle stability.

However, magnesium-based materials have restricted practical utility due to their sluggish adsorption and desorption reaction rates, necessitating higher temperatures for hydrogen release [173]. Additionally, the generation of a very durable magnesium oxide layer decreases the rates at which adsorption occurs [128]. Adding catalysts is often one of the easiest and most efficient methods to improve MgH_2/Mg's hydrogen storage capabilities. Many catalysts, including metals, covalent compounds, metal oxides, metal halides, and carbons, have been added to the MgH_2 system, with notable results. A summary of various common catalysts in MgH_2 is provided in Table 4.8. Ball-milling is the preferred process for creating appropriate catalysts such as Mg/MgH_2–catalyst composite materials to enhance sorption kinetics [174]. There has been a recent effort to reduce the size of nanofibers

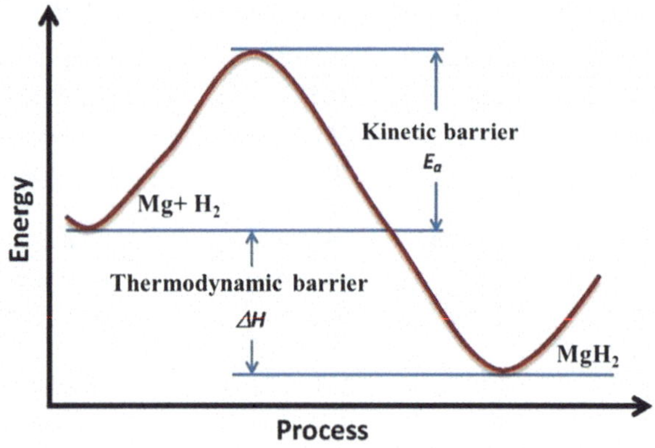

Fig. 4.20 Demonstration of the MgH_2/Mg hydrogenation and dehydrogenation process [32]

using the hydriding chemical vapor deposition process, resulting in a notable rise in the specific surface area [175]. Ball-milled MgH_2 combined with catalysts such as V_2O_5 or Nb_2O_5 exhibit rapid hydrogen absorption and desorption rates of under 2 minutes.

The process of mechanical ball milling can effectively decrease the size of MgH_2 into nano- and micron-sized particles that are significantly enhancing the hydrogen sorption energy [176]. Johnson et al. [177] found that milling MgH_2 for 15 hours at 300 °C under a hydrogen pressure of 10 bar increased the hydrogen capacity from 4.86 wt% to 6.01 wt%. Figure 4.20 depicts the hydrogenation/dehydrogenation process of MgH_2/Mg. Accordingly, enhancing the hydrogen storage capacity of MgH_2/Mg can be achieved by reducing the energy barrier and altering the reaction enthalpy.

Modifying the thermodynamics of the Mg/MgH_2 system could potentially enable the system's large storage capacity to be utilized in less extreme settings. Theoretical models indicate that the thermodynamic properties vary significantly based on the size of magnesium clusters. Particles measuring 0.9 nm and consisting of 20 magnesium atoms are less stable than the bulk substance [178]. This instability leads to a desorption temperature of around 200 °C [179]. These calculations were conducted in a vacuum; therefore, the findings may vary if a more realistic hydrogen atmosphere is utilized.

Recently, melt infiltration was utilized to create magnesium nanoparticles supported by carbon [180]. Magnesium nanoparticles smaller than 2 nm can be produced based on the pore size of the carbon material. Experimental proof of a material achieving the necessary equilibrium pressure of 0.1 MPa at lower temperatures has not been shown. MgH_2 possesses a high energy density of 9 MJ per kilogram of magnesium. Although not ideal for storing hydrogen, this material demonstrates potential for efficiently storing large quantities of thermal energy across a significant temperature range [181].

T. Christensen et al. [203] used the well-researched Al_2O_3 ALD method, which employs water and trimethylaluminum. Although the use of Al-based additives for $Mg(BH_4)_2$ has been limited, they showed that 100 cycles of Al_2O_3 inhibited the

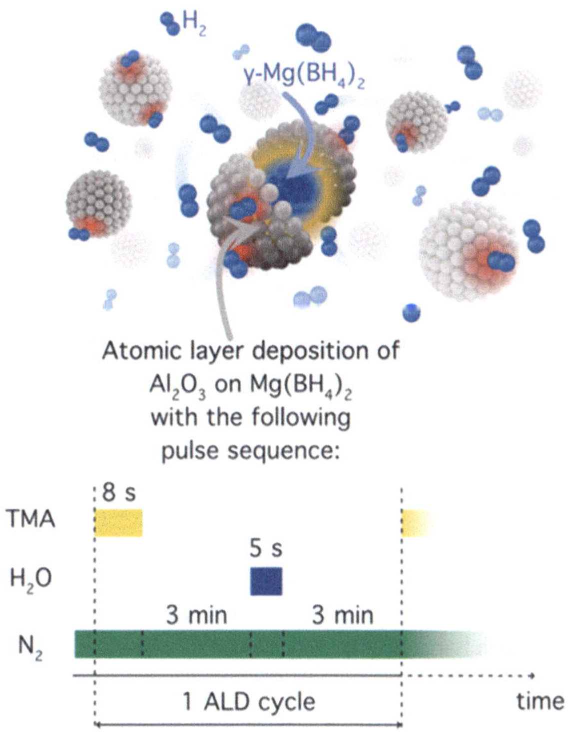

Fig. 4.21 Diagram showing the encapsulation of γ-$Mg(BH_4)_2$ (blue shell) in Al_2O_3 (white shell) with hydrogen molecules present (as tiny blue spheres around the encapsulated Mg $(BH_4)_2$). The precursor's pulse sequence and duration (TMA and H_2O) which are employed in this work to deposit Al_2O_3 via ALD are displayed. The length of the N_2 purging stages and the precursor pulses was increased to accommodate for the porosity of γ-Mg $(BH_4)_2$ [203]

desorption of diborane compared to the bare $Mg(BH_4)_2$, quadrupled the low-temperature hydrogen capacity, and a three-fold increase in the desorption kinetics. They found that the breakdown route was impacted by the use of water and trimethylaluminum (TMA) during the ALD process, and that infiltration dominates the formation of the Al_2O_3 film because of the high porosity of the γ-phase Mg $(BH_4)_2$. These findings suggest that ALD may be used to modify solid-state hydrogen storage materials, and they also provided suggestions for new ALD methods in the future. Figure 4.21 depicts the Al_2O_3 ALD model system with TMA and H_2O as the precursors.

Complex Hydrides

Unlike classical metal hydrides where hydrogen atoms occupy interstitial sites, in complex metal hydrides, hydrogen covalently interacted to a metal or nonmetal atom to create a complex anion [142]. An anion's charge is neutralized by a cation, typically an alkali or alkaline earth ion. The storage capacity is determined by the metals' weight, the number of potential bonded H atoms, and the complex anion charge. High valency light metals, such as boron and beryllium and hydrides with magnesium or lithium as cations, are preferred for their high storage capacity.

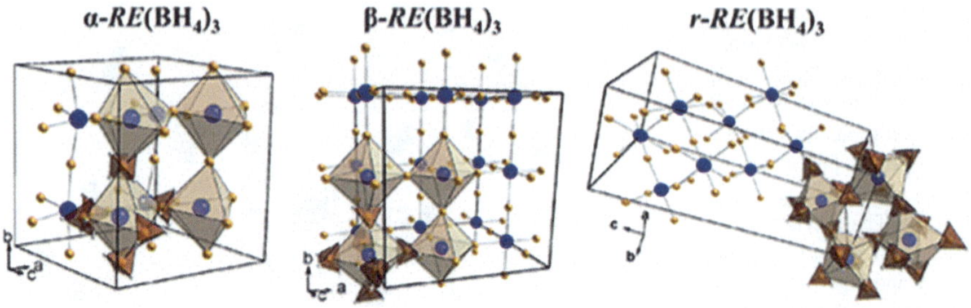

Fig. 4.22 Crystal structural trends of trivalent rare-earth borohydrides. Crystal structures of α-RE $(BH_4)_3$ in *Pa-3*, β-RE$(BH_4)_3$ in *Fm-3m*, and r-RE$(BH_4)_3$ in *R-3c*. Color scheme: RE^{3+} (blue), B (orange), and H (gray) [172]

Various lithium beryllium hydrides have been researched [204], including a compound called $Li_3Be_2H_7$, which has a storage capacity of approximately 8 wt% and an equilibrium pressure of 0.1 MPa at around 250 °C. However, the significant toxicity of beryllium compounds would likely hinder widespread use. The most famous boron hydrides, $NaBH_4$ and $LiBH_4$, are excessively stable. Additionally, boron hydrides are prone to producing volatile boranes as decomposition by-products. Although boranes are only produced in small amounts, their production could result in significant reduction in storage capacity over numerous cycles and harm the fuel cell.

Rare-earth metal borohydrides mostly exist as trivalent $RE(BH_4)_3$, although some like Sm, Eu, and Yb may produce divalent $RE(BH_4)_2$ due to their stable +2 oxidation state. This change can occur directly or by reduction from RE^{3+} to RE^{2+} during heating [205–207]. The crystal structures of divalent $RE(BH_4)_2$ are similar to alkaline earth metal borohydrides due to their comparable ionic radii. Specifically, $RE(BH_4)_2$ (where RE^{2+} = Sm, Eu) have the same structure as $Sr(BH_4)_2$, whereas Yb $(BH_4)_2$ has a structure similar to various forms of $Ca(BH_4)_2$ [208]. Trivalent RE $(BH_4)_3$ may form three distinct crystalline structures: α-, β-, and r-$RE(BH_4)_3$, with space group symmetries Pa-3, Fm-3m, and R-3c, respectively. All of them are connected to the rhenium oxide (ReO_3) structural type. The β-$RE(BH_4)_3$ has the ideal ReO_3 configuration; however, α- and r-$RE(BH_4)_3$ are achieved by tilting the $[RE(BH_4)_6]$ octahedra, as seen in Fig. 4.22.

Table 4.9 illustrates the very high hydrogen density of complex hydrides, which has sparked considerable interest in their characteristics and potential applications. Magnesium nickel hydride is one of the few complex hydrides that can store hydrogen reversibly under moderate conditions. Magnesium iron hydride is exceptional, possessing great volumetric hydrogen density of $\rho_V \sim 150$ g H^2 l^{-1}, which is more than two times that of liquid hydrogen, i.e., $\rho_V = 71$ g H_2 l^{-1} [38].

Researchers have focused on lightweight complicated aluminum hydrides for these reasons. The sodium molecule $NaAlH_4$ is the most extensively researched reversible complex aluminum hydride, with a reversible hydrogen concentration of

Table 4.9 Material properties of complex hydrides

	ρ (g ml^{-1})	ρ_V (g H_2 l^{-1})	ρ_m (wt% H_2)	M (g mol^{-1})	T_{dec} (°C)	ΔH_{dec} (kJ mol^{-1})	wt%	References
$LiAlH_4$	0.92	97	10.6	37.95	~150	10	10.54	[209]
$Mg(BH_4)_2$	0.79	117	14.8	53.99	~350	120	14.82	[210]
$NaBH_4$	1.07	116	10.8	37.83	~500	108	10.57	[211]
$LiBH_4$	0.66	122	18.4	21.78	~400	69	18.36	[212]
Mg_2NiH_4	2.70	98	3.6	111.34	~325	64	–	[213]
Mg_2CoH_5	2.84	126	4.5	111.85	~260	86		[214]
Mg_2FeH_6	2.74	150	5.5	110.50	~400	98	–	[215]
$LiNH_2$	1.17	103	8.8	22.96	~250	66	–	[216]
Na_3AlH_6	1.45	86	5.9	102.00	~275	49	–	[217]
$NaAlH_4$	1.8	93	7.3	54.00	~230	33	–	[218]
Li_3AlH_6	1.02	114	11.2	53.85	~200	25	–	[219]

ρ density, ρ_v volumetric hydrogen density, ρ_m gravimetric hydrogen density, M molecular weight, T_{dec} decomposition temperature, ΔH_{dec} decomposition enthalpy

5.6 wt% H_2. The decomposition occurs in three stages as outlined in Eqs. (4.8), (4.9) and (4.10).

$$3NaAlH_4 \longrightarrow Na_3AlH_6 + 2Al + 3H_2 \tag{4.8}$$

$$Na_3AlH_6 \longrightarrow 3NaH + Al + 1.5H_2 \tag{4.9}$$

$$3NaH \longrightarrow 3Na + 1.5H_2 \tag{4.10}$$

The initial dissociation stage achieves an equilibrium pressure of 0.1 MPa at approximately 308 °C, indicating that the material is similar to a standard low-temperature metal hydride. The hydrogen level in this initial decomposition stage (3.7 wt% H_2) is twice as high as that of a typical low-temperature metal hydride. The second decomposition process yields 1.9% by weight of hydrogen. The equilibrium pressure at this stage is 0.1 MPa at around 130 °C, indicating a common characteristic of a medium-temperature metal hydride. The decomposition temperature for the third step is too elevated for practical usage and cannot serve as a hydrogen source. $NaAlH_4$ exhibits the common structural alterations seen in many complicated aluminum hydrides. In sodium alanate, the Al atom exhibits a tetrahedral coordination, but in the intermediate Na_3AlH_6, it demonstrates an octahedral coordination. Al metal is generated at a different stage during the decomposing process. The segregation process mechanism is currently under debate, with some arguments proposing a transport pathway through volatile AlxHy molecules.

$NaAlH_4$ was developed approximately 50 years ago as a cost-effective alternative to $LiAlH_4$ for reducing reactions in chemical synthesis [220]. In contrast to $LiAlH_4$, $NaAlH_4$ has a substantially lower solubility in diethyl ether [221]. Tetrahydrofuran was the most effective solvent for $NaAlH_4$, but it was not accessible for organic synthesis operations at that time. Complex hydrides like this one were not previously considered viable for hydrogen storage due to the difficulty of rehydrogenation under certain conditions [222]. Recently, it was reported that reducing the size of $NaAlH_4$ by depositing it on a support can enhance its reaction rate and need less severe rehydrogenation conditions.

4.3.3.2 Amides/Imides

Li_3N was first discovered to absorb hydrogen in 1910 [223]. Amides have emerged as highly promising materials for hydrogen storage following the reexamining the hydrogen storage capabilities of lithium imide/lithium amide complexes in 2002 by Chen and colleagues [216]. Lithium nitride can react with two moles of hydrogen to form one mole of lithium amide and two moles of lithium hydride (Li_3NH_4). A total theoretical quantity of 10.4 wt% of hydrogen gas is released or taken in during a two-step process.

$$LiNH_2 + 2LiH \leftrightarrow Li2NH + LiH + H_2 \leftrightarrow Li_3N + 2H_2 \tag{4.11}$$

H_2 can be desorbed at temperatures up to 200 °C with a wt% of 6. To fully release hydrogen, temperatures as high as 420 °C are required. Calorimetric experiments indicate that both desorption processes are endothermic, with a value of around 60 kJ $mole^{-1}$ of H_2 for the second stage. The $LiNH_2$/LiH system is not suited for practical application with PEM fuel cells due to its inability to achieve an equilibrium pressure of 0.1 MPa at temperatures of 250 °C. The elevated decomposition enthalpy associated with the first step leads to an inappropriate rehydrogenation process within PEM fuel cell devices. Another drawback of amide systems is the thermodynamically favored release of ammonia from the pure amide compounds at elevated temperatures [224]. Since the initial report on the hydrogen storage capabilities of amide systems in 2002, other research teams have studied these molecules. The $Mg(NH_2)_2/MgH_2$ system has a high capacity, resulting in 7.4 wt% H_2 evolution during milling. The release of hydrogen occurs through a two-step degradation process, wherein magnesium imide (MgNH) acts as the intermediary. However, the low endothermic dissociation enthalpy of just 3.5 kJ mol^{-1} of H_2 makes these systems impractical for hydrogen storage [225].

4.3.3.3 Metal Nitrides

Metal nitrides, amides, and imides exhibit superior hydrogen storage capabilities at lower temperatures when compared to alternative chemisorption methods. Light metal-nitride compounds have demonstrated promise as a solid-state hydrogen storage material. Lithium nitride is commonly utilized as an initial substance to generate binary or ternary nitrides and can also absorb significant quantities of hydrogen in a reversible manner. The temperature needed to desorb the hydrogen

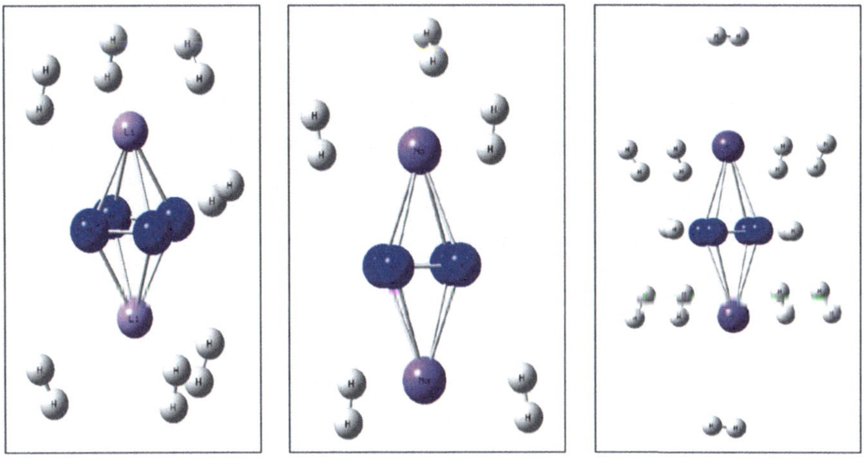

Fig. 4.23 Optimized structures of Li_2N_4 $(H_2)_n$, Na_2N_4 $(H_2)_n$, and K_2N_4 $(H_2)_n$, where $n = 5–12$ [231]

at pressures suitable for use is too high for the current material to be practically applied [216]. Lithium nitride has a hydrogen storage capacity of 11.4% of its weight, which is 50% more than that of magnesium hydride. The storage capacity of most other metal hydrides is generally limited to 2–4% of their weight. This innovative material is not yet suitable for practical use due to the elevated temperatures needed to release the hydrogen, but it indicates a promising direction for developing a practical hydrogen storage material [226]. Figure 4.23 demonstrates the optimized geometry of the three metal nitrides with surrounded hydrogen molecules.

Recently, Takayuki et al. [227] stated that lithium nitrides are an option for reversible hydrogen storage materials due to their ability to absorb and desorb significant amounts of hydrogen. The reaction is as follows:

$$Li_3N + 2H_2 \leftrightarrow LiNH_2 + 2LiH \tag{4.12}$$

Chen et al. [228] have demonstrated that the inclusion of alkaline earth metals like Ca and Mg significantly reduces the temperatures required for hydrogen storage, raises the pressures where the desorption occurs, and enables the achievement of large storage capacities. This outcome significantly advances the metal-N-H system towards fulfilling practical goals. In addition, the synthesis of various metal-N-H systems, including ternary imides such as Li-Mg-N-H and Mg-Na-N-H systems [229], and quaternary imides such as the Mg-Ca-Li-N-H system [230], for the purpose of hydrogen storage, has been accomplished by the reaction of different metal amides with hydrides.

4.3.3.4 Ammonia (NH_3)

Ammonia possesses several benefits that indicate its potential as a medium for hydrogen storage. Initially, it may be converted into a liquid state using moderate conditions. Ammonia's vapor pressure at normal temperature is 9.2 bar, which is approximately equivalent to 121 (psig) [232]. Its physical characteristics closely resemble those of propane. Ammonia can be kept in a basic and cost-effective pressure container. Ammonia contains a high weight percentage of hydrogen. Hydrogen makes up 17.65% of the mass of ammonia [233]. Combining these two parameters results in a liquid with a volumetric hydrogen density approximately 45% more than that of liquid hydrogen. Ammonia has the potential to undergo decomposition using a catalyst, resulting in the production of hydrogen and nitrogen. This process yields a nontoxic and non-greenhouse gas.

Ammonia ranks as the second most frequently manufactured chemical globally [234]. The infrastructure required for its production, transportation, and distribution is extensive. It is proficient in catalytic decomposition and possesses cutting-edge synthesis and distribution technologies. It can be diluted with water and kept in liquid form at room temperature and pressure. The liquid form offers significant hydrogen storage capacities with little cryogenic limitations. Ammonia can be converted into hydrogen without generating any toxic by-products, or it can be blended with current fuels and burned effectively. Ammonia has an advantage over alcohols and hydrocarbons in that it does not emit CO_2 at the conclusion. Advancements in safe ammonia storage involve the utilization of metal ammine complexes (Fig. 4.24).

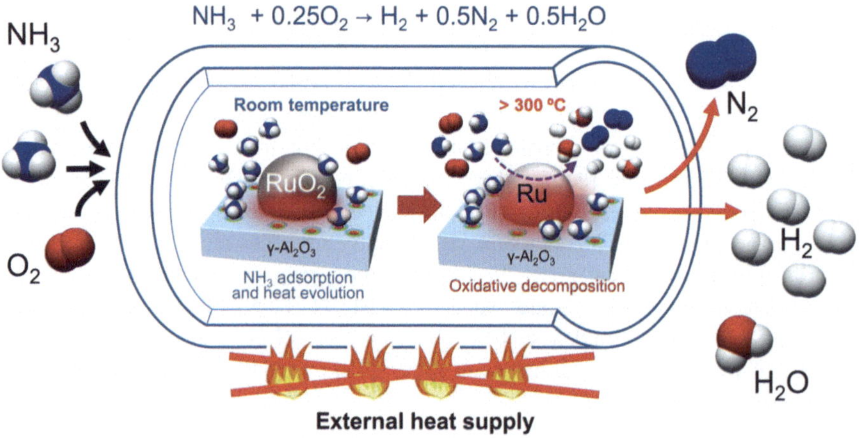

Fig. 4.24 H_2 storage media activated at room temperature using ammonia and oxygen without the need for additional heat input [235]

4.3.3.5 Liquid Organic Hydrogen Carriers (LOHCs)

A non-cryogenic approach was first proposed for LOHCs technology in 1975. These systems prefer using aromatic chemicals like toluene and benzene as hydrogen storage carriers for automotive fuels. The technology behind LOHCs is based on reversible hydrogen storage and release reactions that use corresponding saturates (like decalin, methylcyclohexane, and dodecahydro-N-ethylcarbazole) that serve as hydrogen carriers and unsaturated liquid organics (like naphthalene, toluene, and N-ethylcarbazole) as hydrogen storage actors [236]. Figure 4.25 illustrates the underlying concept of reversible hydrogen storage mechanism of LOHCs technology. In the dehydrogenation device, hydrogen is removed from Hx-LOHCs in the presence of the catalyst. Since the energy needed to activate the C-H bond is higher than the energy needed for hydrogen atom dissociation, the process necessitates constant absorption of external heat [237]. It is essential to address the high energy consumption resulting from temperature variations during the reaction, along with the decrease in catalytic activity. Existing liquid fuel transportation methods including trucks, ships, and pipelines can be used for long-distance transportation of hydrogen carriers. A study has revealed that unsaturated aromatics and their hydrides have the ability to undergo hydrogenation and dehydrogenation processes without causing any changes to the core carbon ring structure [238].

The hydrogenation process is a thermodynamically driven reaction in which the mixture of an organic hydrogen storage liquid and raw hydrogen occurs within a reactor. Next, the system is heated to a certain temperature with the catalyst to produce the matching saturated hydride. The outcomes of the hydrogenation process are referred to as hydrogen carriers, also known as Hx-LOHCs. It is a catalytic

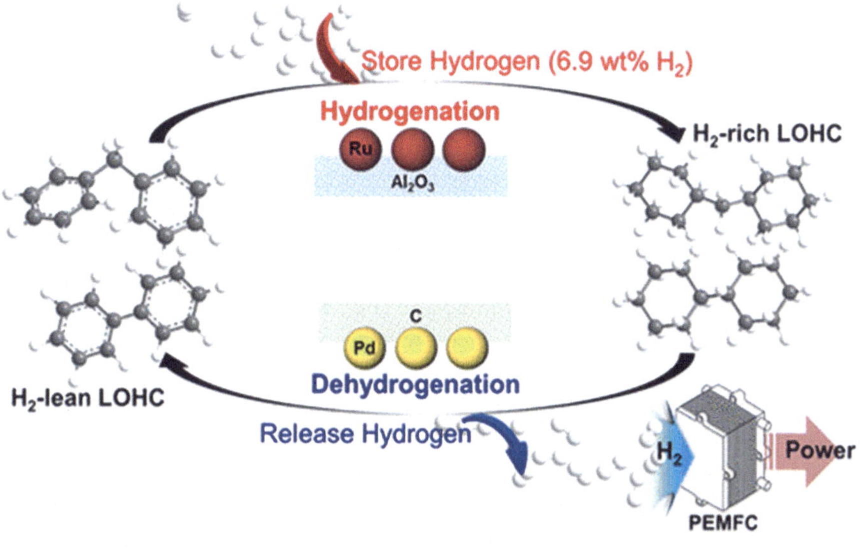

Fig. 4.25 Reversible hydrogen storage mechanism of liquid organic hydrogen carrier [240]

process that utilizes hydrogen to convert unsaturated bonds into saturated ones [239]. Chemical equilibrium favors hydrogenation at low temperatures and high pressures. Dehydrogenation is an endothermic process that is the opposite of hydrogenation.

4.3.3.6 Formic Acid

Formic acid, also referred to as methanoic acid, is the most basic carboxylic acid with the density of 1.22 g/cm^3 making it ideal for convenient transportation, refilling, and handling [241]. It is a crucial compound in chemical reactions and is found naturally in the venom of bees and ants. Researchers are studying formic acid as a hydrogen storage material because the hydrogen produced in the reaction does not contain carbon monoxide. The process involves the use of ruthenium catalysts which dissolve in water to convert HCOOH into hydrogen gas and carbon dioxide gas in a water solution. Enhancing pressure stability (1–600 bar) improves catalytic lifetime and removes CO, making it a feasible hydrogen storage material. The carbon dioxide generated during decomposition can serve as a hydrogen carrier by undergoing hydrogenation to transform it back into formic acid. Formic acid has a concentration of 53 gr L^{-1} of hydrogen, with a wt% of 4.3% at room temperature and atmospheric pressure. Formic acid and its conjugated base, formate, are commonly utilized as a hydrogen source in transfer hydrogenation processes [242]. It decomposes to create solely gaseous compounds such as H_2 and CO_2. The H_2–CO_2 combination may be readily separated under specific circumstances [243].

$$HCOOH \longrightarrow H_2 + CO_2 \quad \Delta G = 48.4\ \mathrm{kJ\ mol^{-1}} \tag{4.13}$$

$$HCOOH \longrightarrow H_2O + CO \quad \Delta G = -28.5\ \mathrm{kJ\ mol^{-1}} \tag{4.14}$$

Formic acid breakdown occurs through two main paths. One pathway leads to the desirable production of CO_2 and H_2 (Eq. 4.13), whereas the other pathway results in the unwanted production of CO and H_2O (Eq. 4.14) [243]. The selective breakdown of formic acid into CO_2 and H_2, a reversible process known as CO_2 hydrogenation, is essential for formic acid-based hydrogen storage. The utilization of carbon dioxide and formic acid as a hydrogen storage system is an attractive and straightforward idea. It involves selectively decomposing formic acid into H_2 and CO_2, and then recycling CO_2 by H_2 reduction to formic acid (Fig. 4.26). CO_2 is plentiful on Earth, inexpensive, and easily accessible. Here, reducing CO_2 emissions involves using CO_2 as a hydrogen transporter.

Fuel cells utilizing formic acid are very suitable for applications involving power production at room temperature and portable electronic devices. The efficiency of a direct formic acid fuel cell is mostly governed by the catalyst applied to the anode for the formic acid oxidation (FAO). Platinum (Pt) serves as a very efficient catalyst for the process of FAO. Hsieh et al. synthesized Pt nanocatalysts on graphene oxide (GO) sheets using ALD. Moreover, They studied the electrocatalytic activity of

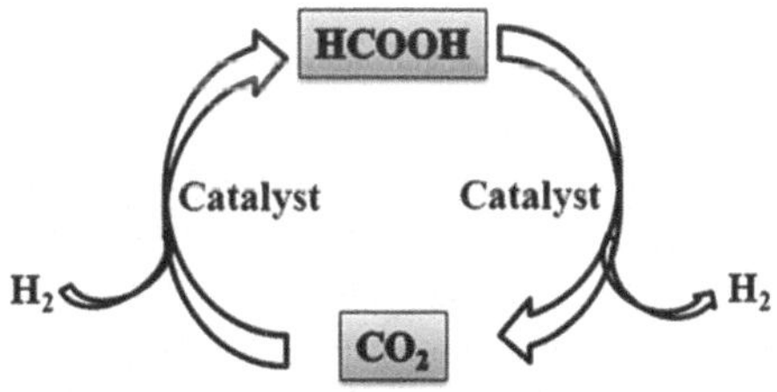

Fig. 4.26 The carbon dioxide-formic acid (CO_2-HCOOH)-based hydrogen storage technology [244]

these catalysts in FAO experiments [245, 246]. The ALD Pt atoms were placed on the margins of the GO sheets, where there were exposed bonds and oxygen groups. This arrangement, as shown in Fig. 4.27, led to remarkable FAO catalytic activity.

4.3.3.7 Carbohydrate

Carbohydrates are compounds having the empirical formula $C_m(H_2O)n$, where m and n can vary. While certain chemical molecules like glycolaldehyde and formaldehyde follow this formula, they are not classified as carbohydrates. Carbohydrates are best defined as polyhydroxy aldehydes or polyhydroxy ketones containing a minimum of three carbon atoms [247]. Carbohydrates, in the form of polymeric $C_6H_{10}O_5$, are a widespread renewable bioresource that have high hydrogen storage capacities in liquid form, without requiring high pressure or extremely low temperatures, and can also be stored as a solid powder. Researchers have recently achieved the production of approximately 12 moles of hydrogen per glucose unit from cellulosic materials and water. Carbohydrates can serve as a high-energy-density hydrogen carrier (14.8 wt%) due to thorough conversion and mild reaction conditions [248]. As depicted in Fig. 4.28, glucose and xylose are the most prevalent monosaccharides globally as they are the primary constituents of plant cell walls. An oligosaccharide is composed of 2 to 10 monosaccharides linked together by glycosidic bonds. When over 10 monosaccharides are linked by glycosidic bonds, it is commonly referred to as a polysaccharide, such as starch (Fig. 4.28c) and cellulose (Fig. 4.28d).

4.3.3.8 Clathrate Hydrates

Hydrogen storage via hydrate formation is a novel approach that involves trapping molecules inside the structure of a crystalline host material, such as water. These cages capture guest molecules by forming VdW interactions between the guest molecules and the surrounding water cage walls. Hydrogen clathrate hydrate is a potential storage media because of its low energy consumption for charging and discharging, low manufacturing costs, safety, and absence of negative environmental effect [250]. Clathrate and semi-clathrate hydrates are currently receiving significant attention as hydrogen storage materials. Hydrates provide advantages such as

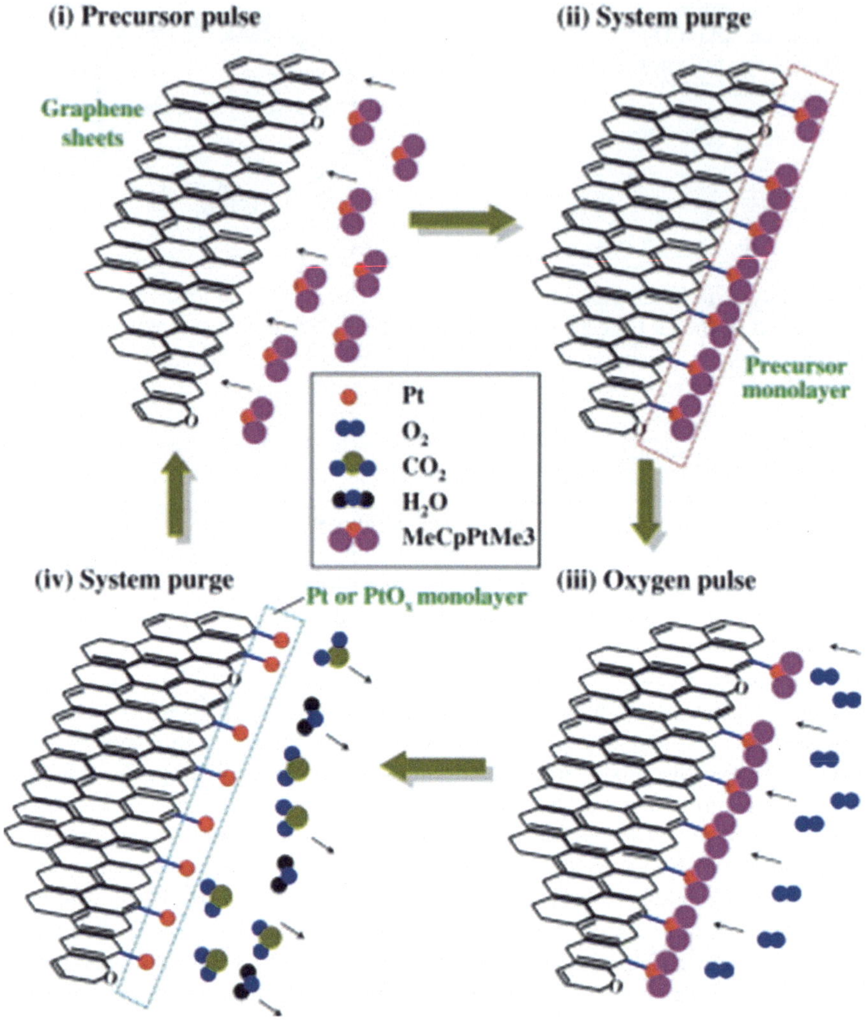

Fig. 4.27 ALD Pt NPs–GO for FOR [245]

reversible formation and dissociation, eco-friendly properties, cost-effectiveness, and reduced fire hazard, positioning them as very promising materials for hydrogen storage. However, high pressure and temperature needed to produce hydrogen hydrates is one of the main issues to hydrogen storage in hydrate crystals. Utilizing promoter molecules to address extreme pressure and temperature issues may shift these materials into a viable storage solution with significant promise.

Research demonstrated that hydrogen hydrates, capable of storing 5.3 wt%, could be synthesized at 249 K and 250 MPa [251]. Additionally, a thorough examination was conducted to evaluate the hydrogen occupancy in hydrate cages and its interaction with molecules of water. Figure 4.29 depicts two-dimensional (2D) three

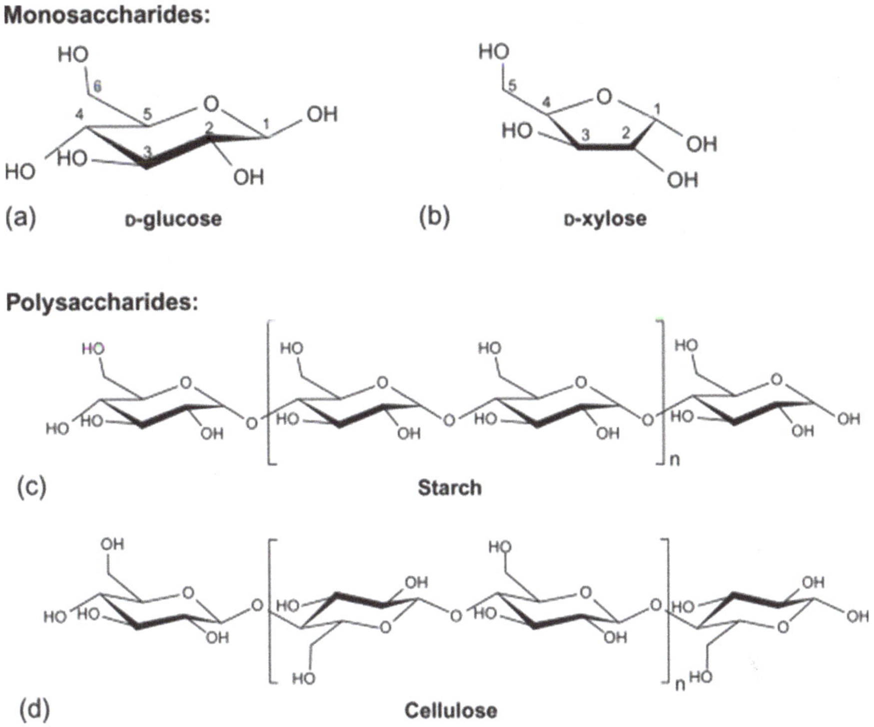

Fig. 4.28 Various carbohydrates differ in their chain length or carbon concentration. The prevalent monosaccharide is d-glucose, while the most abundant pentose is d-xylose. The two predominant polysaccharides are starch and cellulose [249]

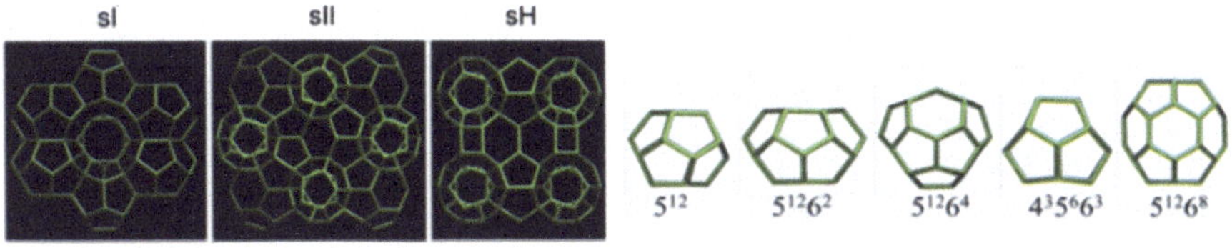

Fig. 4.29 Hydrogen hydrate structures [252]

possible hydrogen clathrate structures: sI, sII, and SH, However, compared to sI, sII, the sH type is far less common. SII hydrogen hydrate crystal promoters consist of tetrahydrofuran, cyclopentane, furan, butane, propane, cyclohexane, tetrahydrothiophene, nitrogen, and argon. Type sI hydrogen hydrate crystal promoters (Fig. 4.30) include cyclopropane, methane, ethane, carbon dioxide, and ethylene oxide. Table 4.10 lists the storage capacities achieved in different investigations using various promoters and/or under varied temperature and pressure conditions.

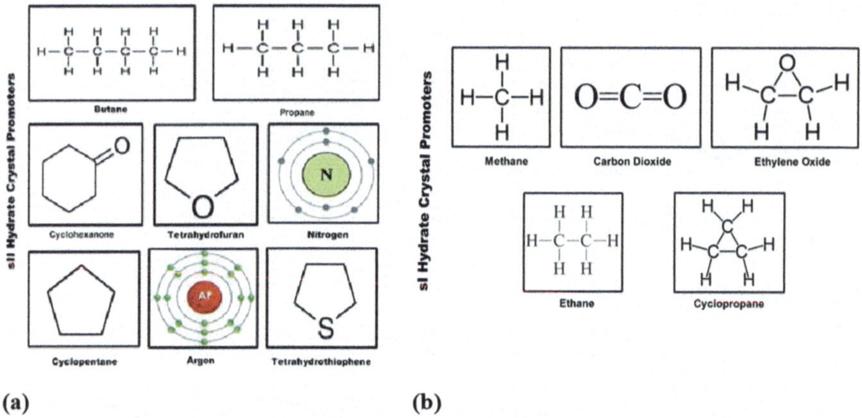

Fig. 4.30 (**a**) Hydrate promoters (Type sII crystal structure). (**b**) Hydrate promoters (Type sI crystal structure) [251]

Table 4.10 Comparison of hydrogen hydrate storage capacity among several promoters

Promoter	Hydrogen storage capacity (wt.%)	Promoter concentration (Mol%)	Operating pressure (MPa)	Operating temperature (K)	References
THF	1.875	3	14.53	273.15	[253]
THF	0.169	3.5	12	279.2	[254]
THF	0.12	5.6	10	276.2	[255]
THP	0.19	5.6	10	272.3	[255]
DHF	0.16	5.6	10	271.5	[255]
DXL	0.36	5.6	10	269.2	[255]
THT	0.5	5.6	10	274.5	[255]
THF/ carbon	0.041	5/70 g	6.4	274	[256]
TBAB	0.052	3.5	12	279.2	[254]
TBAF	0.45	3.4	10	294	[257]
MCH	0.6	1.6	25	274	[258]
TBAOH	0.47	0.0323	20	290	[259]
Ar	1.7	1	25	235	[260]
DMCH	0.85	3	60	275	[261]
MTBE	0.94	5	70	273	[261]
$TBPBH_4$	0.12	2	12.1	285	[262]
Furan	0.23	5	15.5	275.1	[263]

4.4 Hydrogen Storage Measurement Techniques

Various methodologies are employed to quantify the amount of hydrogen absorbed by solid-state materials. The significance of these procedures depends on the specific material and physical characteristics employed. Certain techniques are applicable to chemical hydrides and metal hydrides, while others are applicable to porous carbon materials. In addition, certain equipment is designed to be sensitive in low-pressure conditions, while others are suitable for use in high-pressure environments. However, in terms of precision, certain measurements more precisely quantify the amount of hydrogen uptake [wt %], while others exhibit a degree of ambiguity in their measurements. Hence, this article presents an overview of the methodologies and techniques employed for measuring hydrogen absorption.

4.4.1 Sieverts Technique

This methodology uses volumetric methods to accurately estimate the amount of hydrogen adsorption in terms of weight percentage (wt%). It is the most prevalent, cost-effective, and straightforward procedure among all the methods. The Sieverts apparatus is employed in this method, as shown in Fig. 4.31a. The apparatus comprises a sample volume, a sample reference volume, a gas source for hydrogen gas, and a pressure transducer for measuring pressure. The measurement of hydrogen is conducted incrementally. First, the hydrogen is put into the sample cell. Subsequently, the valve between the sample cell and the reference cell is opened (Fig. 4.31a). The observation of a pressure shift is attributed to the adsorption of hydrogen in the sample cell, which causes a change in volume. Therefore, this approach is centered around the examination of changes in pressure in relation to volume. The manometer will produce signals when the sample adsorbs or desorbs hydrogen atoms in N steps, and these signals correspond to the quantitative measurement of hydrogen atoms in moles. Subsequently, by utilizing the pressure–composition isotherm (PCI) curve, it is possible to quantify the amount of hydrogen or any other experimental gas that is adsorbed by the specific sample [264]. Several research have been conducted to ascertain the hydrogen storage capabilities using the PCI curves [111]. This method may be utilized to investigate the hydrogen adsorption capacity of metal hydrides at low-pressure conditions, namely below 100 bar. In order to achieve precise results, it is essential to calibrate the apparatus correctly and utilize a certain quantity of sample. This ensures that the ratio between the mass of the sample and the volume of the system is appropriately balanced.

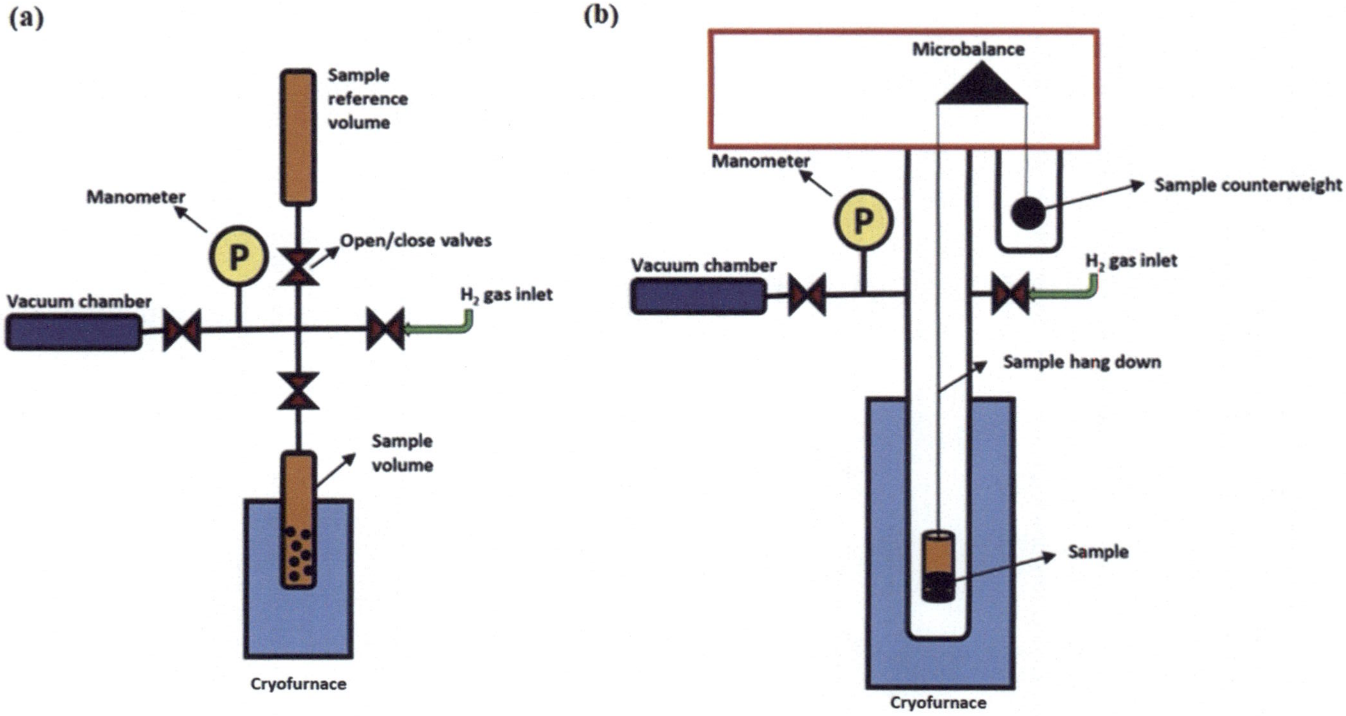

Fig. 4.31 The Sieverts device is used to measure the gas adsorption in the sample cell. (**a**) Gravimetric sorption system for measuring the amount of hydrogen absorbed using a microbalance, expressed as weight percentage. (**b**) The gravimetric sorption device. Open/close valves let hydrogen gas to enter the sample and exit it under vacuum, respectively [76]

4.4.2 Gravimetric Technique

The gravimetric approach is a method that involves applying high pressure. This approach utilizes mass measurement to determine the hydrogen uptake [wt %] in a specific sample. Additionally, the change in mass during hydrogen sorption is determined using pressure variation. To quantify any change in mass value, a very sensitive microbalance with a resolution of micrograms (μg) is necessary (Fig. 4.31b). This approach is particularly well-suited for investigating the process of hydrogen adsorption in metal hydrides and porous materials, which has been widely employed by researchers [265]. In order to obtain a precise measurement, it is necessary to utilize a sample that is both immaculate and free from any volatile substances. A thermogravimetric analyzer may be used to perform gravimetric analysis on smaller mass samples [266]. Nevertheless, the thermogravimetric analyzer's readings have some limitations. Upon introducing warmth to the sample, the alterations caused by friction and buoyancy forces may be identical to those of the gravimetric changes. Occasionally, these modifications contribute to the accurate measurement of the gravimetric weight. Therefore, it is necessary to have a robust calibration in order to address any uncertainties before proceeding with the elimination process.

4.4.3 Electrochemical Method

The electrochemical approach is superior in sensitivity and accuracy compared to the volumetric or gravimetric methodology [267]. Hydrogen is adsorbed at the contact between the electrode and electrolyte in this process. The electrode or electrocatalyst itself is the material that has to be investigated for use in hydrogen storage applications. Hydrogen is adsorbed as hydrogen ions onto the surface of the electrode. The oxidation–reduction reaction could occur at the electrodes during the hydrogen adsorption and desorption process. Nevertheless, the electrolyte composition, the presence of contaminants, and the gases utilized are crucial in this context.

Conventional hydrogen storage methods include physically storing hydrogen molecules across a broad surface area of solid materials. The process of electrochemical hydrogen storage entails the storage of hydrogen ions and atoms at the working electrode, with the counter electrode acting to promote the flow of electric current. Electrochemical hydrogen storage typically involves the process of hydrogen ions being adsorbed onto the surface of the electrode film and then inserted into the bulk materials [268–270]. During the hydrogen loading and release operation, the counter electrode can undergo the oxygen reduction and evolution process. The electrochemical hydrogen storage phenomena have similarities to existing electrochemical energy storage devices, such as supercapacitors and batteries [8, 271, 272]. This approach facilitates the assessment of the hydrogen-storage capabilities of thin-film materials in comparison to the gravimetric and Sieverts techniques

[264]. The electrolyte solution that is most beneficial for electrochemical hydrogen storage is potassium hydroxide (KOH) [273]. KOH utilizes water to generate hydrogen adatoms, which are atoms situated on a crystal surface and employed in the process of hydrogen adsorption. This approach is not appropriate for very corrosive metals.

4.4.4 Secondary Ion Mass Spectroscopy

Secondary ion mass spectrometry (SIMS) is a method employed to characterize and measure the composition of surfaces and subsurfaces with high precision. The setup of SIMS is presented in Fig. 4.32. The SIMS technique involves analyzing the secondary ions generated by bombarding a sample surface with ions. This analysis provides information about the chemical compositions and structural characteristics of the sample. The SIMS method is very sensitive and capable of detecting most elements at levels as low as parts per billion (ppb). It offers lateral resolutions ranging from tens to hundreds of microns and depth resolutions ranging from a few nanometers to microns. Gnaser [274] showed that, given optimal experimental circumstances with a high secondary ion production and little interference from other molecules, it is possible to attain a detection limit of 10^{12} atoms/cm^3.

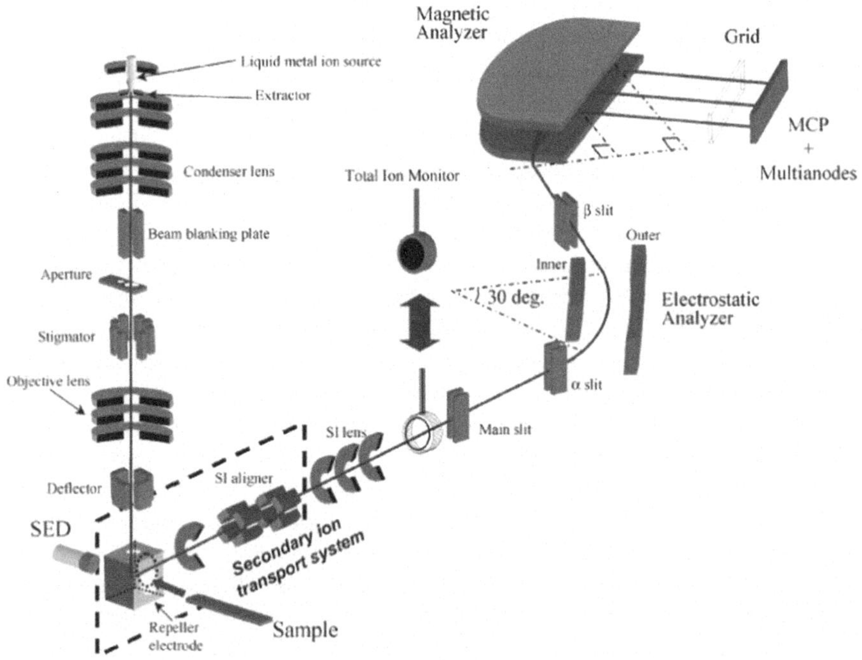

Fig. 4.32 A schematic illustration of SIMS [277]

Nevertheless, the examination of light atoms like hydrogen might be exceedingly difficult because of the substantial background signal. The hydrogen depth profile detection limit of SIMS in a silicon substrate is around 1.3–3 × 10^{18} [275, 276]. By substituting deuterium for hydrogen, the detection limit can be reduced by a factor of ten.

4.4.5 Thermal Desorption Spectroscopy

Thermal desorption spectroscopy (TDS) is a straightforward and nondestructive technology that allows the investigation of the surface kinetics of adsorbates. It involves monitoring the rate at which gas is released in a vacuum or specific gas environment, while controlling the temperature ramping rate. The use of TDS initially focused on analyzing hydrogen content in materials and evaluating the interactions between trapping sites and hydrogen atoms. Additionally, it may be utilized to ascertain the kinetic and thermodynamic characteristics of hydrogen storage alloys. A TDS consists of four main components: heating element, vacuum system, the sample chamber, and detector. The mass spectrometer (MS) is commonly used in thermal desorption spectroscopy (TDS) to achieve a more sensitive detection limit. Figure 4.33 depicts a schematic diagram of Total Dissolved Solids (TDS) with Mass Spectrometry (MS). TDS is used to determine the hydrogen storage capacity of materials with various microstructure characteristics and flaws [278, 279]. By subjecting a substance to various temperature gradients, it is possible to ascertain important data such as the energy required for hydrogen bonding, the rate at which particles diffuse, and the order of the chemical process.

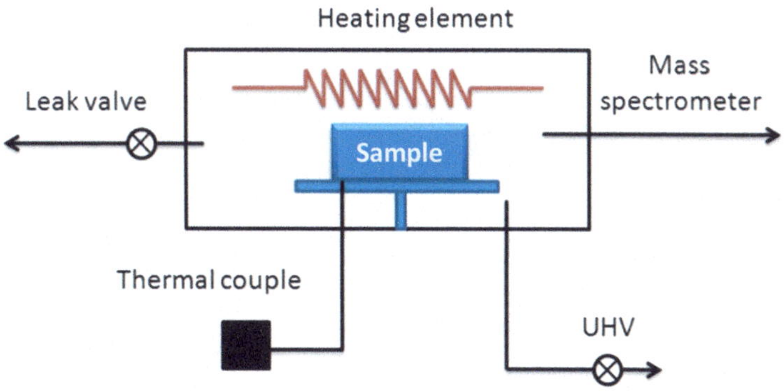

Fig. 4.33 Diagram showing the mass spectrometer-equipped thermal desorption spectroscopy system [264]

4.4.6 Neutron Scattering

In typical structural characterization utilizing X-ray methods, the signals produced by heavy atoms and light atoms are not equivalent, resulting in hydrogen being almost transparent during X-ray diffraction. Neutron scattering closely resembles X-ray diffraction; however, due to the inherent characteristics of neutrons, it enables the examination of light atoms inside the lattice. A neutron is a neutral subatomic particle that is present in the nuclei of atoms, with the exception of hydrogen atoms. The wavelength of this phenomenon is usually around 1 to 10 angstroms (Å), and it can be produced by either nuclear fission or spallation. Neutrons undergo nuclear reactions with atomic nuclei, resulting in the visibility of hydrogen to neutrons. Neutron methods need little sample preparation and may be used on several types of samples, such as zeolites, carbonaceous materials, MOFs, and metal hydride. Due to the low neutron absorption properties of most materials, instruments such as furnaces, pressurized containers, and cryostats may be utilized for conducting in situ investigations. Neutrons can be utilized to gather data about hydride materials, including their crystal structure (using diffraction techniques), geometry, particle size, and fine structures of hydrogenated nanoparticles (using small angle neutron scattering, SANS). Neutrons can also provide information about hydrogen dynamics (using inelastic scattering, INS) and hydrogen diffusion and liberation (using quasi-elastic scattering, QENS). Due to the wide range of methods used in neutron scattering, a detailed treatment of this topic is not included in this study. For a more comprehensive analysis of the uses of neutron scattering on solid-state hydrogen storage materials, readers are advised to refer [280]. Figure 4.34 depicts a schematic representation of neutron scattering.

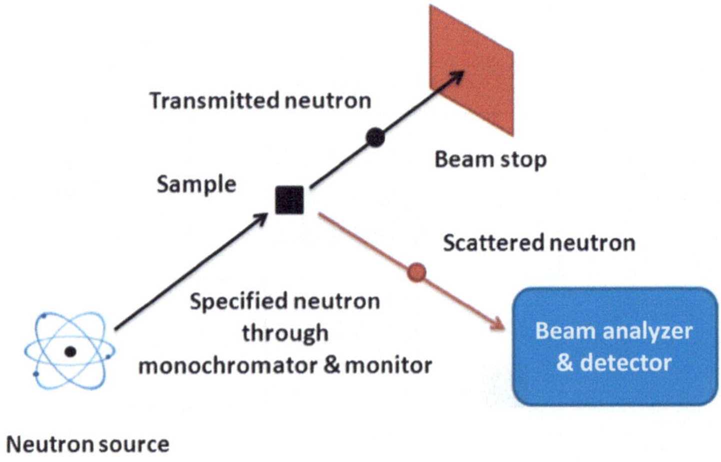

Fig. 4.34 A schematic working mechanism of neutron scattering [264]

4.5 Hydrogen Economy and Storage

Hydrogen is an odorless and colorless gas that is very combustible. Hydrogen generated from natural gas and coal reforming is mostly utilized for petroleum refining and ammonia production. The theoretical hydrogen economy has four main processes: hydrogen production, storage, distribution, and conversion into electrical energy for end consumers. Hydrogen is proposed as the most suitable alternative fuel to substitute fossil fuel-derived fuels in transportation because of its clean energy characteristics, high specific energy density (J/kg), and high-energy conversion efficiency. Despite its appeal as an energy carrier, the hydrogen economy encounters technical challenges such as cost-effective, scalable, and sustainable hydrogen production; absence of high-density hydrogen storage methods; expensive infrastructure for hydrogen distribution; development of economical and long-lasting fuel cells; and safety issues. To address these issues, using carbohydrates like starch or cellulose is recommended as an innovative option.

Common methods of storing hydrogen include compressed hydrogen in cryogenic liquid hydrogen or high-pressure gas cylinders. Compressing and liquefying hydrogen technologies are excessively energy-intensive, and their hydrogen storage capacities are insufficient. Conventional storage techniques for onboard hydrogen in vehicles are not ideal because of limitations in capacity, high expenses, and safety concerns. The storage must be compact, lightweight, safe, and cost-effective. Novel hydrogen storage technologies for automobiles aim to identify materials capable of storing hydrogen at significantly higher levels than compressed or cryogenic hydrogen. Extensive research has been carried out to study and enhance metal hydrides, complex hydrides, and carbon compounds. Their high hydrogen storage capacity and favorable temperature and pressure circumstances make them attractive alternatives to compressing or cryogenically liquefying hydrogen, but releasing hydrogen from these materials is challenging. Figure 4.35 compares the hydrogen storage capacities achieved through various technologies with the target set by the DOE.

4.6 Conclusion and Future Perspectives

Hydrogen is a promising synthetic fuel for the future due to its high heating value and the potential to utilize hydrogen generated from renewable sources in a closed loop system. Just like fossil fuels initiated the industrial age, hydrogen will embark an economic and technical revolution. After investing in renewable energy generation, humans will benefit from a hydrogen-based ecologically friendly energy economy. Hydrogen will be stored differently based on the specific use, such as mobile or stationary applications. Currently, there are several effective and secure methods for storing hydrogen. Nevertheless, numerous novel materials and techniques are being explored for hydrogen storage. The objective in materials science is

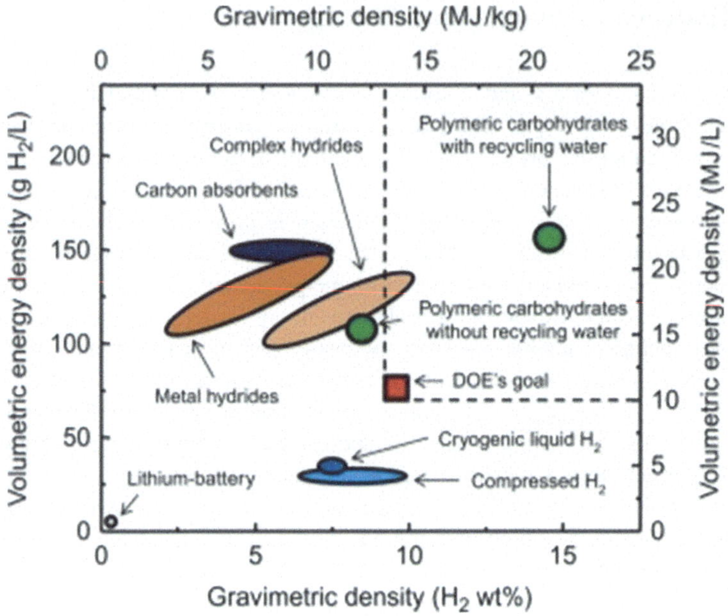

Fig. 4.35 The existing methods of hydrogen storage and the DOE's objective for hydrogen storage

to enhance our comprehension of the electrical behavior when hydrogen interacts with other elements, particularly metals.

Investigation of complex compounds such as $Al(BH_4)_3$ will lead to the discovery of novel compounds involving lightweight metals and hydrogen. A automobile equipped with an internal combustion engine consumes 2.4 kg (3 L) of fuel every 100 km using the most advanced technology available today. This equates to 0.8 kg of hydrogen every 100 km in terms of energy. The automobile requires 4 kg of hydrogen to travel 500 km before needing to refuel the hydrogen storage system. The tank contains 4 kg of hydrogen held in a metal hydride. The tank weighs around 300 kg and has a capacity of roughly 60 L. A storage jar made of steel loaded with metallic powder can serve as a structural component of the automobile. Not all of the weight of the storage system is considered as additional weight to the automobile in this scenario. In addition, more efficient energy conversion devices such as fuel cells will be developed to minimize the quantity of hydrogen required onboard and thereby decrease the weight of the storage system compared to internal combustion engines.

Although certain carbonaceous materials possess exceptional hydrogen storage capability, there are several obstacles to the commercialization of these materials. Here are some of the challenges:

Carbon compounds doped with heteroatoms have demonstrated excellent storage capacity. Consequently, there is a requirement for synthesis techniques that can provide uniform doping on a big scale. Additionally, it is imperative that the process of materials synthesis is capable of being replicated.

Aligned graphene structures have been discovered to possess a higher capacity for hydrogen adsorption. The hydrogen adsorption in aligned structures mostly relies on the distance between the layers. Therefore, it is necessary to adjust the distance between layers in order to increase the storage capacity at normal ambient temperature.

The adsorption mechanism plays a crucial role in the advancement of reversible solid-state materials. Kubas-type adsorption was determined to be more advantageous for actual applications. Therefore, it is crucial to carefully evaluate the various processes of hydrogen adsorption.

The catalyst's involvement is regarded as a crucial factor in hydrogen storage in solid-state materials. It was discovered that carbon-based materials, when combined with an appropriate catalyst, enhance the speed at which hydrogen is absorbed and taken up. Therefore, it is necessary to adjust the kind and concentration of catalysts.

The hydrogen storage capacity in some carbon materials is mostly determined by the size and volume of the pores. Hence, it is crucial to create consistent and homogeneous porous materials that can effectively adsorb hydrogen at normal environmental conditions.

One of the primary obstacles in hydrogen storage technologies is the creation of affordable materials and techniques that provide the necessary kinetics and thermodynamics. The primary objective is to create carbon-based materials that can store hydrogen with a capacity of at least 6.5 wt%, while maintaining the necessary rate of adsorption and desorption at ambient temperature and pressure. Although certain materials have sufficient storage capacity, there are still several obstacles to overcome when it comes to utilizing carbon-based materials in hydrogen storage systems. It is important to prioritize future difficulties, as previously noted.

The significant aspect of ALD is the advancement of nanomaterials or thin films produced using ALD, which is a subject of interest in the realm of energy applications. (1) While posttreatment including substances such as inert gases, air, and H_2O is a simple and promising process for creating ALD thin films, there is limited research on using ALD-derived materials for energy storage systems. Metal oxide thin films and carbon/metal oxide composite films are considered to be highly promising as electrodes for applications connected to energy. The ability to manipulate the size of pores and the amount of carbon in metal cones should be further utilized for the development of membranes/catalysts. (2) Moreover, the incorporation of heteroatoms (such as nitrogen, sulfur, and phosphorus) or the simultaneous incorporation of several heteroatoms into carbon layers may be accomplished by carefully choosing the precursors used in the ALD process.

Therefore, by persistently applying ALD-based methodologies, several solutions may be provided to tackle the existing issues in the field of energy storage and conversion. Further research should be conducted to explore the many uses of ALD in the field of energy storage, in order to fully harness the utmost capabilities of this innovative and adaptable approach.

References

1. M. Sharma, A. Pramanik, G.D. Bhowmick, A. Tripathi, M.M. Ghangrekar, C. Pandey, et al., Premier, progress and prospects in renewable hydrogen generation: A review. Ferment **9**, 537 (2023). https://doi.org/10.3390/FERMENTATION9060537
2. A. Pareek, R. Dom, J. Gupta, J. Chandran, V. Adepu, P.H. Borse, Insights into renewable hydrogen energy: Recent advances and prospects. Mater. Sci. Energy Technol. **3**, 319–327 (2020). https://doi.org/10.1016/J.MSET.2019.12.002
3. M. Yue, H. Lambert, E. Pahon, R. Roche, S. Jemei, D. Hissel, Hydrogen energy systems: A critical review of technologies, applications, trends and challenges. Renew. Sustain. Energy Rev. **146**, 111180 (2021). https://doi.org/10.1016/J.RSER.2021.111180
4. Y. Kojima, H. Miyaoka, T. Ichikawa, Hydrogen storage materials. New Future Dev. Catal. Batter Hydrog. Storage Fuel Cells, 99–136 (2013). https://doi.org/10.1016/B978-0-444-53880-2.00006-5
5. K.L. Lim, H. Kazemian, Z. Yaakob, W.R.W. Daud, Solid-state materials and methods for hydrogen storage: A critical review. Chem. Eng. Technol. **33**, 213–226 (2010). https://doi.org/10.1002/CEAT.200900376
6. C. Altintas, S. Keskin, On the shoulders of high-throughput computational screening and machine learning: Design and discovery of MOFs for H2 storage and purification. Mater. Today Energy **38**, 101426 (2023). https://doi.org/10.1016/J.MTENER.2023.101426
7. E.S. Cho, A.M. Ruminski, S. Aloni, Y.S. Liu, J. Guo, J.J. Urban, Graphene oxide/metal nanocrystal multilaminates as the atomic limit for safe and selective hydrogen storage. Nat. Commun. **7**, 1–8 (2016). https://doi.org/10.1038/ncomms10804
8. S. Karimzadeh, B. Safaei, C. Yuan, T.C. Jen, Emerging atomic layer deposition for the development of high-performance lithium-ion batteries. Electrochem. Energy Rev. **6**, 1–50 (2023). https://doi.org/10.1007/S41918-023-00192-8
9. S. Karimzadeh, B. Safaei, W. Huang, T.C. Jen, Theoretical investigation on niobium doped LiFePO4 cathode material for high performance lithium-ion batteries. J. Energy Storage **67**, 107572 (2023). https://doi.org/10.1016/J.EST.2023.107572
10. L. Zhou, Y. Zhou, A comprehensive model for the adsorption of supercritical hydrogen on activated carbon. Ind. Eng. Chem. Res. **35**, 4166–4168 (1996). https://doi.org/10.1021/IE960275G
11. K. Ledwaba, S. Karimzadeh, T.C. Jen, Enhancement in the hydrogen storage capability of borophene through yttrium doping: A theoretical study. J. Energy Storage **55**, 105500 (2022). https://doi.org/10.1016/J.EST.2022.105500
12. S.K. Bhatia, A.L. Myers, Optimum conditions for adsorptive storage. Langmuir **22**, 1688–1700 (2006). https://doi.org/10.1021/LA0523816
13. J. Andersson, S. Grönkvist, Large-scale storage of hydrogen. Int. J. Hydrog. Energy **44**, 11901–11919 (2019). https://doi.org/10.1016/J.IJHYDENE.2019.03.063
14. J. Ren, N.M. Musyoka, H.W. Langmi, M. Mathe, S. Liao, Current research trends and perspectives on materials-based hydrogen storage solutions: A critical review. Int. J. Hydrog. Energy **42**, 289–311 (2017). https://doi.org/10.1016/J.IJHYDENE.2016.11.195
15. C.X. Guo, Y. Wang, C.M. Li, Hierarchical graphene-based material for over 4.0 wt % physisorption hydrogen storage capacity. ACS Sustain. Chem. Eng. **1**, 14–18 (2013). https://doi.org/10.1021/SC3000306/SUPPL_FILE/SC3000306_SI_001.PDF
16. O.O. Dada, S. Karimzadeh, P.E. Imoisili, T.-C. Jen, Theoretical insights of electrocatalytic hydrogen evolution on MoP nanocrystal. Int. J. Hydrog. Energy (2023). https://doi.org/10.1016/J.IJHYDENE.2023.05.302
17. K. Ledwaba, S. Karimzadeh, T.C. Jen, Emerging borophene two-dimensional nanomaterials for hydrogen storage. Mater. Today Sustain. **22**, 100412 (2023). https://doi.org/10.1016/J.MTSUST.2023.100412

18. R. Prins, Hydrogen spillover. Facts and fiction. Chem. Rev. **112**, 2714–2738 (2012). https://doi.org/10.1021/CR200346Z
19. W.C. Conner, J.L. Falconer, Spillover in heterogeneous catalysis. Chem. Rev. **95**, 759–788 (1995). https://doi.org/10.1021/CR00035A014
20. A. Lueking, R.T. Yang, Hydrogen spillover from a metal oxide catalyst onto carbon nanotubes – implications for hydrogen storage. J. Catal. **206**, 165–168 (2002). https://doi.org/10.1006/JCAT.2001.3472
21. L. Wang, R.T. Yang, Hydrogen storage properties of carbons doped with ruthenium, platinum, and nickel nanoparticles. J. Phys. Chem. C **112**, 12486–12494 (2008). https://doi.org/10.1021/JP803093W
22. J.H. Guo, S.J. Li, Y. Su, G. Chen, Theoretical study of hydrogen storage by spillover on porous carbon materials. Int. J. Hydrog. Energy **45**, 25900–25911 (2020). https://doi.org/10.1016/J.IJHYDENE.2019.12.146
23. C.C. Huang, N.W. Pu, C.A. Wang, J.C. Huang, Y. Sung, G.M. Der, Hydrogen storage in graphene decorated with Pd and Pt nano-particles using an electroless deposition technique. Sep. Purif. Technol. **82**, 210–215 (2011). https://doi.org/10.1016/J.SEPPUR.2011.09.020
24. A. Züttel, P. Sudan, P. Mauron, T. Kiyobayashi, C. Emmenegger, L. Schlapbach, Hydrogen storage in carbon nanostructures. Int. J. Hydrog. Energy **27**, 203–212 (2002). https://doi.org/10.1016/S0360-3199(01)00108-2
25. X. Li, S. Sun, J. Zhang, K. Luo, P. Gao, T. Wu, et al., Hybridization of inorganic CoB noncrystal with graphene and its Kubas-enhanced hydrogen adsorption at room temperature. RSC Adv. **6**, 93238–93244 (2016). https://doi.org/10.1039/C6RA19238F
26. E. Durgun, S. Ciraci, W. Zhou, T. Yildirim, Transition-metal-ethylene complexes as high-capacity hydrogen-storage media. Phys. Rev. Lett. **97**, 226102 (2006). https://doi.org/10.1103/PHYSREVLETT.97.226102/FIGURES/4/MEDIUM
27. S. Nachimuthu, P.J. Lai, E.G. Leggesse, J.C. Jiang, A first principles study on boron-doped graphene decorated by Ni-Ti-Mg atoms for enhanced hydrogen storage performance. Sci. Rep. **5**, 16797 (2015). https://doi.org/10.1038/SREP16797
28. G. Kim, S.H. Jhi, N. Park, S.G. Louie, M.L. Cohen, Optimization of metal dispersion in doped graphitic materials for hydrogen storage. Phys. Rev. B Condens. Matter. Mater. Phys. **78**, 085408 (2008). https://doi.org/10.1103/PHYSREVB.78.085408/FIGURES/5/MEDIUM
29. C.H. Chung, J. Ihm, H. Lee, Recent progress on Kubas-type hydrogen-storage nanomaterials: From theories to experiments. J. Korean Phys. Soc. **66**, 1649–1655 (2015). https://doi.org/10.3938/JKPS.66.1649/METRICS
30. L. Ma, J.M. Zhang, K.W. Xu, V. Ji, Hydrogen adsorption and storage of Ca-decorated graphene with topological defects: A first-principles study. Physica E **63**, 45–51 (2014). https://doi.org/10.1016/J.PHYSE.2014.05.004
31. M. Zarezadeh Mehrizi, J. Abdi, M. Rezakazemi, E. Salehi, A review on recent advances in hollow spheres for hydrogen storage. Int. J. Hydrog. Energy **45**, 17583–17604 (2020). https://doi.org/10.1016/J.IJHYDENE.2020.04.201
32. Y. Wang, Y. Wang, Recent advances in additive-enhanced magnesium hydride for hydrogen storage. Prog. Nat. Sci.-Mater. Int. **27**, 41–49 (2017). https://doi.org/10.1016/J.PNSC.2016.12.016
33. X. Yang, W. Li, J. Zhang, Q. Hou, Hydrogen storage performance of Mg/MgH_2 and its improvement measures: Research progress and trends. Materials (Basel) **16** (2023). https://doi.org/10.3390/MA16041587
34. J. Zhang, S. Yan, L.P. Yu, X.J. Zhou, T. Zhou, P. Peng, Enhanced hydrogen storage properties and mechanisms of magnesium hydride modified by transition metal dissolved magnesium oxides. Int. J. Hydrog. Energy **43**, 21864–21873 (2018). https://doi.org/10.1016/J.IJHYDENE.2018.10.017
35. A. Züttel, Hydrogen storage methods. Naturwissenschaften **91**, 157–172 (2004). https://doi.org/10.1007/S00114-004-0516-X/FIGURES/6

36. J. Liu, M. Zhao, L. Rong, Overview of hydrogen-resistant alloys for high-pressure hydrogen environment: On the hydrogen energy structural materials. Clean Energy **7**, 99–115 (2023). https://doi.org/10.1093/CE/ZKAD009
37. A.I. Osman, N. Mehta, A.M. Elgarahy, M. Hefny, A. Al-Hinai, A.H. Al-Muhtaseb, et al., Hydrogen production, storage, utilisation and environmental impacts: A review. Environ. Chem. Lett. **20**, 153–188 (2021). https://doi.org/10.1007/S10311-021-01322-8
38. A. Züttel, Materials for hydrogen storage. Mater Today **6**, 24–33 (2003). https://doi.org/10.1016/S1369-7021(03)00922-2
39. H. Oesterreicher, Hydrides of intermetallic compounds. Appl. Phys. **24**, 169–186 (1981). https://doi.org/10.1007/BF00899753/METRICS
40. M. Nachtane, M. Tarfaoui, M. Abichou, A. Vetcher, M. Rouway, A. Aâmir, et al., An overview of the recent advances in composite materials and artificial intelligence for hydrogen storage vessels design. J. Compos. Sci. **7**(3), 119 (2023). https://doi.org/10.3390/JCS7030119
41. H. Zheng, X. Zeng, J. Zhang, H. Sun, H. Zheng, X. Zeng, et al., The application of carbon fiber composites in cryotank. Solidification (2018). https://doi.org/10.5772/INTECHOPEN.73127
42. R.N. Muthu, S. Rajashabala, R. Kannan, Hydrogen storage performance of lithium borohydride decorated activated hexagonal boron nitride nanocomposite for fuel cell applications. Int. J. Hydrog. Energy **42**, 15586–15596 (2017). https://doi.org/10.1016/J.IJHYDENE.2017.04.240
43. G. Xia, X. Yu, Y. Guo, Z. Wu, C. Yang, U. Liu, et al., Amminelithium Amidoborane Li(NH3)NH2BH3: A new coordination compound with favorable dehydrogenation characteristics. Chem. A Eur. J. **16**, 3763–3769 (2010). https://doi.org/10.1002/CHEM.200903220
44. A. Leela Mohana Reddy, A.E. Tanur, G.C. Walker, Synthesis and hydrogen storage properties of different types of boron nitride nanostructures. Int. J. Hydrog. Energy **35**, 4138–4143 (2010). https://doi.org/10.1016/J.IJHYDENE.2010.01.072
45. S.M. Aceves, G. Petitpas, F. Espinosa-Loza, M.J. Matthews, E. Ledesma-Orozco, Safe, long range, inexpensive and rapidly refuelable hydrogen vehicles with cryogenic pressure vessels. Int. J. Hydrog. Energy **38**, 2480–2489 (2013). https://doi.org/10.1016/J.IJHYDENE.2012.11.123
46. H. Barthelemy, M. Weber, F. Barbier, Hydrogen storage: Recent improvements and industrial perspectives. Int. J. Hydrog. Energy **42**, 7254–7262 (2017). https://doi.org/10.1016/J.IJHYDENE.2016.03.178
47. S.M. Aceves, G.D. Berry, J. Martinez-Frias, F. Espinosa-Loza, Vehicular storage of hydrogen in insulated pressure vessels. Int. J. Hydrog. Energy **31**, 2274–2283 (2006). https://doi.org/10.1016/J.IJHYDENE.2006.02.019
48. L.G. Scanlon, W.A. Feld, P.B. Balbuena, G. Sandi, X. Duan, K.A. Underwood, et al., Hydrogen storage based on physisorption. J. Phys. Chem. B **113**, 4708–4717 (2009). https://doi.org/10.1021/JP809097V
49. Y. Zhang, S. Wu, L. Wang, X. Zhang, Chemisorption solid materials for hydrogen storage near ambient temperature: A review. Front Energy **17**, 72–101 (2022). https://doi.org/10.1007/S11708-022-0835-7
50. M. Ismail, M.S. Yahya, N.H. Idris, N.S.M. Mustafa, M.F.A.A.H. Yap, Novel materials and technologies for hydrogen storage. New Dimens. Prod. Util. Hydrog., 337–365 (2020). https://doi.org/10.1016/B978-0-12-819553-6.00014-3
51. R.M. Barrer, D.E.W. Vaughan, Solution and diffusion of helium and neon in tridymite and cristobalite. Trans. Faraday Soc. **63**, 2275–2290 (1967). https://doi.org/10.1039/TF9676302275
52. J. Weitkamp, M. Fritz, S. Ernst, Zeolites as media for hydrogen storage. Int. J. Hydrog. Energy **20**, 967–970 (1995). https://doi.org/10.1016/0360-3199(95)00058-L
53. H.W. Langmi, D. Book, A. Walton, S.R. Johnson, M.M. Al-Mamouri, J.D. Speight, et al., Hydrogen storage in ion-exchanged zeolites. J. Alloys Compd. **404–406**, 637–642 (2005). https://doi.org/10.1016/J.JALLCOM.2004.12.193

54. J. Dong, X. Wang, H. Xu, Q. Zhao, J. Li, Hydrogen storage in several microporous zeolites. Int. J. Hydrog. Energy **32**, 4998–5004 (2007). https://doi.org/10.1016/J.IJHYDENE.2007.08.009
55. D.J. Collins, H.C. Zhou, Hydrogen storage in metal–organic frameworks. J. Mater. Chem. **17**, 3154–3160 (2007). https://doi.org/10.1039/B702858J
56. J.L.C. Rowsell, O.M. Yaghi, Strategies for hydrogen storage in metal–organic frameworks. Angew. Chem. Int. Ed. **44**, 4670–4679 (2005). https://doi.org/10.1002/ANIE.200462786
57. R. Li, X. Han, Q. Liu, A. Qian, F. Zhu, J. Hu, et al., Enhancing hydrogen adsorption capacity of metal organic frameworks M(BDC)TED0.5 through constructing a bimetallic structure. ACS Omega **7**, 20081–20091 (2022). https://doi.org/10.1021/ACSOMEGA.2C01914/ASSET/IMAGES/MEDIUM/AO2C01914_M004.GIF
58. H. Zhou, J. Zhang, J. Zhang, X.F. Yan, X.P. Shen, A.H. Yuan, Spillover enhanced hydrogen storage in Pt-doped MOF/graphene oxide composite produced via an impregnation method. Inorg. Chem. Commun. **54**, 54–56 (2015). https://doi.org/10.1016/J.INOCHE.2015.02.001
59. N.L. Rosi, J. Eckert, M. Eddaoudi, D.T. Vodak, J. Kim, M. O'Keeffe, et al., Hydrogen storage in microporous metal-organic frameworks. Science **300**, 1127–1129 (2003). https://doi.org/10.1126/SCIENCE.1083440
60. J.L.C. Rowsell, A.R. Millward, K.S. Park, O.M. Yaghi, Hydrogen sorption in functionalized metal-organic frameworks. J. Am. Chem. Soc. **126**, 5666–5667 (2004). https://doi.org/10.1021/JA049408C/SUPPL_FILE/JA049408CSI20040324_112257.PDF
61. H. Furukawa, M.A. Miller, O.M. Yaghi, Independent verification of the saturation hydrogen uptake in MOF-177 and establishment of a benchmark for hydrogen adsorption in metal–organic frameworks. J. Mater. Chem. **17**, 3197–3204 (2007). https://doi.org/10.1039/B703608F
62. A.G. Wong-Foy, A.J. Matzger, O.M. Yaghi, Exceptional H_2 saturation uptake in microporous metal-organic frameworks. J. Am. Chem. Soc. **128**, 3494–3495 (2006). https://doi.org/10.1021/JA058213H/SUPPL_FILE/JA058213HSI20060202_065946.PDF
63. S.S. Han, W.A.G. Iii, Lithium-doped metal-organic frameworks for reversible H_2 storage at ambient temperature. J. Am. Chem. Soc. **129**, 8422–8423 (2007). https://doi.org/10.1021/JA072599
64. X. Zou, M.H. Cha, S. Kim, M.C. Nguyen, G. Zhou, W. Duan, et al., Hydrogen storage in Ca-decorated, B-substituted metal organic framework. Int. J. Hydrog. Energy **35**, 198–203 (2010). https://doi.org/10.1016/J.IJHYDENE.2009.10.054
65. S.S. Han, J.L. Mendoza-Cortés, W.A. Goddard, Recent advances on simulation and theory of hydrogen storage in metal–organic frameworks and covalent organic frameworks. Chem. Soc. Rev. **38**, 1460–1476 (2009). https://doi.org/10.1039/B802430H
66. K.K. Gangu, S. Maddila, S.B. Mukkamala, S.B. Jonnalagadda, Characteristics of MOF, MWCNT and graphene containing materials for hydrogen storage: A review. J. Energy Chem. **30**, 132–144 (2019). https://doi.org/10.1016/J.JECHEM.2018.04.012
67. L. Wang, R.T. Yang, Hydrogen storage properties of carbons doped with ruthenium, platinum, and nickel nanoparticles. J. Phys. Chem. C **112**, 12486–12494 (2008). https://doi.org/10.1021/JP803093W/ASSET/IMAGES/LARGE/JP-2008-03093W_0006.JPEG
68. S.P. Shet, S. Shanmuga Priya, K. Sudhakar, M. Tahir, A review on current trends in potential use of metal-organic framework for hydrogen storage. Int. J. Hydrog. Energy **46**, 11782–11803 (2021). https://doi.org/10.1016/J.IJHYDENE.2021.01.020
69. A.P. Côté, A.I. Benin, N.W. Ockwig, M. O'Keeffe, A.J. Matzger, O.M. Yaghi, Chemistry: Porous, crystalline, covalent organic frameworks. Science **310**, 1166–1170 (2005). https://doi.org/10.1126/SCIENCE.1120411/SUPPL_FILE/COTE.SOM.PDF
70. H.M. El-Kaderi, J.R. Hunt, J.L. Mendoza-Cortés, A.P. Côté, R.E. Taylor, M. O'Keeffe, et al., Designed synthesis of 3D covalent organic frameworks. Science **316**, 268–272 (2007). https://doi.org/10.1126/SCIENCE.1139915/SUPPL_FILE/EL-KADERI.SOM.REVISED.PDF

71. E. Klontzas, E. Tylianakis, G.E. Froudakis, Designing 3D COFs with enhanced hydrogen storage capacity. Nano Lett. **10**, 452–454 (2010). https://doi.org/10.1021/NL903068A/SUPPL_FILE/NL903068A_SI_001.PDF
72. H. Kobayashi, M. Yamauchi, H. Kitagawa, Y. Kubota, K. Kato, M. Takata, On the nature of strong hydrogen atom trapping inside Pd nanoparticles. J. Am. Chem. Soc. **130**, 1828–1829 (2008). https://doi.org/10.1021/JA7102372/SUPPL_FILE/JA7102372_SI.PDF
73. S.B. Kalidindi, H. Oh, M. Hirscher, D. Esken, C. Wiktor, S. Turner, et al., Metal@COFs: Covalent organic frameworks as templates for Pd nanoparticles and hydrogen storage properties of Pd@COF-102 hybrid material. Chem. A Eur. J. **18**, 10848–10856 (2012). https://doi.org/10.1002/CHEM.201201340
74. B. Karthikeyan, Novel synthesis and optical properties of Sm3+ doped Au- polyvinyl alcohol nanocomposite films. Chem. Phys. Lett. **432**, 513–517 (2006). https://doi.org/10.1016/J.CPLETT.2006.10.113
75. I. Cabasso, Y. Yuan, *FY 2008 Annual Progress Report* (U.S. DOE, Washington, DC, 2008)
76. E. Boateng, A. Chen, Recent advances in nanomaterial-based solid-state hydrogen storage. Mater. Today Adv. **6**, 100022 (2020). https://doi.org/10.1016/J.MTADV.2019.100022
77. T. Rimza, S. Saha, C. Dhand, N. Dwivedi, S.S. Patel, S. Singh, et al., Carbon-based sorbents for hydrogen storage: Challenges and sustainability at operating conditions for renewable energy. ChemSusChem **15**, e202200281 (2022). https://doi.org/10.1002/CSSC.202200281
78. R. Ströbel, J. Garche, P.T. Moseley, L. Jörissen, G. Wolf, Hydrogen storage by carbon materials. J. Power Sources **159**, 781–801 (2006). https://doi.org/10.1016/J.JPOWSOUR.2006.03.047
79. P. Chen, X. Wu, J. Lin, K.L. Tan, High H2 uptake by alkali-doped carbon nanotubes under ambient pressure and moderate temperatures. Science **285**, 91–93 (1999). https://doi.org/10.1126/SCIENCE.285.5424.91/ASSET/F5A52D4F-0F6B-42FC-AD53-1B1E13889218/ASSETS/GRAPHIC/SE2597628002.JPEG
80. V. Gayathri, R. Geetha, Hydrogen adsorption in defected carbon nanotubes. Adsorption **13**, 53–59 (2007). https://doi.org/10.1007/S10450-007-9002-Z/METRICS
81. X. Li, H. Zhu, L. Ci, C. Xu, Z. Mao, B. Wei, et al., Hydrogen uptake by graphitized multi-walled carbon nanotubes under moderate pressure and at room temperature. Carbon N. Y. **39**, 2077–2079 (2001). https://doi.org/10.1016/S0008-6223(01)00183-X
82. C. Liu, Y. Chen, C.Z. Wu, S.T. Xu, H.M. Cheng, Hydrogen storage in carbon nanotubes revisited. Carbon N. Y. **48**, 452–455 (2010). https://doi.org/10.1016/J.CARBON.2009.09.060
83. C. Park, M.A. Keane, Controlled growth of highly ordered carbon nanofibers from Y zeolite supported nickel catalysts. Langmuir **17**, 8386–8396 (2001). https://doi.org/10.1021/LA0110390
84. C. Liu, Y.Y. Fan, M. Liu, H.T. Cong, H.M. Cheng, M.S. Dresselhaus, Hydrogen storage in single-walled carbon nanotubes at room temperature. Science **286**, 1127–1129 (1999). https://doi.org/10.1126/SCIENCE.286.5442.1127/ASSET/3583E91A-2068-4C36-AF1E-E45671795BB4/ASSETS/GRAPHIC/SE4397962003.JPEG
85. V. Tozzini, V. Pellegrini, Prospects for hydrogen storage in graphene. Phys. Chem. Chem. Phys. **15**, 80–89 (2012). https://doi.org/10.1039/C2CP42538F
86. A.C. Dillon, M.J. Heben, Hydrogen storage using carbon adsorbents: Past, present and future. Appl. Phys. A Mater. Sci. Process **72**, 133–142 (2001). https://doi.org/10.1007/S003390100788/METRICS
87. A.C. Dillon, K.M. Jones, T.A. Bekkedahl, C.H. Kiang, D.S. Bethune, M.J. Heben, Storage of hydrogen in single-walled carbon nanotubes. Nature, 386, 377–389 (1997). https://doi.org/10.1038/386377a0
88. A. Chambers, C. Park, R.T.K. Baker, N.M. Rodriguez, Hydrogen storage in graphite nanofibers. J. Phys. Chem. B **102**, 4253–4256 (1998). https://doi.org/10.1021/JP980114L
89. D.P. Broom, C.J. Webb, K.E. Hurst, P.A. Parilla, T. Gennett, C.M. Brown, et al., Outlook and challenges for hydrogen storage in nanoporous materials. Appl. Phys. A **122**, 1–21 (2016). https://doi.org/10.1007/S00339-016-9651-4

90. B.J. Kim, S.J. Park, Influence of surface treatments on micropore structure and hydrogen adsorption behavior of nanoporous carbons. J. Colloid Interface Sci. **311**, 619–621 (2007). https://doi.org/10.1016/J.JCIS.2007.03.049
91. S. Brunauer, P.H. Emmett, E. Teller, Adsorption of Gases in Multimolecular Layers. J. Am. Chem. Soc. **60**, 309–319 (1938). https://doi.org/10.1021/JA01269A023/ASSET/JA01269A023.FP.PNG_V03
92. B.J. Kim, Y.S. Lee, S.J. Park, Preparation of platinum-decorated porous graphite nanofibers, and their hydrogen storage behaviors. J. Colloid Interface Sci. **318**, 530–533 (2008). https://doi.org/10.1016/J.JCIS.2007.10.018
93. Z.M. Ao, T.T. Tan, S. Li, Q. Jiang, Molecular hydrogen storage in Al-doped bulk graphite with wider layer distances. Solid State Commun. **149**, 1363–1367 (2009). https://doi.org/10.1016/J.SSC.2009.05.022
94. F.I. Desai, M.N. Uddin, M.M. Rahman, R. Asmatulu, A critical review on improving hydrogen storage properties of metal hydride via nanostructuring and integrating carbonaceous materials. Int. J. Hydrog. Energy **48**, 29256–29294 (2023). https://doi.org/10.1016/J.IJHYDENE.2023.04.029
95. W. Li, C. Liang, W. Zhou, J. Qiu, Z. Zhou, G. Sun, et al., Preparation and characterization of multiwalled carbon nanotube-supported platinum for cathode catalysts of direct methanol fuel cells. J. Phys. Chem. B **107**, 6292–6299 (2003). https://doi.org/10.1021/JP022505C
96. R. Geetha, V. Gayathri, Comparative study on gas adsorption in defected carbon and boron nitride nanotube. Curr. Nanosci. **6**, 131–136 (2010). https://doi.org/10.2174/157341310790945731
97. R.K. Ahluwalia, J.K. Peng, T.Q. Hua, Sorbent material property requirements for on-board hydrogen storage for automotive fuel cell systems. Int. J. Hydrog. Energy **40**, 6373–6390 (2015). https://doi.org/10.1016/J.IJHYDENE.2015.03.025
98. Y. Zhao, Y.H. Kim, A.C. Dillon, M.J. Heben, S.B. Zhang, Hydrogen storage in novel organometallic buckyballs. Phys. Rev. Lett. **94**, 155504 (2005). https://doi.org/10.1103/PHYSREVLETT.94.155504/FIGURES/4/MEDIUM
99. Q. Sun, Q. Wang, P. Jena, Y. Kawazoe, Clustering of Ti on a C60 surface and its effect on hydrogen storage. J. Am. Chem. Soc. **127**, 14582–14583 (2005). https://doi.org/10.1021/JA0550125/ASSET/IMAGES/MEDIUM/JA0550125N00001.GIF
100. Q. Sun, P. Jena, Q. Wang, M. Marquez, First-principles study of hydrogen storage on Li12C60. J. Am. Chem. Soc. **128**, 9741–9745 (2006). https://doi.org/10.1021/JA058330C/ASSET/IMAGES/MEDIUM/JA058330CN00001.GIF
101. K.R.S. Chandrakumar, S.K. Ghosh, Alkali-metal-induced enhancement of hydrogen adsorption in C60 fullerene: An ab initio study. Nano Lett. **8**, 13–19 (2008). https://doi.org/10.1021/NL071456I/ASSET/IMAGES/MEDIUM/NL071456IN00001.GIF
102. Q. Wang, Q. Sun, P. Jena, Y. Kawazoe, Theoretical study of hydrogen storage in Ca-Coated fullerenes. J. Chem. Theory Comput. **5**, 374–379 (2009). https://doi.org/10.1021/CT800373G
103. M. Yoon, S. Yang, C. Hicke, E. Wang, D. Geohegan, Z. Zhang, Calcium as the superior coating metal in functionalization of carbon fullerenes for high-capacity hydrogen storage. Phys. Rev. Lett. **100**, 206806 (2008). https://doi.org/10.1103/PHYSREVLETT.100.206806/FIGURES/3/MEDIUM
104. S. Niaz, T. Manzoor, A.H. Pandith, Hydrogen storage: Materials, methods and perspectives. Renew. Sustain. Energy Rev. **50**, 457–469 (2015). https://doi.org/10.1016/J.RSER.2015.05.011
105. N.H. Phan, S. Rio, C. Faur, L. Le Coq, P. Le Cloirec, T.H. Nguyen, Production of fibrous activated carbons from natural cellulose (jute, coconut) fibers for water treatment applications. Carbon N. Y. **44**, 2569–2577 (2006). https://doi.org/10.1016/J.CARBON.2006.05.048
106. M. Jordá-Beneyto, D. Lozano-Castelló, F. Suárez-García, D. Cazorla-Amorós, Á. Linares-Solano, Advanced activated carbon monoliths and activated carbons for hydrogen storage. Microporous Mesoporous Mater. **112**, 235–242 (2008). https://doi.org/10.1016/J.MICROMESO.2007.09.034

107. M. Sharon, T. Soga, R. Afre, D. Sathiyamoorthy, K. Dasgupta, S. Bhardwaj, et al., Hydrogen storage by carbon materials synthesized from oil seeds and fibrous plant materials. Int. J. Hydrog. Energy **32**, 4238–4249 (2007). https://doi.org/10.1016/J.IJHYDENE.2007.05.038
108. H. Marsh, E.A. Heintz, F. Rodríguez-Reinoso, *Introduction to carbon technologies* (University of Alicante, Alicante, 1997)
109. S. Hynek, Hydrogen storage by carbon sorption. Int. J. Hydrog. Energy **22**, 601–610 (1997). https://doi.org/10.1016/S0360-3199(96)00185-1
110. I. Rossetti, G. Ramis, A. Gallo, A. Di Michele, Hydrogen storage over metal-doped activated carbon. Int. J. Hydrog. Energy **40**, 7609–7616 (2015). https://doi.org/10.1016/J.IJHYDENE.2015.04.064
111. W. Zhao, V. Fierro, C. Zlotea, M.T. Izquierdo, C. Chevalier-César, M. Latroche, et al., Activated carbons doped with Pd nanoparticles for hydrogen storage. Int. J. Hydrog. Energy **37**, 5072–5080 (2012). https://doi.org/10.1016/J.IJHYDENE.2011.12.058
112. S.Y. Lee, S.J. Park, Effect of platinum doping of activated carbon on hydrogen storage behaviors of metal-organic frameworks-5. Int. J. Hydrog. Energy **36**, 8381–8387 (2011). https://doi.org/10.1016/J.IJHYDENE.2011.03.038
113. Y. Wang, K. Wang, C. Guan, Z. He, Z. Lu, T. Chen, et al., Surface functionalization-enhanced spillover effect on hydrogen storage of Ni–B nanoalloy-doped activated carbon. Int. J. Hydrog. Energy **36**, 13663–13668 (2011). https://doi.org/10.1016/J.IJHYDENE.2011.08.049
114. W. Zhao, V. Fierro, N. Fernández-Huerta, M.T. Izquierdo, A. Celzard, Impact of synthesis conditions of KOH activated carbons on their hydrogen storage capacities. Int. J. Hydrog. Energy **37**, 14278–14284 (2012). https://doi.org/10.1016/J.IJHYDENE.2012.06.110
115. W.C. Xu, K. Takahashi, Y. Matsuo, Y. Hattori, M. Kumagai, S. Ishiyama, et al., Investigation of hydrogen storage capacity of various carbon materials. Int. J. Hydrog. Energy **32**, 2504–2512 (2007). https://doi.org/10.1016/J.IJHYDENE.2006.11.012
116. M. Hirscher, B. Panella, Nanostructures with high surface area for hydrogen storage. J. Alloys Compd. **404–406**, 399–401 (2005). https://doi.org/10.1016/J.JALLCOM.2004.11.109
117. R. Ströbel, L. Jörissen, T. Schliermann, V. Trapp, W. Schütz, K. Bohmhammel, et al., Hydrogen adsorption on carbon materials. J. Power Sources **84**, 221–224 (1999). https://doi.org/10.1016/S0378-7753(99)00320-1
118. M. Zieliński, R. Wojcieszak, S. Monteverdi, M. Mercy, M.M. Bettahar, Hydrogen storage on nickel catalysts supported on amorphous activated carbon. Cat. Com. **6**, 777–783 (2005). https://doi.org/10.1016/J.CATCOM.2005.07.001
119. Y. Jia, L. Cheng, N. Pan, J. Zou, G. Lu, X. Yao, Catalytic de/hydrogenation in Mg by Co-Doped Ni and VOx on active carbon: Extremely fast kinetics at low temperatures and high hydrogen capacity. Adv. Energy Mater. **1**, 387–393 (2011). https://doi.org/10.1002/AENM.201000025
120. H.Y. Tian, C.E. Buckley, D.A. Sheppard, M. Paskevicius, N. Hanna, A synthesis method for cobalt doped carbon aerogels with high surface area and their hydrogen storage properties. Int. J. Hydrog. Energy **35**, 13242–13246 (2010). https://doi.org/10.1016/J.IJHYDENE.2010.09.018
121. M. Zieliński, R. Wojcieszak, S. Monteverdi, M. Mercy, M.M. Bettahar, Hydrogen storage in nickel catalysts supported on activated carbon. Int. J. Hydrog. Energy **32**, 1024–1032 (2007). https://doi.org/10.1016/J.IJHYDENE.2006.07.004
122. L. Wang, R.T. Yang, Hydrogen storage on carbon-based adsorbents and storage at ambient temperature by hydrogen spillover. Catal. Rev. **52**, 411–461 (2010). https://doi.org/10.1080/01614940.2010.520265
123. B. Duret, A. Saudin, Microspheres for on-board hydrogen storage. Int. J. Hydrog. Energy **19**, 757–764 (1994). https://doi.org/10.1016/0360-3199(94)90240-2

124. D.K. Kohli, R.K. Khardekr, R. Singh, P.K. Gupta, Glass micro-container based hydrogen storage scheme. Int. J. Hydrog. Energy **33**, 417–422 (2008). https://doi.org/10.1016/J.IJHYDENE.2007.07.044
125. L.N. Smith, P.S. Midha, Computer simulation of morphology and packing behaviour of irregular particles, for predicting apparent powder densities. Comput. Mater. Sci. **7**, 377–383 (1997). https://doi.org/10.1016/S0927-0256(97)00003-7
126. M.L. Schmitt, J.E. Shelby, M.M. Hall, Preparation of hollow glass microspheres from sol–gel derived glass for application in hydrogen gas storage. J. Non Cryst. Solids **352**, 626–631 (2006). https://doi.org/10.1016/J.JNONCRYSOL.2005.11.057
127. G. Wicks, L.K. Heung, R.F. Schumacher, Microspheres and microworlds. Am. Ceram. Soc. Bull. **87**(6) (2008). https://www.agriculturedefensecoalition.org/
128. N.K. Zhevago, V.I. Glebov, Hydrogen storage in capillary arrays. Energ. Conver. Manage. **48**, 1554–1559 (2007). https://doi.org/10.1016/J.ENCONMAN.2006.11.017
129. N. Wadnerkar, V. Kalamse, A. Chaudhari, Hydrogen uptake capacity of C_2H_4Sc and its ions: A density functional study. J. Comput. Chem. **31**, 1656–1661 (2010). https://doi.org/10.1002/JCC.21449
130. V. Kalamse, N. Wadnerkar, A. Chaudhari, Hydrogen storage in C_2H_4V and $C_2H_4V^+$ organometallic compounds. J. Phys. Chem. C **114**, 4704–4709 (2010). https://doi.org/10.1021/JP910614N/SUPPL_FILE/JP910614N_SI_001.PDF
131. N. Wadnerkar, V. Kalamse, A. Chaudhari, VC3H3 organometallic compound: A possible hydrogen storage material. Int. J. Hydrog. Energy **36**, 664–670 (2011). https://doi.org/10.1016/J.IJHYDENE.2010.09.081
132. W.Q. Deng, X. Xu, W.A. Goddard, New alkali doped pillared carbon materials designed to achieve practical reversible hydrogen storage for transportation. Phys. Rev. Lett. **92**, 166103 (2004). https://doi.org/10.1103/PHYSREVLETT.92.166103/FIGURES/5/MEDIUM
133. M.H. Park, Y.S. Lee, Hydrogen adsorption on 3d transition-metal-doped organosilica complexes. Chem. Phys. Lett. **488**, 7–9 (2010). https://doi.org/10.1016/J.CPLETT.2010.01.070
134. S. Niaz, T. Manzoor, N. Islam, A.H. Pandith, Theoretical investigations on C2H4Nb complex as a potential hydrogen storage system, using moller–plesset (MP2) and density functional theory. Int J Quantum Chem **114**, 449–457 (2014). https://doi.org/10.1002/QUA.24587
135. Y. Yürüm, A. Taralp, T.N. Veziroglu, Storage of hydrogen in nanostructured carbon materials. Int. J. Hydrog. Energy **34**, 3784–3798 (2009). https://doi.org/10.1016/J.IJHYDENE.2009.03.001
136. A.K. Kandalam, B. Kiran, PJ., Multidecker organometallic complexes for hydrogen storage. J. Phys. Chem. C **112**, 6181–6185 (2008). https://doi.org/10.1021/jp074759+
137. J. Niu, B.K. Rao, P. Jena, Binding of hydrogen molecules by a transition-metal ion. Phys. Rev. Lett. **68**, 2277 (1992). https://doi.org/10.1103/PhysRevLett.68.2277
138. F. Yang, J. Wang, Y. Zhang, Z. Wu, Z. Zhang, F. Zhao, et al., Recent progress on the development of high entropy alloys (HEAs) for solid hydrogen storage: A review. Int. J. Hydrog. Energy **47**, 11236–11249 (2022). https://doi.org/10.1016/J.IJHYDENE.2022.01.141
139. S.L. Suib, New and future developments in catalysis: Batteries, hydrogen storage and fuel cells. New Future Dev. Catal. Batter Hydrog. Storage Fuel Cells, 1–535 (2013). https://doi.org/10.1016/C2010-0-68687-1
140. J. Reedijk, *Reference Module in Chemistry, Molecular Sciences and Chemical Engineering* (Elsevier, 2013)
141. S. Srinivasan, P.C. Sharma, Development of novel polymer nanostructures and nanoscale complex hydrides for reversible hydrogen storage, in *Hydrogen Storage*, (2012). https://doi.org/10.5772/50171
142. F. Sclüth, B. Bogdanović, M. Felderhoff, Light metal hydrides and complex hydrides for hydrogen storage. Chem. Commun., 2249–2258 (2004). https://doi.org/10.1039/B406522K
143. H.I. Schlesinger, H.C. Brown, A.E. Finholt, J.R. Gilbreath, H.R. Hoekstra, E.K. Hyde, Sodium borohydride, its hydrolysis and its use as a reducing agent and in the generation of

hydrogen. J. Am. Chem. Soc. **75**, 215–219 (1953). https://doi.org/10.1021/JA01097A057/ASSET/JA01097A057.FP.PNG_V03
144. Y. Kojima, K.I. Suzuki, K. Fukumoto, Y. Kawai, M. Kimbara, H. Nakanishi, et al., Development of 10 kW-scale hydrogen generator using chemical hydride. J. Power Sources **125**, 22–26 (2004). https://doi.org/10.1016/S0378-7753(03)00827-9
145. R.M.M. Michael Strizki. US Patent 6 534 033, assigned to Millennium Cell, n.d.
146. S.C. Abrahams, A.P. Ginsberg, K. Knox, Transition metal-hydrogen compounds. II. The crystal and molecular structure of potassium rhenium hydride, K2ReH9. Inorg. Chem. **3**, 558–567 (1964). https://doi.org/10.1021/IC50014A026/ASSET/IC50014A026.FP.PNG_V03
147. W. Grochala, P.P. Edwards, Thermal decomposition of the non-interstitial hydrides for the storage and production of hydrogen. Chem. Rev. **104**, 1283–1315 (2004). https://doi.org/10.1021/CR030691S/ASSET/CR030691S.FP.PNG_V03
148. D.G. Westlake, Site occupancies and stoichiometries in hydrides of intermetallic compounds: Geometric considerations. J. Less-Common Met. **90**, 251–273 (1983). https://doi.org/10.1016/0022-5088(83)90075-9
149. H. Iba, E. Akiba, Hydrogen absorption and modulated structure in Ti–V–Mn alloys. J. Alloys Compd. **253–254**, 21–24 (1997). https://doi.org/10.1016/S0925-8388(96)03072-1
150. Metal hydrides: The key to sustainable and infinite hydrogen storage (n.d.). https://mincatec-energy.com/en/technology/. Accessed 8 Mar 2024
151. B. Sakintuna, F. Lamari-Darkrim, M. Hirscher, Metal hydride materials for solid hydrogen storage: A review. Int. J. Hydrog. Energy **32**, 1121–1140 (2007). https://doi.org/10.1016/J.IJHYDENE.2006.11.022
152. D. Mori, N. Kobayashi, T. Shinozawa, T. Matsunaga, H. Kubo, K. Toh, et al., Research on in-vehicle hydrogen storage systems using hydrogen storage materials high-pressure hydrogen storage alloy tank. J. Japan Inst. Met. **69**, 308–311 (2005). https://doi.org/10.2320/JINSTMET.69.308
153. A.A. Akunets, N.G. Basov, V.S. Bushuev, V.M. Dorogotovtsev, A.I. Gromov, A.I. Isakov, et al., Super-high-strength microballoons for hydrogen storage. Int. J. Hydrog. Energy **19**, 697–700 (1994). https://doi.org/10.1016/0360-3199(94)90157-0
154. A. Szajek, M. Jurczyk, I. Okońska, K. Smardz, E. Jankowska, L. Smardz, Electrochemical and electronic properties of nanocrystalline Mg-based hydrogen storage materials. J. Alloys Compd. **436**, 345–350 (2007). https://doi.org/10.1016/J.JALLCOM.2006.07.043
155. R.K. Jain, A. Jain, S. Agarwal, N.P. Lalla, V. Ganesan, D.M. Phase, et al., Hydrogenation behaviour of Ce-based AB5 intermetallic compounds. J. Alloys Compd. **440**, 84–88 (2007). https://doi.org/10.1016/J.JALLCOM.2006.08.326
156. M. Balcerzak, M. Jurczyk, Influence of gaseous activation on hydrogen sorption properties of TiNi and Ti_2Ni alloys. J. Mater. Eng. Perform. **24**, 1710–1717 (2015). https://doi.org/10.1007/S11665-015-1445-X/TABLES/1
157. F. Cuevas, M. Latroche, A. Percheron-Guégan, Relationship between polymorphism and hydrogenation properties in $Ti_{0.64}Zr_{0.36}Ni$ alloy. J. Alloys Compd. **404–406**, 545–549 (2005). https://doi.org/10.1016/J.JALLCOM.2005.02.072
158. M. Matsushita, I. Tajima, M. Abe, H. Tokuyama, Experimental study of porosity and effective thermal conductivity in packed bed of nano-structured FeTi for usage in hydrogen storage tanks. Int. J. Hydrog. Energy **44**, 23239–23248 (2019). https://doi.org/10.1016/J.IJHYDENE.2019.07.037
159. K.D. Ćirić, A. Kocjan, A. Gradišek, V.J. Koteski, A.M. Kalijadis, V.N. Ivanovski, et al., A study on crystal structure, bonding and hydriding properties of Ti–Fe–Ni intermetallics – behind substitution of iron by nickel. Int. J. Hydrog. Energy **37**, 8408–8417 (2012). https://doi.org/10.1016/J.IJHYDENE.2012.02.047
160. W. Liu, K.F. Aguey-Zinsou, Low temperature synthesis of $LaNi_5$ nanoparticles for hydrogen storage. Int. J. Hydrog. Energy **41**, 1679–1687 (2016). https://doi.org/10.1016/J.IJHYDENE.2015.10.128

161. L. Pickering, J. Li, D. Reed, A.I. Bevan, D. Book, Ti–V–Mn based metal hydrides for hydrogen storage. J. Alloys Compd. **580**, S233–S237 (2013). https://doi.org/10.1016/J.JALLCOM.2013.03.208
162. H.I. Schlesinger, H.C. Brown, A.E. Finholt, The preparation of sodium borohydride by the high temperature reaction of sodium hydride with borate esters. J. Am. Chem. Soc. **75**, 205–209 (1953). https://doi.org/10.1021/JA01097A054/ASSET/JA01097A054.FP.PNG_V03
163. S.I. Orimo, Y. Nakamori, J.R. Eliseo, A. Züttel, C.M. Jensen, Complex hydrides for hydrogen storage. Chem. Rev. **107**, 4111–4132 (2007). https://doi.org/10.1021/CR0501846/ASSET/CR0501846.FP.PNG_V03
164. R. Aiello, M.A. Matthews, D.L. Reger, J.E. Collins, Production of hydrogen gas from novel chemical hydrides. Int. J. Hydrog. Energy **23**, 1103–1108 (1998). https://doi.org/10.1016/S0360-3199(98)00001-9
165. A. El Kharbachi, E. Pinatel, I. Nuta, M. Baricco, A thermodynamic assessment of $LiBH_4$. Calphad **39**, 80–90 (2012). https://doi.org/10.1016/J.CALPHAD.2012.08.005
166. U.B. Demirci, P. Miele, Chemical hydrogen storage: 'material' gravimetric capacity versus 'system' gravimetric capacity. Energ. Environ. Sci. **4**, 3334–3341 (2011). https://doi.org/10.1039/C1EE01612A
167. D. Bae, U. Jung, H. Lee, H. Yoo, S.Y. Moon, K.H. Lee, et al., Synthesis of double-walled boron nitride nanotubes from ammonia borane by thermal plasma methods. ACS Omega **8**, 21514–21521 (2023). https://doi.org/10.1021/ACSOMEGA.3C00498/ASSET/IMAGES/LARGE/AO3C00498_0007.JPEG
168. P. Wang, S. Orimo, T. Matsushima, H. Fujii, G. Majer, Hydrogen in mechanically prepared nanostructured h-BN: A critical comparison with that in nanostructured graphite. Appl. Phys. Lett. **80**, 318–320 (2002). https://doi.org/10.1063/1.1432447
169. M. Oertel, W. Weirich, B. Kügler, L. Lücke, M. Pietsch, U. Winkelmann, The lithium-lithium hydride process for the production of hydrogen: Comparison of two concepts for 950 and 1300°C HTR helium outlet temperature. Int. J. Hydrog. Energy **12**, 211–217 (1987). https://doi.org/10.1016/0360-3199(87)90024-3
170. T.K. Nielsen, A. Karkamkar, M. Bowden, F. Besenbacher, T.R. Jensen, T. Autrey, Methods to stabilize and destabilize ammonium borohydride. Dalton Trans. **42**, 680–687 (2012). https://doi.org/10.1039/C2DT31591B
171. Z. Xiong, C.K. Yong, G. Wu, P. Chen, W. Shaw, A. Karkamkar, et al., High-capacity hydrogen storage in lithium and sodium amidoboranes. Nat. Mater. **7**(2), 138–141 (2007). https://doi.org/10.1038/nmat2081
172. E.M. Dematteis, M.B. Amdisen, T. Autrey, J. Barale, M.E. Bowden, C.E. Buckley, et al., Hydrogen storage in complex hydrides: Past activities and new trends. Prog. Energy **4**, 032009 (2022). https://doi.org/10.1088/2516-1083/AC7499
173. L. Schlapbach, A. Züttel, Hydrogen-storage materials for mobile applications. Nature **414**, 353–358 (2001). https://doi.org/10.1038/35104634
174. H. Reule, M. Hirscher, A. Weißhardt, H. Kronmüller, Hydrogen desorption properties of mechanically alloyed MgH_2 composite materials. J. Alloys Compd. **305**, 246–252 (2000). https://doi.org/10.1016/S0925-8388(00)00710-6
175. I. Saita, T. Toshima, S. Tanda, T. Akiyama, Hydriding chemical vapor deposition of metal hydride nano-fibers. Mater. Trans. **47**, 931–934 (2006). https://doi.org/10.2320/MATERTRANS.47.931
176. J. Huot, G. Liang, S. Boily, A. Van Neste, R. Schulz, Structural study and hydrogen sorption kinetics of ball-milled magnesium hydride. J. Alloys Compd. **293–295**, 495–500 (1999). https://doi.org/10.1016/S0925-8388(99)00474-0
177. S.R. Johnson, P.A. Anderson, P.P. Edwards, I. Gameson, J.W. Prendergast, M. Al-Mamouri, et al., Chemical activation of MgH_2; a new route to superior hydrogen storage materials. Chem. Commun. **36**, 2823–2825 (2005). https://doi.org/10.1039/B503085D

178. Y. Liu, Y. Guo, Y. Jiang, L. Feng, Y. Sun, Y. Wang, Recent progress in thermodynamic and kinetics modification of magnesium hydride hydrogen storage materials. Mater. Rep. Energy **4**, 100252 (2024). https://doi.org/10.1016/J.MATRE.2024.100252
179. I. Zavaliy, V. Berezovets, R. Denys, O. Kononiuk, V. Yartys, Hydrogen absorption-desorption properties and hydrolysis performance of MgH_2-$Zr_3V_3O_{0.6}H_x$ and MgH_2-$Zr_3V_3O_{0.6}H_x$-C composites. J. Energy Storage **65**, 107245 (2023). https://doi.org/10.1016/J.EST.2023.107245
180. P.E. De Jongh, R.W.P. Wagemans, T.M. Eggenhuisen, B.S. Dauvillier, P.B. Radstake, J.D. Meeldijk, et al., The preparation of carbon-supported magnesium nanoparticles using melt infiltration. Chem. Mater. **19**, 6052–6057 (2007). https://doi.org/10.1021/CM702205V/SUPPL_FILE/CM702205V-FILE002.PDF
181. B. Bogdanović, A. Ritter, B. Spliethoff, Active MgH_2-Mg systems for reversible chemical energy storage. Angew. Chem. Int. Ed. Engl. **29**, 223–234 (1990). https://doi.org/10.1002/ANIE.199002233
182. J.L. Bobet, B. Chevalier, M.Y. Song, B. Darriet, J. Etourneau, Reactive mechanical grinding of magnesium in hydrogen and the effects of additives. Mater. Manuf. Process **17**, 351–361 (2002). https://doi.org/10.1081/AMP-120005381
183. G. Liang, J. Huot, S. Boily, A. Van Neste, R. Schulz, Catalytic effect of transition metals on hydrogen sorption in nanocrystalline ball milled MgH_2–Tm (Tm=Ti, V, Mn, Fe and Ni) systems. J. Alloys Compd. **292**, 247–252 (1999). https://doi.org/10.1016/S0925-8388(99)00442-9
184. A. Denis, E. Sellier, C. Aymonier, J.L. Bobet, Hydrogen sorption properties of magnesium particles decorated with metallic nanoparticles as catalyst. J. Alloys Compd. **476**, 152–159 (2009). https://doi.org/10.1016/J.JALLCOM.2008.09.016
185. J.L. Bobet, E. Akiba, Y. Nakamura, B. Darriet, Study of Mg-M (M=Co, Ni and Fe) mixture elaborated by reactive mechanical alloying – Hydrogen sorption properties. Int. J. Hydrog. Energy **25**, 987–996 (2000). https://doi.org/10.1016/S0360-3199(00)00002-1
186. R.A. Varin, S. Li, C. Chiu, L. Guo, O. Morozova, T. Khomenko, et al., Nanocrystalline and non-crystalline hydrides synthesized by controlled reactive mechanical alloying/milling of Mg and Mg–X (X = Fe, Co, Mn, B) systems. J. Alloys Compd. **404–406**, 494–498 (2005). https://doi.org/10.1016/J.JALLCOM.2004.12.176
187. H. Gasan, O.N. Celik, N. Aydinbeyli, Y.M. Yaman, Effect of V, Nb, Ti and graphite additions on the hydrogen desorption temperature of magnesium hydride. Int. J. Hydrog. Energy **37**, 1912–1918 (2012). https://doi.org/10.1016/J.IJHYDENE.2011.05.086
188. W. Oelerich, T. Klassen, R. Bormann, Comparison of the catalytic effects of V, V2O5, VN, and VC on the hydrogen sorption of nanocrystalline Mg. J. Alloys Compd. **322**, L5–L9 (2001). https://doi.org/10.1016/S0925-8388(01)01173-2
189. X. Xu, C. Song, Improving hydrogen storage/release properties of magnesium with nano-sized metal catalysts as measured by tapered element oscillating microbalance. Appl. Catal. A. Gen. **300**, 130–138 (2006). https://doi.org/10.1016/J.APCATA.2005.10.062
190. A. Zaluska, L. Zaluski, J.O. Ström-Olsen, Nanocrystalline magnesium for hydrogen storage. J. Alloys Compd. **288**, 217–225 (1999). https://doi.org/10.1016/S0925-8388(99)00073-0
191. M. Waiblinger, K. Lips, W. Harneit, A. Weidinger, E. Dietel, A. Hirsch, Corrected article: Thermal stability of the endohedral fullerenes. Phys. Rev. B **64**, 159901 (2001). https://doi.org/10.1103/PhysRevB.64.159901
192. M.P. Pitt, M. Paskevicius, C.J. Webb, D.A. Sheppard, C.E. Buckley, E.M. Gray, The synthesis of nanoscopic Ti based alloys and their effects on the MgH_2 system compared with the MgH_2 + $0.01Nb_2O_5$ benchmark. Int. J. Hydrog. Energy **37**, 4227–4237 (2012). https://doi.org/10.1016/J.IJHYDENE.2011.11.114
193. S.A. Jin, J.H. Shim, Y.W. Cho, K.W. Yi, Dehydrogenation and hydrogenation characteristics of MgH_2 with transition metal fluorides. J. Power Sources **172**, 859–862 (2007). https://doi.org/10.1016/J.JPOWSOUR.2007.04.090

194. S. Milošević, Ž. Rašković-Lovre, S. Kurko, R. Vujasin, N. Cvjetićanin, L. Matović, et al., Influence of VO_2 nanostructured ceramics on hydrogen desorption properties from magnesium hydride. Ceram. Int. **39**, 51–56 (2013). https://doi.org/10.1016/J.CERAMINT.2012.05.091
195. M. Polanski, J. Bystrzycki, Comparative studies of the influence of different nano-sized metal oxides on the hydrogen sorption properties of magnesium hydride. J. Alloys Compd. **486**, 697–701 (2009). https://doi.org/10.1016/J.JALLCOM.2009.07.042
196. K.S. Jung, D.H. Kim, E.Y. Lee, K.S. Lee, Hydrogen sorption of magnesium hydride doped with nano-sized TiO_2. Catal. Today **120**, 270–275 (2007). https://doi.org/10.1016/J.CATTOD.2006.09.028
197. Z. Jalil, A. Rahwanto, H. Sofyan, M. Usman, E. Handoko, The use of Silica from beach sand as catalyst in Magnesium based hydrides for Hydrogen storage materials. IOP Conf. Ser. Earth Environ. Sci. **105**, 012093 (2018). https://doi.org/10.1088/1755-1315/105/1/012093
198. G. Barkhordarian, T. Klassen, R. Bormann, Kinetic investigation of the effect of milling time on the hydrogen sorption reaction of magnesium catalyzed with different Nb_2O_5 contents. J. Alloys Compd. **407**, 249–255 (2006). https://doi.org/10.1016/J.JALLCOM.2005.05.037
199. A. Jubair, M.A. Rahman, M.M.R. Khan, M.W. Rahman, Catalytic role of binary oxides (CuO and Al_2O_3) on hydrogen storage in MgH_2. Environ. Prog. Sustain. Energy **42**, e13940 (2023). https://doi.org/10.1002/EP.13940
200. A. Eslami, S.A. Lachini, M. Shaterian, M. Karami, M. Enhessari, Synthesis, characterization, and hydrogen storage capacity of $MgMn_2O_4$ spinel nanostructures. Inorg. Chem. Commun. **154**, 110875 (2023). https://doi.org/10.1016/J.INOCHE.2023.110875
201. Z.G. Huang, Z.P. Guo, A. Calka, D. Wexler, C. Lukey, H.K. Liu, Effects of iron oxide (Fe_2O_3, Fe_3O_4) on hydrogen storage properties of Mg-based composites. J. Alloys Compd. **422**, 299–304 (2006). https://doi.org/10.1016/J.JALLCOM.2005.11.074
202. M.Y. Song, J.L. Bobet, B. Darriet, Improvement in hydrogen sorption properties of Mg by reactive mechanical grinding with Cr_2O_3, Al_2O_3 and CeO_2. J. Alloys Compd. **340**, 256–262 (2002). https://doi.org/10.1016/S0925-8388(02)00019-1
203. N. Leick, N.A. Strange, A. Schneemann, V. Stavila, K. Gross, N. Washton, et al., Al_2O_3 atomic layer deposition on nanostructured γ-$Mg(BH_4)_2$ for H_2 storage. ACS Appl. Energy Mater. **4**, 1150–1162 (2021). https://doi.org/10.1021/ACSAEM.0C02314/SUPPL_FILE/AE0C02314_SI_001.PDF
204. A. Zaluska, L. Zaluski, J.O. Ström-Olsen, Lithium–beryllium hydrides: The lightest reversible metal hydrides. J. Alloys Compd. **307**, 157–166 (2000). https://doi.org/10.1016/S0925-8388(00)00883-5
205. T.D. Humphries, M.B. Ley, C. Frommen, K.T. Munroe, T.R. Jensen, B.C. Hauback, Crystal structure and in situ decomposition of $Eu(BH_4)_2$ and $Sm(BH_4)_2$. J. Mater. Chem. A **3**, 691–698 (2014). https://doi.org/10.1039/C4TA04080E
206. J.E. Olsen, C. Frommen, M.H. Sørby, B.C. Hauback, Crystal structures and properties of solvent-free $LiYb(BH_4)_4{-}xCl_x$, $Yb(BH_4)_3$ and $Yb(BH_4)_2{-}xCl_x$. RSC Adv. **3**, 10764–10774 (2013). https://doi.org/10.1039/C3RA40435H
207. B. Richter, J.B. Grinderslev, K.T. Møller, M. Paskevicius, T.R. Jensen, From metal hydrides to metal borohydrides. Inorg. Chem. **57**, 10768–10780 (2018). https://doi.org/10.1021/ACS.INORGCHEM.8B01398/SUPPL_FILE/IC8B01398_SI_001.PDF
208. J.B. Grinderslev, K.T. Møller, M. Bremholm, T.R. Jensen, Trends in synthesis, crystal structure, and thermal and magnetic properties of rare-earth metal borohydrides. Inorg. Chem. **58**, 5503–5517 (2019). https://doi.org/10.1021/ACS.INORGCHEM.8B03258/SUPPL_FILE/IC8B03258_SI_001.PDF
209. J. Block, A.P. Gray, The thermal decomposition of lithium aluminum hydride. Inorg. Chem. **4**, 304–305 (1965). https://doi.org/10.1021/IC50025A009/ASSET/IC50025A009.FP.PNG_V03
210. O. Zavorotynska, A. El-Kharbachi, S. Deledda, B.C. Hauback, Recent progress in magnesium borohydride $Mg(BH_4)_2$: Fundamentals and applications for energy storage. (2016). https://doi.org/10.1016/j.ijhydene.2016.02.015

211. P. Martelli, R. Caputo, A. Remhof, P. Mauron, A. Borgschulte, Z�ttel A., Stability and decomposition of $NaBH_4$. J. Phys. Chem. C **114**, 7173–7177 (2010). https://doi.org/10.1021/JP909341Z/ASSET/IMAGES/MEDIUM/JP-2009-09341Z_0005.GIF
212. M.B. Smith, G.E. Bass, Heats and free energies of formation of the alkali aluminum hydrides and of cesium hydride: Heat of formation of aluminum chloride in hydrochloric acid. J. Chem. Eng. Data **8**, 342–346 (1963). https://doi.org/10.1021/JE60018A020/ASSET/JE60018A020.FP.PNG_V03
213. R. Martínez-Coronado, M. Retuerto, B. Torres, M.J. Martínez-Lope, M.T. Fernández-Díaz, J.A. Alonso, High-pressure synthesis, crystal structure and cyclability of the Mg_2NiH_4 hydride. Int. J. Hydrog. Energy **38**, 5738–5745 (2013). https://doi.org/10.1016/J.IJHYDENE.2013.02.108
214. P. Zolliker, K. Yvon, P. Fischer, J. Schefer, Dimagnesium Cobalt(I) Pentahydride, Mg_2CoH_5, containing square-pyramidal CoH_5^4 anions. Inorg. Chem. **24**, 4177–4180 (1985). https://doi.org/10.1021/IC00218A039/SUPPL_FILE/IC00218A039_SI_001.PDF
215. S.S. Sai Raman, D.J. Davidson, J.L. Bobet, O.N. Srivastava, Investigations on the synthesis, structural and microstructural characterizations of Mg-based K_2PtCl_6 type (Mg_2FeH_6) hydrogen storage material prepared by mechanical alloying. J. Alloys Compd. **333**, 282–290 (2002). https://doi.org/10.1016/S0925-8388(01)01729-7
216. P. Chen, Z. Xiong, J. Luo, J. Lin, T.K. Lee, Interaction of hydrogen with metal nitrides and imides. Nature **420**, 302–304 (2002). https://doi.org/10.1038/nature01210
217. E. Rönnebro, D. Noréus, K. Kadir, A. Reiser, B. Bogdanovic, Investigation of the perovskite related structures of $NaMgH_3$, $NaMgF_3$ and Na_3AlH_6. J. Alloys Compd. **299**, 101–106 (2000). https://doi.org/10.1016/S0925-8388(99)00665-9
218. M. Afzal, P. Sharma, Design of a large-scale metal hydride based hydrogen storage reactor: Simulation and heat transfer optimization. Int. J. Hydrog. Energy **43**, 13356–13372 (2018). https://doi.org/10.1016/J.IJHYDENE.2018.05.084
219. H.W. Brinks, B.C. Hauback, The structure of Li_3AlD_6. J. Alloys Compd. **354**, 143–147 (2003). https://doi.org/10.1016/S0925-8388(02)01348-8
220. J.W. Lauher, D. Dougherty, P.J. Herley, IUCr., Sodium tetrahydroaluminate. Acta Cryst. **35**, 1454–1456 (1979). https://doi.org/10.1107/S0567740879006701
221. C. Pratthana, K.F. Aguey-Zinsou, Surfactant induced synthesis of $LiAlH_4$ and $NaAlH_4$ nanoparticles for hydrogen storage. Appl. Sci. **12**, 4742 (2022). https://doi.org/10.3390/APP12094742/S1
222. Slattery DK, Hampton MD. Complex Hydrides for Hydrogen Storage n.d.
223. J. Sangster, A.D. Pelton, The Li-N (lithium-nitrogen) system. J. Ph. Equilibr. **13**, 291–296 (1992). https://doi.org/10.1007/BF02667557/METRICS
224. S. Hino, T. Ichikawa, N. Ogita, M. Udagawa, H. Fujii, Quantitative estimation of NH_3 partial pressure in H_2 desorbed from the Li–N–H system by Raman spectroscopy. Chem. Commun., 3038–3040 (2005). https://doi.org/10.1039/B503356J
225. J. Hu, G. Wu, Y. Liu, Z. Xiong, P. Chen, K. Murata, et al., Hydrogen release from $Mg(NH_2)_2$–MgH_2 through mechanochemical reaction. J. Phys. Chem. B **110**, 14688–14692 (2006). https://doi.org/10.1021/JP061279U
226. K. Murata, K. Kaneko, F. Kokai, K. Takahashi, M. Yudasaka, S. Iiji-Ma, et al., Ternary imides for hydrogen storage. Adv. Mater. **16**, 1522–1525 (2004). https://doi.org/10.1002/ADMA.200400571
227. T. Ichikawa, S. Isobe, N. Hanada, H. Fujii, Lithium nitride for reversible hydrogen storage. J. Alloys Compd. **365**, 271–276 (2004). https://doi.org/10.1016/S0925-8388(03)00637-6
228. Z. Xiong, J. Hu, G. Wu, P. Chen, Hydrogen absorption and desorption in Mg–Na–N–H system. J. Alloys Compd. **395**, 209–212 (2005). https://doi.org/10.1016/J.JALLCOM.2004.10.062

229. Z. Xiong, J. Hu, G. Wu, P. Chen, W. Luo, K. Gross, et al., Thermodynamic and kinetic investigations of the hydrogen storage in the Li–Mg–N–H system. J. Alloys Compd. **398**, 235–239 (2005). https://doi.org/10.1016/J.JALLCOM.2005.02.010
230. Y. Liu, Z. Xiong, J. Hu, G. Wu, P. Chen, K. Murata, et al., Hydrogen absorption/desorption behaviors over a quaternary Mg–Ca–Li–N–H system. J. Power Sources **159**, 135–138 (2006). https://doi.org/10.1016/J.JPOWSOUR.2006.04.006
231. T. He, Q. Liu, H. Fan, Y. Yang, H. Wang, S. Zhang, et al., Exploring the effect of ion concentrations on the electrode activity and stability for direct alkaline seawater electrolysis. Int. J. Hydrog. Energy **48**, 19385–19395 (2023). https://doi.org/10.1016/J.IJHYDENE.2023.01.321
232. Thomas, G., Parks, G. *Potential Roles of Ammonia in a Hydrogen Economy a Study of Issues Related to the Use Ammonia for On-Board Vehicular Hydrogen Storage* (U.S. Department of Energy, 2006)
233. S. Chatterjee, R.K. Parsapur, K.W. Huang, Limitations of ammonia as a hydrogen energy carrier for the transportation sector. ACS Energy Lett. **6**, 4390–4394 (2021). https://doi.org/10.1021/ACSENERGYLETT.1C02189/ASSET/IMAGES/LARGE/NZ1C02189_0001.JPEG
234. K.S. Chong, M.A. Hamid, D.K.S. Ng, Systematic decision-making framework for evaluation of process alternatives for sustainable ammonia (NH_3) production. Mater. Today Proc. **64**, A6–A17 (2022). https://doi.org/10.1016/J.MATPR.2022.07.054
235. Discovery of a facile process for H_2 production using ammonia as a carrier (n.d.). https://phys.org/news/2017-04-discovery-facile-h2-production-ammonia.html. Accessed 10 Mar 2024
236. P. Preuster, C. Papp, P. Wasserscheid, Liquid organic hydrogen carriers (LOHCs): Toward a hydrogen-free hydrogen economy. Acc. Chem. Res. **50**, 74–85 (2017). https://doi.org/10.1021/ACS.ACCOUNTS.6B00474/ASSET/IMAGES/MEDIUM/AR-2016-00474U_0009.GIF
237. D. Forberg, T. Schwob, M. Zaheer, M. Friedrich, N. Miyajima, R. Kempe, Single-catalyst high-weight% hydrogen storage in an N-heterocycle synthesized from lignin hydrogenolysis products and ammonia. Nat. Commun. **7**, 1–6 (2016). https://doi.org/10.1038/ncomms13201
238. M. Byun, A. Lee, S. Cheon, H. Kim, H. Lim, Preliminary feasibility study for hydrogen storage using several promising liquid organic hydrogen carriers: Technical, economic, and environmental perspectives. Energ. Conver. Manage. **268**, 116001 (2022). https://doi.org/10.1016/J.ENCONMAN.2022.116001
239. L. Li, M. Yang, Y. Dong, P. Mei, H. Cheng, Hydrogen storage and release from a new promising Liquid Organic Hydrogen Storage Carrier (LOHC): 2-methylindole. Int. J. Hydrog. Energy **41**, 16129–16134 (2016). https://doi.org/10.1016/J.IJHYDENE.2016.04.240
240. M. Jang, Y.S. Jo, W.J. Lee, B.S. Shin, H. Sohn, H. Jeong, et al., A high-capacity, reversible liquid organic hydrogen carrier: H_2-release properties and an application to a fuel cell. ACS Sustain. Chem. Eng. **7**, 1185–1194 (2019). https://doi.org/10.1021/ACSSUSCHEMENG.8B04835/SUPPL_FILE/SC8B04835_SI_001.PDF
241. I. Gordiy, L. Steinbach, I. Frank, Ab-initio molecular dynamics simulation of condensed-phase reactivity: The electrolysis of ammonia and ethanimine in aquatic carbon dioxide solutions. Energies **14**, 6510 (2021). https://doi.org/10.3390/EN14206510/S1
242. Y.F. Wei, Y. Liang, R. Luo, L. Ouyang, Recent advances of Cp*Ir complexes for transfer hydrogenation: Focus on formic acid/formate as hydrogen donors. Org. Biomol. Chem. **21**, 7484–7497 (2023). https://doi.org/10.1039/D3OB01034A
243. A. Malouche, *Metal Doped MIL-101(Cr) for Hydrogen Storage. Material Chemistry* (Université Paris-Est Créteil Val de Marne (UPEC), 2019)
244. M. Yadav, Q. Xu, Liquid-phase chemical hydrogen storage materials. Energ. Environ. Sci. **5**, 9698–9725 (2012). https://doi.org/10.1039/C2EE22937D
245. H.C. Te, W.Y. Chen, D.Y. Tzou, A.K. Roy, H.T. Hsiao, Atomic layer deposition of Pt nanocatalysts on graphene oxide nanosheets for electro-oxidation of formic acid. Int. J. Hydrog. Energy **37**, 17837–17843 (2012). https://doi.org/10.1016/J.IJHYDENE.2012.08.139

246. H.C. Te, H.T. Hsiao, D.Y. Tzou, P.Y. Yu, P.Y. Chen, B.S. Jang, Electro-oxidation of methanol and formic acid on platinum nanoparticles with different oxidation levels. Mater. Chem. Phys. **149–150**, 359–367 (2015). https://doi.org/10.1016/J.MATCHEMPHYS.2014.10.030
247. J. Clayden, S. Nick Greeves, *Organic Chemistry* (OUP, Oxford, 2012)
248. Y.H.P. Zhang, A sweet out-of-the-box solution to the hydrogen economy: Is the sugar-powered car science fiction? Energ. Environ. Sci. **2**, 272–282 (2009). https://doi.org/10.1039/B818694D
249. J.E. Kim, Y.H.P. Zhang, Use of carbohydrates for hydrogen storage. Compend Hydrog. Energy Hydrog. Storage, Distrib. Infrastruct. **2**, 219–241 (2016). https://doi.org/10.1016/B978-1-78242-362-1.00009-2
250. Y.H. Hu, E. Ruckenstein, Clathrate hydrogen hydrate – A promising material for hydrogen storage. Angew. Chem. Int. Ed. **45**, 2011–2013 (2006). https://doi.org/10.1002/ANIE.200504149
251. T. Saikia, S. Patil, A. Sultan, Hydrogen hydrate promoters for gas storage – A review. Energies **16**, 2667 (2023). https://doi.org/10.3390/EN16062667/S1
252. S. Liang, P.G. Kusalik, Communication: Structural interconversions between principal clathrate hydrate structures. J. Chem. Phys. **143**, 011102 (2015). https://doi.org/10.1063/1.4923465/566985
253. J. Cai, Y.Q. Tao, N. von Solms, C.G. Xu, Z.Y. Chen, L.X. Sen, Experimental studies on hydrogen hydrate with tetrahydrofuran by differential scanning calorimeter and in-situ Raman. Appl. Energy **243**, 1–9 (2019). https://doi.org/10.1016/J.APENERGY.2019.03.179
254. H.P. Veluswamy, W.I. Chin, P. Linga, Clathrate hydrates for hydrogen storage: The impact of tetrahydrofuran, tetra-n-butylammonium bromide and cyclopentane as promoters on the macroscopic kinetics. Int. J. Hydrog. Energy **39**, 16234–16243 (2014). https://doi.org/10.1016/J.IJHYDENE.2014.01.054
255. P. Di Profio, V. Canale, R. Germani, S. Arca, A. Fontana, Reverse micelles enhance the formation of clathrate hydrates of hydrogen. J. Colloid Interface Sci. **516**, 224–231 (2018). https://doi.org/10.1016/J.JCIS.2018.01.059
256. Y.J. Fang, Y.M. Xie, X.F. Zhou, L. Yan, S.H. Wang, Influence of activated carbon to the hydrogen storage characteristics of THF hydrate. Adv. Mat. Res. **887–888**, 493–496 (2014). https://doi.org/10.4028/WWW.SCIENTIFIC.NET/AMR.887-888.493
257. A.T. Trueba, I.R. Radović, J.F. Zevenbergen, C.J. Peters, M.C. Kroon, Kinetic measurements and in situ Raman spectroscopy study of the formation of TBAF semi-hydrates with hydrogen and carbon dioxide. Int. J. Hydrog. Energy **38**, 7326–7334 (2013). https://doi.org/10.1016/J.IJHYDENE.2013.03.154
258. N.I. Papadimitriou, I.N. Tsimpanogiannis, A.T. Papaioannou, A.K. Stubos, Monte Carlo study of sII and sH argon hydrates with multiple occupancy of cages. Mol. Simul. **34**, 1311–1320 (2008). https://doi.org/10.1080/08927020802101734
259. A.A. Karimi, O. Dolotko, D. Dalmazzone, Hydrate phase equilibria data and hydrogen storage capacity measurement of the system H_2 + tetrabutylammonium hydroxide + H_2O. Fluid Ph. Equilibr. **361**, 175–180 (2014). https://doi.org/10.1016/J.FLUID.2013.10.043
260. R.K. Zhdanov, K.V. Gets, R.V. Belosludov, O.S. Subbotin, Y.Y. Bozhko, V.R. Belosludov, Theoretical modeling of the thermodynamic properties and the phase diagram of binary gas hydrates of argon and hydrogen. Fluid Ph. Equilib. **434**, 87–92 (2017). https://doi.org/10.1016/J.FLUID.2016.11.030
261. Á. Martín, C.J. Peters, Hydrogen storage in sH clathrate hydrates: Thermodynamic model. J. Phys. Chem. B **113**, 7558–7563 (2009). https://doi.org/10.1021/JP8074578
262. O. Dolotko, D. Dalmazzone, A.A. Karimi, The phase behaviours in mixed G + tetra-n-butylphosphonium borohydride and G + tetra-n-butylammonium hydroxide hydrates (where G = H2, N2). (2011), 10

263. T. Tsuda, K. Ogata, S. Hashimoto, T. Sugahara, M. Moritoki, K. Ohgaki, Storage capacity of hydrogen in tetrahydrothiophene and furan clathrate hydrates. Chem. Eng. Sci. **64**, 4150–4154 (2009). https://doi.org/10.1016/J.CES.2009.06.018
264. T.Y. Wei, K.L. Lim, Y.S. Tseng, S.L.I. Chan, A review on the characterization of hydrogen in hydrogen storage materials. Renew. Sustain. Energy Rev. **79**, 1122–1133 (2017). https://doi.org/10.1016/J.RSER.2017.05.132
265. D. Gygi, E.D. Bloch, J.A. Mason, M.R. Hudson, M.I. Gonzalez, R.L. Siegelman, et al., Hydrogen storage in the expanded pore metal-organic frameworks M2(dobpdc) (M = Mg, Mn, Fe, Co, Ni, Zn). Chem. Mater. **28**, 1128–1138 (2016). https://doi.org/10.1021/ACS.CHEMMATER.5B04538/SUPPL_FILE/CM5B04538_SI_007.CIF
266. M. Mehrabi, P. Parvin, A. Reyhani, S.Z. Mortazavi, Hybrid laser ablation and chemical reduction to synthesize Ni/Pd nanoparticles decorated multi-wall carbon nanotubes for effective enhancement of hydrogen storage. Int. J. Hydrog. Energy **43**, 12211–12221 (2018). https://doi.org/10.1016/J.IJHYDENE.2018.04.144
267. Y. Honarpazhouh, F.R. Astaraei, H.R. Naderi, O. Tavakoli, Electrochemical hydrogen storage in Pd-coated porous silicon/graphene oxide. Int. J. Hydrog. Energy **41**, 12175–12182 (2016). https://doi.org/10.1016/J.IJHYDENE.2016.05.241
268. S.K. Konda, A. Chen, Palladium based nanomaterials for enhanced hydrogen spillover and storage. Mater. Today **19**, 100–108 (2016). https://doi.org/10.1016/J.MATTOD.2015.08.002
269. A. Eftekhari, B. Fang, Electrochemical hydrogen storage: Opportunities for fuel storage, batteries, fuel cells, and supercapacitors. Int. J. Hydrog. Energy **42**, 25143–25165 (2017). https://doi.org/10.1016/J.IJHYDENE.2017.08.103
270. S. Karimzadeh, B. Safaei, T.-C. Jen, Investigation on electrochemical performance of striped, β12 and χ3 Borophene as anode materials for lithium-ion batteries. J. Mol. Graph. Model. **120**, 108423 (2023). https://doi.org/10.1016/J.JMGM.2023.108423
271. F. Wang, X. Wu, X. Yuan, Z. Liu, Y. Zhang, L. Fu, et al., Latest advances in supercapacitors: From new electrode materials to novel device designs. Chem. Soc. Rev. **46**, 6816–6854 (2017). https://doi.org/10.1039/c7cs00205j
272. M. Kaur, K. Pal, Review on hydrogen storage materials and methods from an electrochemical viewpoint. J. Energy Storage **23**, 234–249 (2019). https://doi.org/10.1016/J.EST.2019.03.020
273. Y. Dabaki, C. Khaldi, N. Fenineche, O. ElKedim, M. Tliha, J. Lamloumi, Electrochemical studies on the Ca-based hydrogen storage alloy for different milling times. Met. Mater. Int. **27**, 1005–1024 (2021). https://doi.org/10.1007/S12540-019-00496-9/FIGURES/16
274. H. Gnaser, SIMS detection in the 1012 atoms cm^{-3} range. Surf. Interface Anal. **25**, 737–740 (1997). https://doi.org/10.1002/(SICI)1096-9918(199709)25:10%3C737::AID-SIA294%3E3.0.CO;2-M
275. F.A. Stevie, C. Zhou, M. Hopstaken, M. Saccomanno, Z. Zhang, A. Turansky, SIMS measurement of hydrogen and deuterium detection limits in silicon: Comparison of different SIMS instrumentation. J. Vac. Sci. Technol. B Nanotechnol. Microelectron. Mater. Process. Meas. Phenom. **34** (2016). https://doi.org/10.1116/1.4940151/593420
276. Z. Zhu, V. Shutthanandan, M. Engelhard, An investigation of hydrogen depth profiling using ToF-SIMS. Surf. Interface Anal. **44**, 232–237 (2012). https://doi.org/10.1002/SIA.3826
277. M. Nojima, M. Toi, A. Maekawa, B. Tomiyasu, T. Sakamoto, M. Owari, et al., Evaluation of the nano-beam SIMS apparatus. Appl. Surf. Sci. **231–232**, 930–935 (2004). https://doi.org/10.1016/J.APSUSC.2004.03.176
278. C. Zlotea, Y. Oumellal, S.J. Hwang, C.M. Ghimbeu, P.E. De Jongh, M. Latroche, Ultrasmall MgH_2 nanoparticles embedded in an ordered microporous carbon exhibiting rapid hydrogen sorption kinetics. J. Phys. Chem. C **119**, 18091–18098 (2015). https://doi.org/10.1021/ACS.JPCC.5B05754/SUPPL_FILE/JP5B05754_SI_001.PDF

279. K. Alsabawi, T.A. Webb, E.M. Gray, C.J. Webb, The effect of C60 additive on magnesium hydride for hydrogen storage. Int. J. Hydrog. Energy **40**, 10508–10515 (2015). https://doi.org/10.1016/J.IJHYDENE.2015.06.110
280. D.K. Ross, Neutron scattering studies for analysing solid-state hydrogen storage. Solid-State Hydrog. Storage Mater. Chem., 135–173 (2008). https://doi.org/10.1533/9781845694944.2.135

Chapter 5
Analysis and Characterization

Procedures for identifying, isolating, or quantifying substances or materials as well as describing their physical characteristics are known as analysis and characterization. Characterization, in the context of materials science, is the exhaustive and all-inclusive process of measuring and probing the structure and properties of a material. Without it, it would be impossible to develop a scientific understanding of engineering materials. It is a decisive step in the mechanism of studying materials. Characterization techniques include examining a material's intrinsic composition or determining its physical characteristics. To enlarge or view interior structure and apprehend the allotment and synergies of elements within the specimen, sophisticated microscopy or elemental analysis are frequently used. We can better predict how a material will function in a given application or how it might be modified to suit a particular usage by fully comprehending a material's qualities at the elemental level [1, 2]. A wide range of properties, including mechanical strength, thermal stability, electrical conductivity, and corrosion resistance, can be examined using advanced material characterization techniques. Understanding the material properties of various components in a system allows engineers to optimize their designs to increase performance and dependability [3].

Recently, Atomic layer deposition (ALD) technology has been utilized to synthesize a series of tandem catalysts with several metal-oxide contacts and high-capacity metal-based catalysts, single crystal atoms, etc., for hydrogen (H_2) generation, purification, and storage research. These materials are often analyzed and characterized to probe the process viability and synthesizing materials' structure and properties using analytical means such as surface morphology, growth and uniformity, structural characterization, materials' density, materials' composition, and electrical and optical properties.

P. O. Oviroh et al., *Emerging Atomic Layer Deposition for Hydrogen Energy*,
Springer Series in Materials Science 348,
https://doi.org/10.1007/978-3-031-67774-8_5

5.1 Surface Morphology

Surface morphology is a subset of analytical imaging, which employs high-end microscopes to create pictures of goods, materials, and objects that are invisible to the unaided eye. Analytical imaging is an advanced kind of high spatial resolution imaging. These photos are from the sample or product's exposed surface [4–6]. By studying the small morphological structure on a substance's surface, electron microscopy instruments are frequently used to assess dispersibility, sizes of particle, and distribution of the materials. Field Emission Scanning Electron Microscopy (FE-SEM), Scanning Electron Microscopes (SEM), high-resolution transmission electron microscopes (HRTEM), and scanning tunnelling microscopes (STM), and others are often used for surface morphology characterization of atomic layer deposition (ALD)-prepared samples [7].

5.1.1 Scanning Electron Microscopes (SEM)/Field Emission Scanning Electron Microscopy (FE-SEM)

A beam of electrons is used in the topographic method of SEM to reveal the surface shape of objects. Materials' structural characteristics are revealed via the electron's interactions with samples [8]. Figure 5.1 provides an illustration of the SEM's core idea. An object's surface is scanned by the main electron beam, which is generated at the apex of the microscope under high vacuum via heating a metallic element. A surface picture or an elemental composition of the specimen's surface can be created by detecting certain signals that are produced when the electrons impact the sample. The backscattered electrons, X-rays, and secondary electrons are the three signals in SEM that yield the most data [6].

The top surface layer's atoms release secondary electrons that create a picture of the surface. The surface morphology influences the image's contrast. Because the main electron beam's diameter is so tiny, a high-resolution picture may be produced. Primary beam electrons that are backscattered are those that are mirrored by the atoms in the surface covers. The atomic number of the elements in the facet layers

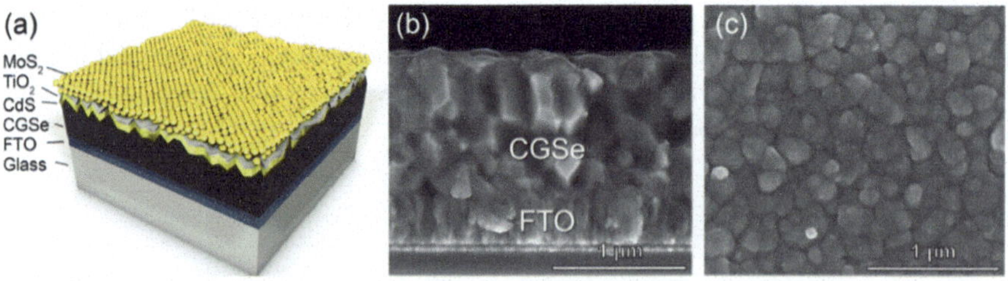

Fig. 5.1 SEM image of MoS_2/TiO_2/CdS/CGSe photocathode: (**a**) schematic, (**b**) cross-sectional SEM image, and (**c**) top view. (Reproduced with permission from Ref. [17])

affects the contrast of the picture that is created. Thus, the dispensation of various chemical phases on the samples' surface will be depicted in the photograph. The imaging resolution for backscattered electrons is typically not as excellent as for secondary electrons since they are emitted from a depth in the material. An X-ray is released as a result of electron shell transitions that are brought on by the primary beam's contact with the specimen's atoms. The energy of the emitted X-ray is consistent with that of its parent element [9–11]. For sample examination in SEM, backscattered electrons and secondary electrons are utilized. In classical SEM, the covering of samples with conductive layers is favored. Recent SEM methods, nevertheless, employ low-intensity electron beams and do not require coating of samples [8]. It should be emphasized that signal detection relies heavily on the gaseous atmosphere within the specimen chamber [11–13]. The secondary electrons strike the gas molecules as they go toward the positively charged detector. An enhanced cascade of electrons travels to the detector as a result of each such collision producing an extra daughter electron. Positive ions are further generated along with extra daughter electrons. The charge buildup at the surface of nonconductive specimens like wood is therefore counteracted by these positive ions as they migrate down toward the specimen. Because of this, it is not necessary to cover nonconductive specimens with a metallic coating to avert charging the components and the ensuing impairment of picture quality.

As a result of the development of Field Emission Microscopy (FEM) by a German physicist called Erwin Muller in 1936, a new technique known as FE-SEM emerged as a result of advances in the development of technologies utilized in SEM, a form of electron microscope [14]. The operation of FE-SEM is similar to that of conventional SEM. The process involves focusing an electron beam by electromagnetic lenses that scan a specimen's surface, and the interacting and reflected electrons generate an image of the specimen's surfaces and topography [14, 15]. The electron-producing procedures is the primary disparity between FE-SEM and SEM, yet FE-SEM largely follows similar principles as SEM. FE-SEMs use a Field Emission Gun (FEG) as their electron source, while thermionic emission is utilized in SEM; FEGs discharge the electron beam using a potential gradient [16].

The surface morphology of prepared catalyst by ALD can be examined effectively by the utilization of SEM. Hellstern et al. (2019) described the SEM micrograph of prepared molybdenum disulfide (MoS_2) catalytic coatings by solar production of hydrogen via copper gallium diselenide (CGSe) photocathodes; they reported that MoS_2/TiO_2 can function as an efficient catalytic mechanism for CdS/CGSe photocathodes. They concluded that the advancement of potential tandem methods for unsupported water splitting would be facilitated by increases in the performance and longevity of CdS/CGSe photocathodes (Fig. 5.1) [17]. In a similar investigation, Scheffe et al. (2011) published SEM micrograph of cobalt ferrites ($Co_xFe_{3-x}O_4$) and iron oxide (γ-Fe_2O_3) on high surface-area (50 m^2 g^{-1}) m-ZrO_2 supports created by ALD, for hydrogen production through chemical looping redox cycles. They confirm the steadiness of the m-ZrO_2 microstructure at ALD processes, with no pore clogging (Fig. 5.2) [18].

Fig. 5.2 SEM image of bulk m-ZrO_2 supports (**a**) at low magnification and (**b**) within the bulk support. (Reproduced with permission from Ref. [18])

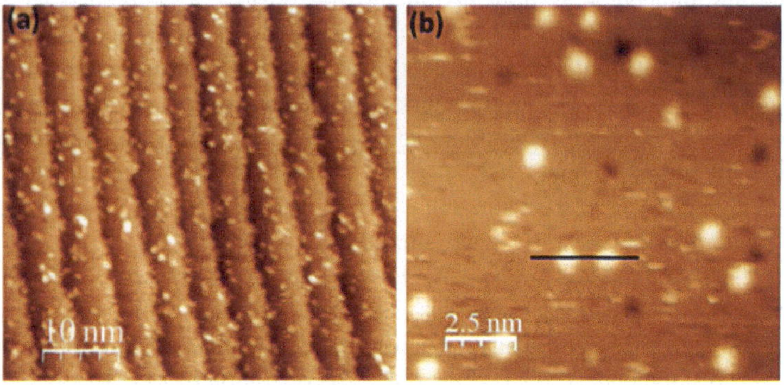

Fig. 5.3 STM images of Pt (111) after exposure to 15 L $FeCp_2$ at 298 K followed by annealing in 1×10^{-6} mbar O_2 at 473 K, at scanning parameters of (**a**) 0.76 V and 0.32 nA and (**b**) 0.77 V and 0.33 nA. (Reproduced with permission from Ref. [19])

5.1.2 Scanning Tunnelling Microscopes (STM)

STM is a characterization technique that has significantly influenced nanoscience and nanotechnology. Since its development by Rohrer and Binnig, who were co-awarded the 1986 Nobel Prize in Physics, STM has undergone significant changes that have made it possible to see nanostructures at atomic size for the first time [16–18]. The foundation of STM is a quantum mechanical process known as tunnelling. In particles of tiny size, electrons can "penetrate" potential barriers because they have wave-like properties in quantum mechanics, which makes quantum mechanical probability processes based on classical Newtonian mechanics impossible. STM typically entails bringing a particularly pointed conductive tip to a conductive sample's interface configuration that is within a tunneling proximity (sub-nanometer) [16].

Using a unique ALD technique, Wang et al. (2022) successfully created an inverted single-site $Fe_1(OH)_X$/Pt (111) model catalyst for the selective oxidation of CO in H_2. STM (Fig. 5.3) verified that the proposed single-crystal surface had been

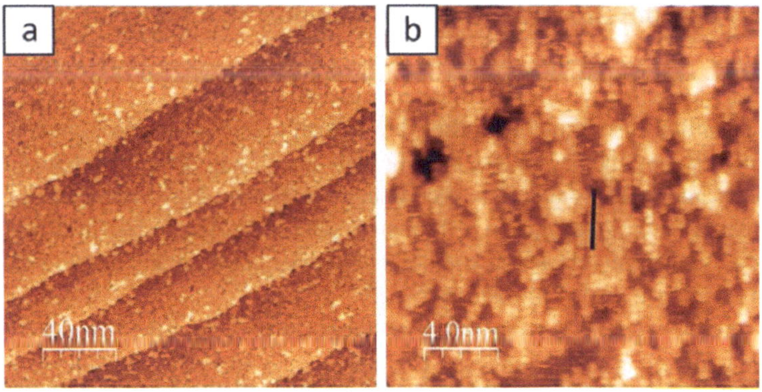

Fig. 5.4 STM images of Pd (111) surfaces after exposure to TMA at 200 °C for ~3000 L: (**a**) low and (**b**) high magnifications. (With permission from Ref. [20])

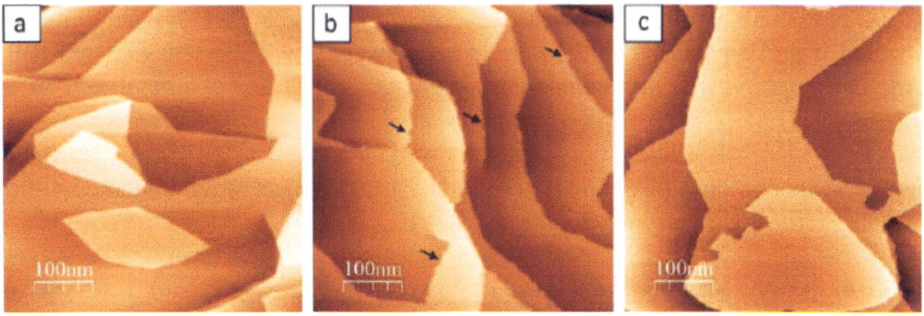

Fig. 5.5 STM images of Cu (111) surface: (**a**) clean, (**b**) after exposing to TMA, (**c**) after sequentially exposing water and TMA for 200 and 3000 L at 200 °C, respectively, $V = -1$ V, $I = 1$ nA. The black arrows emphasize the corrugated step edges. (With permission from Ref. [20])

obtained. STM was used to examine the ferrocene ($FeCp_2$) adsorption mechanisms on Pt (111) at progressively higher coverages. Their findings suggest that such single interfacial sites exhibit greater intrinsic capacity than typical metal–oxide interfaces because of the remarkable flexibility of the single-atom iron catalyst in changing states of oxidation, which was not detected for iron oxide nanoparticles under comparable reaction circumstances [19].

In another study using the ALD procedure, Lu et al. (2014) reported the combined measurements from in situ quartz crystal microbalance (QCM), quadrupole mass spectroscopy (QMS), and STM (Figs. 5.4 and 5.5) to examine Al_2O_3 utilizing trimethylaluminum (TMA) and water on Pd, Ir, Pt, and Cu surfaces. They reported that clean transition metal surfaces are susceptible to TMA dissociative chemisorption of to create Al* or $AlCH_3$* and the ensuing full hydration into $Al(OH)_3$* during water exposures (the asterisk (*) designates a surface species) [20].

5.1.3 Transmission Electron Microscopes (TEM)

One of the most effective microscopic techniques is called TEM, and it is frequently used in scientific disciplines including biological, chemical sciences, nanotechnology, materials, and medicinal research. It made use of a high-energy electron to see the samples and collect data on their composition, morphology, and crystallography [5]. Imagery of atomic resolution is frequently obtained using TEM. In TEM, a plane electron wave that travels across the object changes in phase and amplitude illuminates an electron-transparent material.

The imaging lens system's lens aberrations have a significant impact on the subsequent image creation process. The specimen's structure, thickness, and elemental makeup, as well as the selected imaging settings, all have a significant role in the possible picture contrast [21]. Very tiny objects are needed for high-resolution (HR-TEM) imaging, and depending on their characteristics, they may only slightly alter the electron wave's phase while leaving the amplitude mostly unaltered [22]. Phase contrast often referred to as weak-phase objects (WPOs) may only be attained if an extra phase shift is produced between the dispersed scattered and unshattered portions of the component wave—ideally, a phase shift of $\pi/2$. In a traditional TEM, a phase shift like this is produced by taking advantage of the imaging system's lens aberrations. Equation (5.1)'s phase contrast transfer function (PCTF) explains how phase information data in the object wave underneath the sample is converted into image intensities [23].

$$\mathrm{PCTF} \sim \sin\left({}^{\pi}\!/_{2}\left(2\lambda\,\Delta f u^2 + \lambda^3 C_s u^4 + \varphi_{\mathrm{pp}}\right)\right) \tag{5.1}$$

The PCTF must be adjusted to the object's spatial frequency (SF) spectrum to maximize phase contrast, and is dependent on the electron wavelength λ, the SF μ, the objective lens's spherical aberration coefficient C_s, and the defocus value Δf. In microscopes lacking a physical phase-plate (PP), the extra phase shift φ_{PP} equals zero and is addressed below. The phase shift that the imaging lens system's aberrations cause to the object wave is described by the sine-function argument in the PCTF. The argument must consequently be close to $\pm\ \pi/2$ in order to produce phase contrast of WPOs, which results in PCTF values close to ± 1 and substantial contrasting transmission [21–23]. To achieve the highest contrast, HR-TEM imaging frequently uses Scherzer-defocus Δf_S [24–26], which substantially corrects for the effects of spherical aberration and generates a wide range of SF, where significant phase variation is found.

Kim et al. (2021) used ALD methods to guarantee the deposition of a conformal coating of Al_2O_3 onto Ni nanoparticles anchored on SiO_2, resulting in the enhancement of stable Ni-based dried reforming methane catalysts. They reported that the Ni nanoparticles were partially enclosed by an amorphous (grown-ALD) layer and partially immersed in the support. The amount of active (and available) Ni active sites will therefore decrease as the thickness of the grown-ALD layers of Al_2O_3

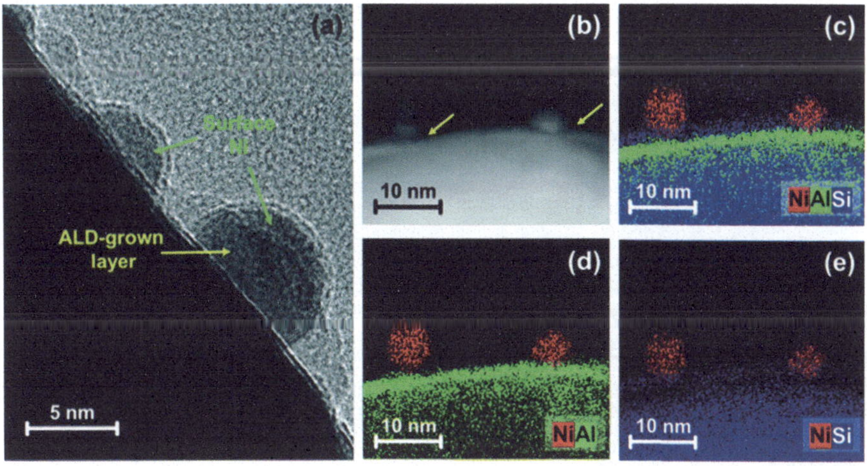

Fig. 5.6 Electron microscopy-based morphological characterization: (**a**) HR-TEM and (**b**) HAADF-STEM images with (**c–e**) EDX mapping for reduced Al (4.0 nm)–Ni/SiO_2. The arrows represent the ALD-grown layer. (With permission from Ref. [19])

increases from initial 0.4 nm to a thickness of 4.0 nm (Fig. 5.6), which is consistent with a drop in the amount of chemisorbed H_2 as the layer's thickness increases [27].

In another study, Ramesh et al. (2019) prepared a 3D Ni foam using the ALD process on a molybdenum nitride (MoN_x) thin film as effective catalysts for the hydrogen evolution reaction (HER). They reported that by sequentially exposing NH_3 and $Mo(CO)_6$ at 225 °C, thin films of MoN_x is proficiently produced on an nickel foam substrate via the ALD process. The characterization analysis revealed that the prepared MoN_x thin film was primarily amorphous, couple with MoN nanocrystalline grains (4 nm) distributed thoroughly, according to TEM examination in Fig. 5.7. Additionally, a rough outer surface of the MoN_x thin film with an aggregate chemical makeup of $Mo_{0.59}N_{0.41}$ is seen in the high-resolution (HR)-TEM examination [28].

5.1.4 Atomic Force Microscopy (AFM)

The adaptable microscopic method known as AFM was invented in 1986 by Gerd Binning, Calvin Quate, and Christoph Gerber [29]. This technology will not only enable nanometer-scale surface visualization of the material but also a variety of surface measurements [30]. In furtherance of facilitating the examination of solid–solid interaction at an unheard-of microscopic dimension, the AFM offers a universal technique for performing nondestructive surface profilometry at a higher than 10 nm resolution and potentially lower to the atomic level [31].

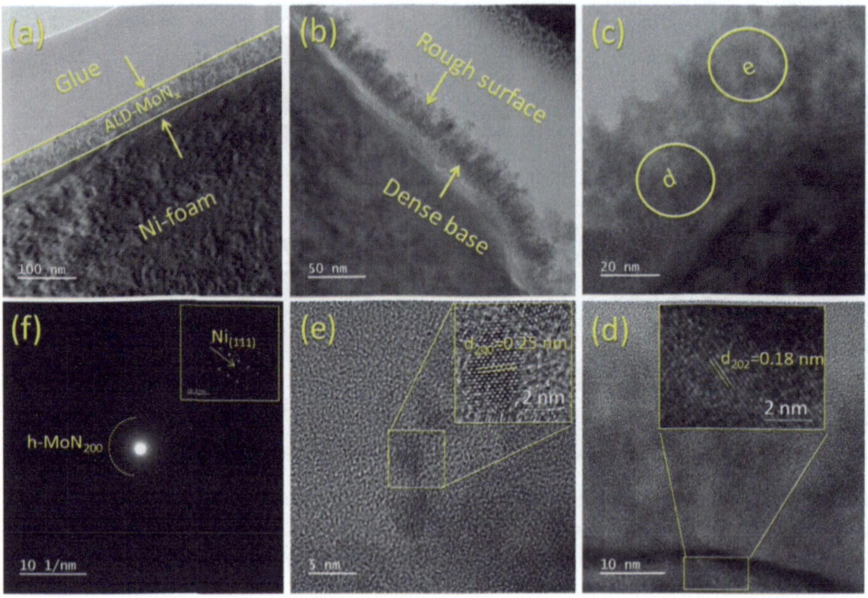

Fig. 5.7 (**a** − **c**) Cross-sectional view of MoN_x/NF (700) composite TEM bright-field images at different magnifications; (**d**, **e**) HR images at location; (**c**, **f**) its selected-area electron diffraction patterns. The insets in (**d**, **e**) are the magnified HR images to show the nanocrystalline grains, and the inset in (**f**) is the selected-area electron diffraction of the NF substrate. (With permission from Ref. [28]).

One of the most alluring aspects of the AFM approach is that practically any specimen, from flexible polymers, solid ceramic, metallic objects, or delicate biomolecules, may be employed for surface research. It also offers the benefit of probing the sample's surface without the need for specific preparation, which is necessary with related methods like SEM and TEM [30]. An AFM instrument is made up of the scanner, a laser beam, a position sensitive photodiode (PSPD), and a sharp tip/probe cantilever [32]. A pointed tip or probe is used to scan the surface of the material in an AFM, which functions similarly to a scanning probe microscope (SPM) [32]. The cantilever experiences a tiny bending or deflection as the probe moves in a raster pattern over the suppose sample, interacting with it in magnetic, electrostatic, capillary, and Van der Waals ways. A laser beam detects this defection, which is accurately recorded by the PSPD, and a three-dimensional surface profile of the suppose sample is produced [33].

Hussain et al. (2020) utilized ALD and magnetron sputtering to create a highly durable Niobium (Nb)-doped titanium oxide (TiO_2) thin film as a capable support for Pt catalyst. Fabricated Pt/Nb-TiO_2 catalyst was characterized by means of AFM as shown in Fig. 5.8. They reported a granular structure of the coating on the glassy carbon substrate and observed an island-like morphology of the deposited catalyst on titanium (Ti)-substrate electrodes. However, the measured root-mean square (RMS) roughness values were obtained as 3.9 and 4.6 nm, which did not differ

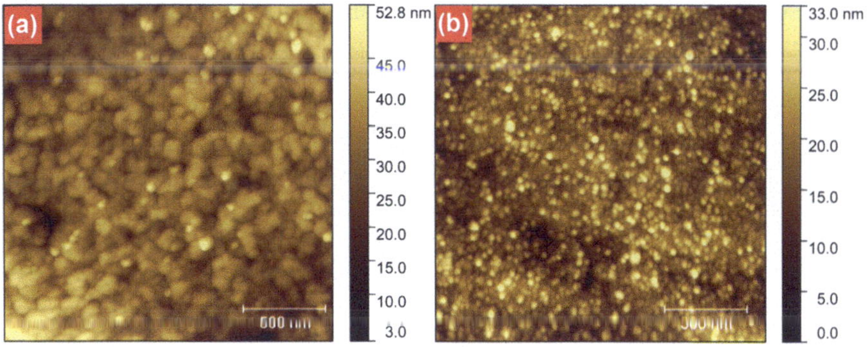

Fig. 5.8 AFM images of (**a**) Pt/Nb-TiO_2(Ti) and (**b**) Pt/Nb-TiO_2(GC) samples. (With permission from Ref [34])

significantly regardless of the differences in the thin film structure between the Ti-substrates and glassy carbon [34].

5.2 Growth and Uniformity

ALD process is based on successive self-saturated surface reactions, resulting in the molecularly controlled layer-by-layer development of thin films. When the highly reactive precursors are kept apart from one another in the gas phase, self-saturated growth is achieved [35]. The self-saturating growth mechanism produces all desirable film features, such as conformality, film quality, and thickness uniformity. Because the uniform distribution of the active sites on the substrate is necessary for uniformity, thickness uniformity becomes independent of the method used to introduce the precursors into the substrate [36]. Although ALD films typically exhibit great uniformity, there are a number of factors that could cause the uniformity to decline as the film is being deposited. Overlaying material pulses, uneven gas dispersion, uneven precursor dispersion in the conduit gas, and thermal precursor self-decomposition are the three most frequent causes. The impact of gaseous by-products on ALD growth is one understudied source of nonuniformity. Gaseous by-products arise in the surface reaction because ALD growth often results from exchange reactions between precursors and the surface reactive sites; these by-products may become a significant factor if they react with the surface sites [35–37].

In a single cycle, there are four sequential processes in film growth using ALD. To promote the adsorption of chemical over the substrate, the first step involves pulsed heating of a vaporized metal precursor within a reactor chamber with a harsh vacuum condition. The second stage is a purging procedure that clears the reactor compartment of any excess or unreacted precursor. A coreactant containing oxygen, such as water (H_2O), ozone (O_3), or oxygen (O_2), is pulsed in the third stage to create oxides, which then react with the precursor and adsorbed on the substrate to create a metal oxide coating. The last step is to remove excess and leftover amounts of both pulsed reactant substances [38, 39]. The ALD process offers superior thickness control, conformality, film quality, and thickness uniformity than other deposition methods thanks to its distinctive, typically self-limiting, growth mechanism. To measure the overall thickness of a thin film deposited by the ALD process, the spectroscopic ellipsometer is utilized [35, 40].

5.2.1 Spectroscopic Ellipsometer

A potent method for measuring thin and ultra-thin films (down to <1 Å) with a single or few distinct layers is spectroscopic ellipsometry (SE). This method can be used to analyze almost any transparent thin film and is incredibly sensitive to film thickness and homogeneity. Layers as thin as the wavelengths of the probe light itself, less than a single atomic layer, can be measured beyond the thickness limits of equivalent ellipsometry and reflectometry-based techniques and is especially helpful for ultra-thin film applications (<100 nm) [41–44].

Ellipsometry (also known reflection ellipsometry) investigates the change in the state of polarizations of the light that is reflected from a surface. Ellipsometry is sensitive to submonolayer surface covering because variations are assessed rather than the absolute intensity of light. In situ research can benefit from using ellipsometry because it is a nondestructive, noninvasive technology that just needs a low-power light source and does not interfere with most processes [45]. In this procedure, a light source generates electromagnetic radiation that is then linearly polarized by a polarizer, passed via a quarter-wave plate or another alternative compensator, and then irradiated the specimen. After reflection, the radiation passes via an analyzer and a second polarizer (optional) prior to being detected [43, 46–48].

The optical properties of several materials, including their extinction coefficient (k) and refractive index (n) or alternatively, their absorption coefficient, were identified using the earliest ellipsometry measurements. The capability to precisely gauge the thickness of some thin films utilizing a single-wave length nulling ellipsometry measurement was discovered in the 1940s. Since then, single wave-length ellipsometry has developed into the industry standard for measuring thickness, notably in the semiconductor sector. Ellipsometry experimentations generate assessments that are not helpful on their own; instead, useful variables like the optical properties or thickness of thin film for the materials must be obtained using computers. Modern computers have evolved to the enhancement of a number of spectroscopic ellipsometers and more realistic analytical tools that are necessary to comprehend SE data [49–52]. One of the crucial instruments for measuring thickness of thin film fabricated by ALD is the SE. Previous studies have reported the characterization of ALD-prepared thin films using SE.

In a study, Müller et al. (2019) fabricated ultrathin Bi_2O_3 films by utilizing [Bi $(tmhd)_3$] as a precursor for bismuth, via thermal and plasma-enhanced ALD, for

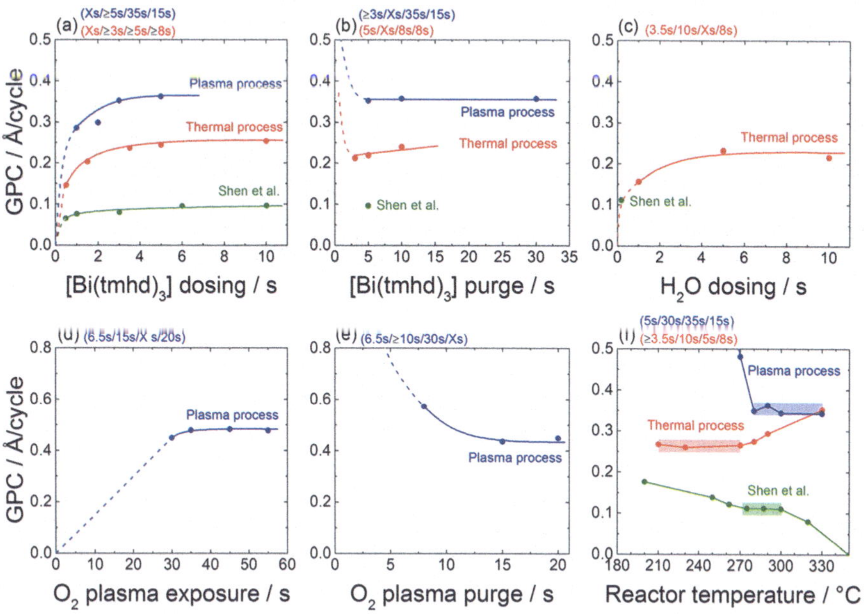

Fig. 5.9 The GPC of Bi_2O_3 films as a function of the ALD process parameters of the plasma-enhanced process (blue), the thermal process (red), and as reported by Shen et al. 28 (green): (**a**) [Bi (tmhd)3] exposure time, (**b**) [Bi(tmhd)3] purge time, (**c**) water exposure time, (**d**) oxygen plasma exposure time, (**e**) oxygen plasma purge time, and (**f**) deposition temperature (temperature windows are indicated by the semitransparent rectangles). (Reproduced with permission from Ref [53])

photocatalytic water splitting. The ultrathin film fabrication growth per cycle (GPC) was monitored with real-time via SE. They reported that, as the temperature window shifts from 210 °C–270 °C to 270 °C–330 °C and the GPC increases from 0.24 to 0.35 when an O_2 plasma is used as the oxidant instead of water. The oxygen-rich, thick, amorphous, conformally formed films that were deposited have bandgaps that fall between 2.7 and 2.9 eV. The enhanced GPC was explained by real-time SE, which demonstrates the plasma's improved reactivity toward the Bi $(tmhd)_x$ surface group [53]. Figure 5.9 shows the GPC of Bi_2O_3 films as a function of the ALD process parameters, the initial 20 cycles of the plasma and thermal-enhanced ALD processes are depicted in Fig. 5.10 demonstrating the progression of the apparent thin film thickness.

In a similar study, Kilic et al. (2020) discuss how in situ ellipsometer with a few seconds time resolution was used to look at how the characteristics of transition metal oxides thin films changed during the ALD growing processes. They reported that with the ability to distinguish the evolution of surface roughness and layer thickness throughout distinct phases of specific cycles, in situ ellipsometer enables the time-dependent monitoring of precursor and surface interactions in plasma-enhanced ALD of TMO films [54]. In a recent study, Labbe et al. (2022) utilized the SE techniques to characterize the air-stable and inexpensive precursor,

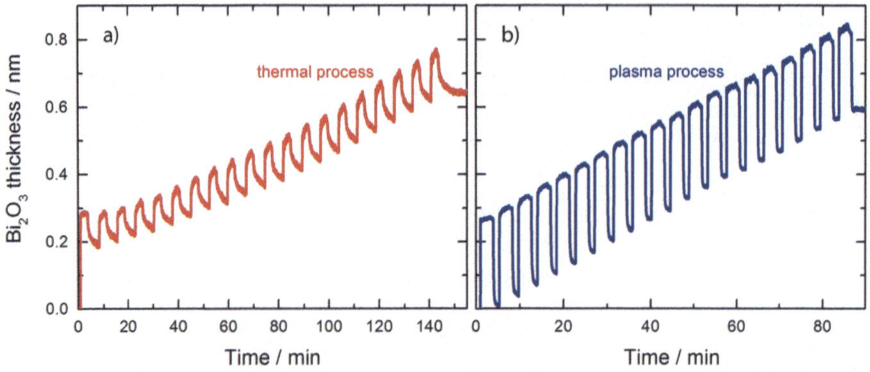

Fig. 5.10 Bi_2O_3 thickness determined during (**a**) thermal- and (**b**) plasma-enhanced ALD growth. (Reproduced with permission from Ref. [53])

$FeC_5H_4C_2H_5$ (ethylferrocene), used in the ALD fabrication of FeO_x thin films. They reported that the FeOx deposition showed a distinctive development process in which numerous layers of islands gradually nucleated, and grew into a continuous thin film with pinholes. They concluded that after prolonged ALD cycles, linear growth was seen with a mixture of substrate-inhibited and enhanced growth [55].

5.3 Structural Characterization

Characterization, in the context of materials science, is the exhaustive and comprehensive procedure of probing and assessing a material's properties and structure. Devoid of this, it would be impossible to develop a scientific understanding of how materials are created. It is a vital phase in the progression of studying materials [56]. The crucial technique like X-Ray Diffraction (XRD) and Raman spectroscopy can all be used to determine the crystallinity, interplanar distance, particle sizes, percentage crystallinity phase, molecular interactions, and other structural characteristics of materials in a nondestructive manner.

5.3.1 X-Ray Diffraction (XRD)

XRD is a flexible nondestructive method of examination used to study the physical properties of powder, and solid materials, notably their crystal structure, orientation, and phase composition [57].

XRD analysis determines the short-range hierarchy of noncrystalline materials and the long-range order (i.e., structure) of materials that are crystalline. Information obtained can be used to deduce the crystallinity level, phases, typical grain size, lattice constants, and crystal defects. Details of crystalline symmetry, strain, electron density, and roughness can also be obtained by enhanced XRD [58]. Additionally, XRD makes it possible to analyze how humidity, pressure, and temperature affect the specimens, adding to the technique's adaptability by extending its range of applications and creating new research opportunities [59].

Max von Laue and Paul Ewalds first proposed the concept of XRD in 1912 via an experiment that demonstrated the "wave nature" of X-rays, and the presence of "space lattices" in crystals [60]. Max von Laue published his research on XRD of crystals as in a 3-D diffraction grating on June 8, 1912 [61]. Similarly, William Henry Bragg and Lawrence Bragg used their findings to support Barlow's hypothetical rock salt model, which is how the XRD investigation of a single crystal was started [62]. Building an accurate X-ray spectrometer and creating the ionization chamber for a more user-friendly quantitative estimation of X-ray intensity were two other key contributions made by William Henry Bragg to the advancement of XRD [63]. A monograph on the hypothesis of XRD for discovering novel structures was first released in 1955 by Alexander Kitaigorodsky and continues to be in use today. Every year, a sizable number of new structures are discovered, creating a vast database of materials' databases for quantum chemistry, crystal chemistry, crystallography, and additional procedures intended to comprehend and describe the atomic structure of substances [61–63]. While it is simple to produce X-rays with wavelengths that match the size of crystal unit cells, diffraction investigations on crystals can also be performed using electrons or neutrons with the proper energy. The fact that single cells have several Å dimensions demonstrates that crystals with larger micron sizes are produced by the periodic repetition of billions of cell units in all 3-D. Because XRD is strongly dependent on the quality of the diffraction properties and the firmly ordered periodicity of the atoms, any deviancy from the ideal order will be seen in the XRD pattern. A toms can be replaced by other atoms, slightly deviate geometrically from their ideal positions due to internal strain, or bigger 2- or 3-D aberrations can cause deviations from the ideal order [64].

Although powder X-ray diffraction (PXRD) provides a wider range of applications than single crystal analysis (SCXRD), PXRD of multicomponent materials results in significantly greater peak overlap, which obscures the location and intensities of the diffraction maxima [65, 66]. When X-rays are impacted on a large number of tiny crystallites in powder, the individual beams of intensity split into cones, and the detector's tiny spots produce rings of varied intensities. To identify crystals, we compare the peak intensity and d-spacing against the JCPDS collection [58]. The position of the diffraction peaks, which are determined by Bragg angle (θ) or interplanar spacing d_{hkl}, and their intensity are unique to each molecule, making the XRD a fingerprint of the atomic dispersion of a given substance [60]. Researchers regularly use XRD because it is able to quickly identify the texture, crystallite size, level of stress, crystal structure, and degree of crystallization of materials. XRD, as a characterization tool, has been utilized in microelectronics, forensic science, geology, glass manufacturing, corrosion analysis [68], geopolymers [69, 70], energy generation [67], environment [71–73], pharmaceutics [68], biopolymers [67, 69–71], materials science and catalysis [72–75].

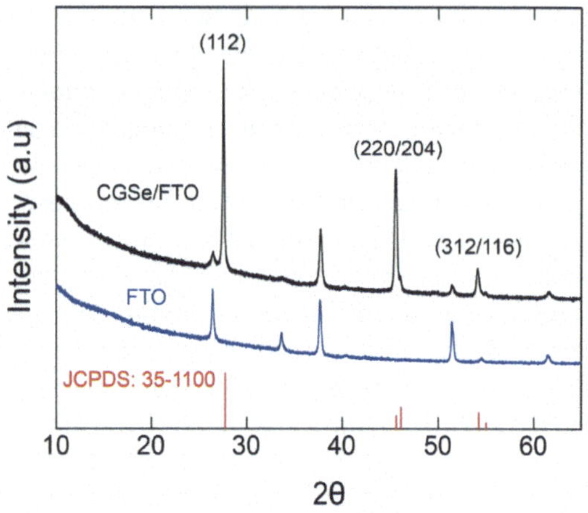

Fig. 5.11 X-ray diffractograms of CGSe/FTO (black) and FTO (blue). (Reproduced with permission from Ref. [17])

Hansen et al. (2011) prepared a substoichiometric TiO_2 25 nm thin film grown via ALD, as a catalyst support for H_2 fuel cells. They characterized the prepared thin films by means of XRD and reported a Magnéli-phase Ti_4O_7 with excellent electrical conductivity that makes them desirable as prospective support materials for fuel cells [76]. In a similar study, Lee et al. (2011) prepared a H_2 gas sensor using arrays of TiO_2 nanotubes using anodic aluminum oxide (AAO) as a template in a template-assisted technique via ALD. The phase composition and crystal structure of the prepared TiO_2 nanotubes were characterized using XRD. They reported that XRD analysis of as-prepared TiO_2 showed only an amorphous halo phase and TiO_2 nanotubes' crystalline anatase phase was observed after an annealing process was carried out [77].

Hellstern et al. 2019 [17] reported the XRD micrograph of prepared molybdenum disulfide (MoS_2) catalytic coatings for solar H_2 production via copper gallium diselenide (CGSe) photocathodes using the ALD process. The chalcopyrite crystal phase in prepared CGSe electrodes was confirmed by XRD analysis, as illustrated in Fig. 5.11. In the preparation process, the substrate was soda lime glass covered with 400 nm fluorine-doped tin oxide (FTO). They reported that MoS_2/TiO_2 can aid as an effectual catalytic strategy for CdS/CGSe photocathodes and concluded that the development of promising tandem methods for unsupported H_2O splitting would be facilitated by increases in the activity and longevity of CdS/CGSe photocathodes [17].

5.3.2 *Raman Spectroscopy*

A nondestructive technique for chemical analysis called Raman spectroscopy (RS) provides comprehensive information on chemical structure, polymorphy, crystallization phase, and molecular interactions. It revolves around the way light and chemical reactions in a substance interact. RS is a subfield of vibrational spectroscopy that makes it possible to identify small amounts of substances structurally and with great sensitivity using their distinct vibrational fingerprints [78]. In order to provide multidimensional analytical knowledge on the sample's surface, RS is a useful imaging tool. The main attributes of this method are good selectivity, simple specimen preparation needs, and low sensitivity to water. When a photon interacts with a molecule's chemical bonds, vibrational vibrations are created that cause the photon to be radiated in all directions. This process is known as light scattering [60].

By simultaneously collecting spectra and laser scanning the specimen's surface, RS pictures are created. Line scan and point-by-point scan are two categories for scanning techniques. By gathering the Raman spectra at every square that divides the region of interest, the point-by-point approach of data cube creation produces maps at high resolution [60]. However, this method achieves spatial resolutions between 2 m and 25 m and spectral resolutions between 1 cm^{-1} and 4 cm^{-1}. Additionally, the apparatus operates in confocal mode to minimize secondary scattering effects and fluorescence background. The line scan approach, on the other hand, uses drawing of either vertical or horizontal lines through cylinder optics or rotating mirrors to scan the region of concern using the incident laser. The quantity of rows, breadth, and height of the area of concern determine the spatial resolution because each scan line generates a spectrum [79–81]. Utilizing a wide aperture optical filter and a light source that illuminates the entire sample, wide-field techniques examine the discrete frequencies of the light across time. Liquid crystal imaging spectrometers (LCIS) often have spectral and spatial resolutions that are competitive with single-stage dispersive monochromators and are diffraction limited. LCIS has a wide-open spectral range, a strong out-of-passband rejection performance, and can operate in the 400–2500 nm spectrum [60].

To conduct nondestructive and noninvasive topographic investigations, RS relies on power and wavelength rather than touch [82, 83]. RS can be used to evaluate chemicals and biological processes in watery conditions because this spectroscopy method only produces minimal interference from water. Additionally, this method involves little to no sample preparation beforehand and only a little amount of material. The Raman instrument is also quite flexible [85–88]. Nevertheless, a weak Raman effect can cause RS to have limited sensitivity and lengthy acquisition times. Due to fluorochromes' interference with the Raman effect, the user is unable to analyze fluorescence in samples in this regard [35, 36]. Despite the aforementioned drawbacks, advanced Raman methods, such as tip-enhanced RS (TERS), surface-enhanced RS (SERS), and shifted subtracted RS (SSRS), respectively, sensitivity, extended spatial resolution, and high fluorescence compatibility to circumvent the limitations of the conventional Raman [82, 84].

Formo et al. (2011) utilize surface-enhanced RS (SERS) to analyze the catalyst structure in situ while it is working at high temperatures and in different atmospheres. By using ALD, strong SERS substrates were created by covering silver nanowires with an extremely thin film layers of protective alumina. They revealed how to use the SERS method to gather catalyst structural data on-site at high temperatures in varied settings. This success was made possible by the use of sturdy

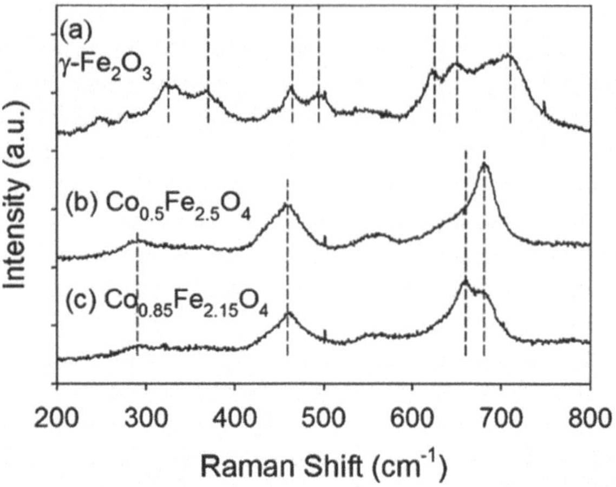

Fig. 5.12 Raman spectra of (**a**) γ-Fe_2O_3 deposited on m-ZrO_2, (**b**) $Co_{0.5}$ $Fe_{2.5}O_4$ deposited on m-ZrO_2, and (**c**) $Co_{0.85}Fe_{2.15}O_4$ deposited on m-ZrO_2. (Reproduced with permission from Ref. [18])

SERS substrates created by the deposition of an incredibly thin alumina layer upon silver nanowires [85].

In another study, Scheffe et al. (2011), ALD process was used to create a high surface area (50 m_2 g^{-1}), thin films of iron oxide (Fe_2O_3) and cobalt ferrite ($CoFe_2O_4$) on m-ZrO_2 support. To identify the chemical makeup of the deposits, phonon modes in the films were examined using a Raman microscope (Fig. 5.12). They reported for ALD iron oxide films, the strongest Raman modes are found at 325, 370, 465, 495, 625, 650, and 710 cm^{-1}, which are denotative of maghemite (γ-Fe_2O_3). They reported in their RS characterization that because no modes suggestive of m-ZrO_2 are evident, the analysis gave indirect indication for the conformal coverage of the ALD films on the m-ZrO_2 support and concluded that their findings point the way to reliable materials for thermal gas splitting and chemical looping cycles. Furthermore, they suggested that ALD films, which retain their reactivity at such temperatures in disparity to oxide powders that are deactivated by frittage and grain formation, can operate as a useful platform for investigating the surface chemistry of these techniques [18].

5.4 Material Density

It is becoming more and more important to characterize thin films and surfaces with exceptionally high sensitivity as the physical characteristics of cutting-edge thin-film structures continue to get narrower. Achieving the desired film qualities requires careful management of the film interface roughness, thickness, and density since thin films' properties are closely correlated to their structure [86]. But utilizing other methods, the film density is a characteristic that can only be nondestructively and indirectly accessed. Solution sample and/or alteration film porosity measurements are typically used to determine the change in thickness and mass density in the instance of alteration layers [87].

Understanding thin film properties and evaluating potential technological applications requires a thorough understanding of surface and structure studies. Neutron reflectometry, ellipsometry, and medium-energy ion scattering spectrometry are techniques for measuring thickness. Density is a special feature that is important for the functional and mechanical effectiveness of the films but is quite difficult to study. By taking different estimates of thickness and mass per unit area, density is frequently inferred indirectly. X-ray fluorescence, forward-recoil spectrometry or Rutherford backscattering, and neutron activation analysis can all be used to calculate mass per unit area. The X-ray reflectivity (XRR) of films can be utilized to calculate their density [88].

5.4.1 X-Ray Reflectivity (XRR)

For almost a century, XRR has been utilized to scrutinize the electron density profile of numerous materials. XRR is an effective, nondestructive, parameter-free method of measuring the mass density of different kinds of films up to the nanoscale thickness range [89]. The vast spectrum of electromagnetic waves includes XRR. Electrons can be accelerated or decelerated in a vacuum (a synchrotron) or in metallic targets (tubes) to produce X-rays. The Cu K α radiation is the type of X-ray that is most frequently employed in material research, with a typically wavelength (λ) of 154 nm [90]. The following traits also apply to the XRR method: It can be used to research amorphous, polycrystalline, or single-crystalline substance. It can be utilized to nondestructively assess surface roughness and interface width (caused by roughness and interdiffusion). Under visible light, XRR can be utilized to examine an opaque film. It can be used to figure out whether a multilayer or single-layer film has layers. It can also be employed to gauge films from a few to a thousand nm thickness [91].

In a conventional XRR experiment, specimen-goniometer, incorrect alignment, which manifests as scanning the specimen in a non-specular (off-specular) state, is the most consequential constraining factor in precisely determining the density [10]. Non-specular data are mistakenly thought to be specular, which results in error. One can cleverly extrapolate the true density (at infinite wavelength) by employing an energy-dispersive detector and making use of the complete energy spectrum of the X-ray source [10]. The drawback to this is that entry-level commercial reflectometers might not be able to use the detector because it is a nontypical, pricey piece of equipment [88].

ALD has been utilized too at low temperatures to explore Iridium (Ir) and iridium (IV) oxide (IrO_2) films, and the resultant films have been analyzed via XRR. Between 100 °C and 180 °C, IrO_2 films were successfully produced utilizing ozone and (MeCp) Ir(CHD) as Ir precursor; however, at 120 °C and below, the density of the films significantly decreased. IrO_2 went from being crystalline to

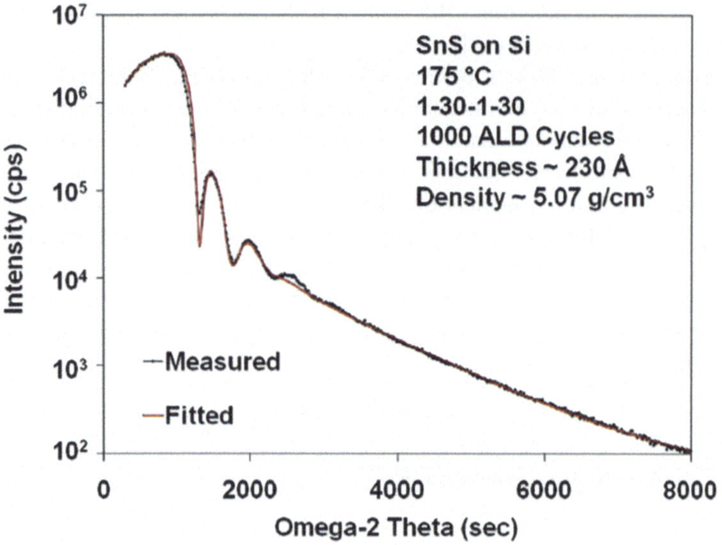

Fig. 5.13 XRR scan showing X-ray intensity vs angle for 1000 cycles of SnS ALD on silicon at 175 °C using a pulse sequence of (1-30-1-30). (Reproduced with permission from Ref. [87])

amorphous as a result of the density loss [92]. In a similar study by Kim and George (2010), hydrogen sulfide (H_2S) and tin (II) 2,4-pentanedionate ($Sn(acac)_2$) were sequentially exposed in the ALD process to grow tin monosulfide (SnS). In this investigation, the hot wall, viscous flow ALD reactor was also utilized to produce SnS thin films on silicon substrates. XRR measurements were utilized to characterize the prepared thin film after 1000 cycles at 175 °C. They reported a film thickness of 230 Å was recorded according to the XRR measurements, and the surface roughness is ~30 Å. The 0.22–0.24 Å growth rate per cycle for SnS at 175 °C and the SnS ALD thickness of 230 Å after 1000 cycles are shown in Fig. 5.13 [87]. In their XRR investigation, they came to the conclusion that the obtained roughness of 30 Å is more than the statistical roughening that is anticipated for a CVD procedure. The surface roughness could be caused by either a film's crystallinity or issues with nucleation. The reported SnS thin film's density of 5.07 g/cm^3 was likewise supported by XRR analysis. This density is quite similar to the value reported in previous literature, 5.0 g/cm^3 [93].

5.5 Material Composition

The identification and measuring of the elements existing in a material is regarded as composition measurement. To assess the quality or purity of raw materials, intermediates, or finished goods, it is crucial to measure the chemical composition of gases, liquids, and solids. This may entail searching for trace elements for contamination or doing a bulk chemical analysis. Both the conventional semiquantitative X-ray fluorescence methodology and other approaches can be utilized to estimate the chemical composition of materials.

5.5.1 X-Ray Fluorescence (XRF)

The electromagnetic radiation known as X-rays is a high energy, short wavelength, high frequency kind that occurs between gamma rays and ultraviolet light. X-ray diffraction (XRD), wavelength dispersive spectroscopy (WDS), and X-ray spectroscopy (XRS) are a few more experimental methodologies that utilize electron and X-ray interactions with samples and depend on similar basic principles [94]. Scientist Wilhelm K. Röntgen (1845–1923) of Germany discovered X-rays for the first time, and for this, he was given the Nobel Prize in 1901. Whereas X-rays have been employed for industrial elemental analysis since the 1950s, the origins of XRS date back to 1909, while Charles G. Barkla established a relationship between the number of X-rays released by a sample and its atomic weight [95].

Henry G. J. Moseley made a significant contribution to X-ray-based elemental identification in 1913 when he noticed that the K line transitions in an X-ray spectrum travelled by identical quantity every time the atomic number improved by one [96]. He has been accredited with replacing the atomic weight-based periodic tables with ones that depend on atomic numbers. He set the groundwork for the identification of elements in XRS by eventually discovering a link between frequency (energy) and atomic number, which is a fundamental of XRS. The method's promise was swiftly realized, and from 1914 to 1924, advancements in X-ray technology received half of the physics Nobel Prizes. The application of electrons as the excitation source for XRS initially created substantial hurdles due to high vacuum, the problem of sample volatility, and the requirement for electrically conducting specimens. To subdue these obstacles, the sample's fluorescent secondary X-ray emission was stimulated using an X-ray source attached with a metal target. This method of sample excitation caused problems because it had a lower photon excitation efficiency and necessitated equipment with complicated detecting parts. Notwithstanding these limitations, the most useful tool for the analyst utilizing profitable instruments would be the fluorescence emission of X-rays [95].

One of the simplest, most precise, and cost-effective analytical techniques for figuring out the chemical make-up of a variety of materials is X-ray fluorescence (XRF) technology. It is trustworthy and nondestructive, stipulates minimal or no preparation of samples, and works with solid, liquid, and powdered samples. Depending on the analyzer, the range of elements covered for qualitative and quantitative analyses begins with carbon (6) and fluorine (9), respectively. Americium (108, 108) is the element with the highest covered atomic number. XRF has detection limits at the sub-ppm level and can quickly and simultaneously assess concentrations up to 100% [94]. Since it can quickly offer a high-resolution evaluation of relative alterations in elemental compositions, XRF spectrometry is particularly useful [98]. The wavelength-dispersive principle, which is the foundation of XRF spectrometry, asserts that specific atoms release a relative abundance of X-ray photons of wavelength or energy that may be calculated.

The analysis of materials from a variety of contexts and habitats has used this technique. Due to its ability to give accuracy and higher precision than competing approaches, XRF has a broad span of applications. The exceptional synergy of benefits that X-ray examination of archaeological artifacts offers in terms of both use and cost includes:

(i) ***Nondestructive***. The vast majority of the time, exposure to X-rays does not destroy or alter the samples being analyzed. As a result, they can be archived for future use or applied to potentially harmful tests like obsidian hydration analyses [95].

(ii) ***Very little preparation***. The majority of obsidian artifacts do not require much prior preparation for study. The dissolution processes required by many other techniques are timewasting and expensive due to the necessity of acids and other reagents. Although it is preferable to wash off any sediments off the specimens. It has been demonstrated that if the amount of dirt is little and the artifact has not been heated to a point where some sediment matrix has melted onto the specimen, a thorough cleaning is not required [99]. Most obsidian artifacts do not require much prior preparation for study. Many other procedures need dissolution processes that are time-consuming and expensive in terms of the acids or other chemicals required. It has been shown that if the amount of dirt is little and the artifact has not been heated to the point where any sediment matrix has melted onto the specimen, a complete cleaning is not necessary, even if it is ideal to wash any sediments off archaeological objects. This is mostly because X-rays can penetrate beyond the surface in the mid-Z X-ray area, and while this includes any contamination on the surface, it usually does not matter if some soil is still present in the flake scars. In comparison to any surface adulteration, the analyzed volume is relatively enormous. This is not the situation with the vast majority of metals, where chemical and weathering patination may drastically modify the composition at the surface, resulting in incorrect conclusions [89].

(iii) ***Rapid analysis.*** Chemical compositions can be identified in a matter of seconds thanks to fast X-ray spectroscopy. The analysis of elements such as Ti-Nb using the Berkeley Spectrace and Thermo desktop instruments at 200 live seconds per specimen requires approximately 5–6 minutes for each specimen, based on mass [95].

(iv) ***Simple to use.*** Modern equipment is computer-controlled, and the measurement setup and result processing are handled by efficient software. Today, talented students may do fully automated tasks that traditionally needed a trained analyst's continual observation necessitate the continual monitoring from a professional analyst [101, 102].

(v) ***Economical.*** By eliminating the more involved preparation of samples necessary for the majority of XRF and all destructive experiments, the amount per sample is substantially decreased. This indicates that XRF will resolve all of our predicaments, but it does not constitute the infinitely knowledgeable black box that we would prefer it to be [103].

Sample size restrictions restricted elemental collection, and failure to define tiny components are some of XRF's drawbacks. The best samples for XRF examinations are those that have the smallest dimension of >10 mm and a thickness of >2 mm [96–103].

5.5.2 *Energy-Dispersive X-Ray Spectroscopy (EDS, EDX, or XEDS)*

The analytical method of EDS, EDX, or XEDS enables the elemental analysis and chemical characterization of materials. A core-shell electron is unbound from a substance after been stimulated by a source of energy, such as the electron beam of an electron microscope, assisting in releasing a portion of the energy that has previously been engrossed. A higher energy outer-shell electron goes in to take its position, and the disparity in energy is then emitted as an X-ray with a specific spectrum based on its parent atom. As a consequence, it is feasible to determine the chemical makeup of an energy source-excited specimen volume. While the strength of the signal represents the element's concentration, the placement of the peaks in the spectrum identifies the element [104]. The spectrometer's ability to get results quickly makes the approach incredibly adaptable. The device is small, reliable, steady, specimen to use, and its output may be rapidly analyzed. The study is founded on the detection of the distinctive X-rays generated by the interaction of the electron beam with the specimen. The data might be gathered at extremely specific local locations or over the entire specimen. Thus, using EDXS and electron microscopy, respectively, it is possible to determine the material's chemical composition and morphology [105].

Knowing the theoretical underpinnings of the approach is crucial for providing an appropriate interpretation of the collected data by the instrument. In a procedure known as inner-shell ionization, the specimen's atoms generate the distinctive X-rays [105]. A vacancy in the shell is created when an electron from the inner shell is eliminated by an electron from the beam. The atom is currently ionized for 10^{-14} s before an outer-shell electron replaces the space left by the inner-shell electron. A photon with the chemical element's characteristic energy is released during this transition, ionizing its shell [106]. The emitted photons are designated as K, L, and M lines by the type of ionization that filled the inner shell, and α, β, γ by the outer shell that corresponds to the released electron. It is crucial to remember that some transitions are prohibited for atoms with higher atomic numbers. The IUPAC and/or Manne Siegbahn rules for notation and the quantum selection rules for permissible transitions can both be applied to notation [105]. Another X-ray source produced during the interaction between the beam and the specimen is bremsstrahlung radiation, commonly known as continuous X-rays. This radiation is what causes the electron beam to slow down in the Coulombic fields of the atoms. The electromagnetic spectrum appears as background in the EDX spectrum when the electrons break apart, releasing photons of any energy [105, 106].

X-rays identified as satellite peaks are produced when multiple electrons are simultaneously expelled from an atom under conditions of extreme overvoltage [107]. Due to the simultaneous electron ejection's effect on the energy levels' overall structure, X-rays with somewhat lesser energies than those generated by lone electron ionization appear close to the distinctive peaks. The Auger process is another significant producer of satellite peaks. Finally, variations between peaks generator by a purer element and those produced by that same element when it is present in a compound can be seen in the X-ray spectra [108]. This fluctuation happens as a result of the outer valence electrons' substantial influence on the electrical configuration of an atom's inner shells. When contrasting metals with the respective oxides and halides, these alterations stand out the most. As a result, it is not desirable to analyze oxide compounds using a metallic standard material. Metals can be utilized as reference parameters for measurement in the case of EDX analysis because the peak shifts are undetected [105, 106]. The sample's whole output of X-rays is detected, and the EDX spectrum shows them all. The ability to accurately identify the chemical components still present in the specimen thanks to their identification in the spectrum allows for a more precise quantification to be produced. The most significant X-rays among all those shown in an EDX spectrum are known as characteristic X-rays [105].

The method used to acquire the data is a crucial factor to take into account when using the EDX. The three components that make up the specimen's X-ray processing are the detector, the electronic processor, and the multichannel analyzer display. EDX main steps are as follows: the detector produces a charge pulse inversely proportional to the X-ray energy. The generated pulse is first transformed to a voltage, after which it is amplified by a field effect transistor, separated from other pulses, and further amplified. Finally, it is electrically identified as being the product of an X-ray with a particular energy. The designated multichannel analyzer's channel for that energy then stores a digital signal. This process moves so quickly that the spectrum appears to be created concurrently with the simultaneous detection of the entire spectrum of X-ray energy [109, 110]. It is crucial to remember that the intensity and peak broadening are taken into account along with absorption correction (A), the atomic number effect (Z), and characteristic fluorescence correction (F) when calculating the chemical composition [111].

The XRF and EDX techniques have been used for characterization of ALD-prepared catalysts. Feng et al. (2010) utilized the ALD process for the preparation of highly distributed, uniform Pd nanoparticles on Al_2O_3 and ZnO-coated mesoporous silica gel. Their catalytic activities of these samples were examined in the methanol degradation reaction. The amount of Pd supported on the Al_2O_3 or ZnO-coated silica gel was evaluated using XRF, while EDX mapping was used to characterize the Pd species spatial distribution within the silica gel particles as shown in Fig. 5.14 [112]. In another study, titanium dioxide was deposited in atomic layers

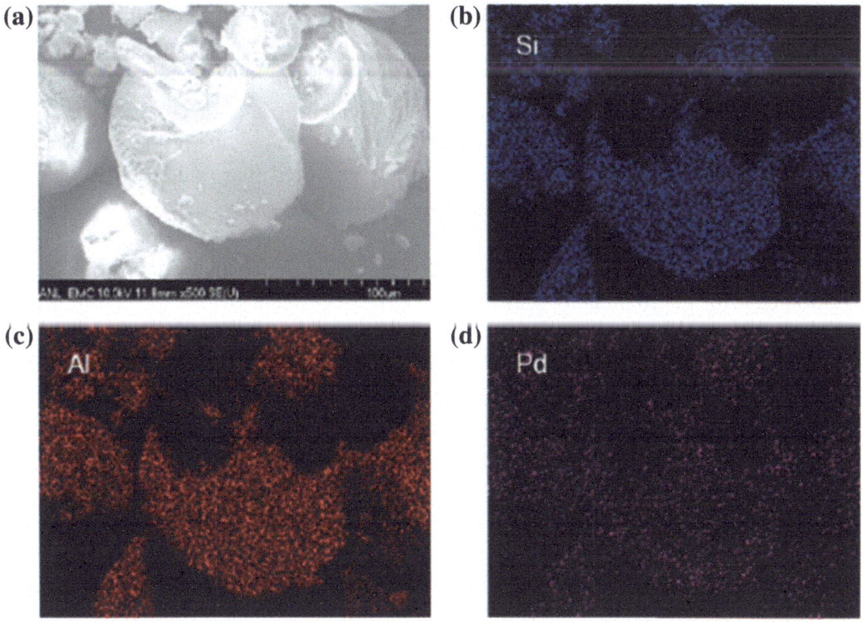

Fig. 5.14 EDX mapping (combined with the SEM image) of the cross-section of a cleaved silica gel particle: (**a**) SEM image of the cross-section, (**b**) EDX mapping of Si, (**c**) EDX mapping of Al, and (**d**) EDX mapping of Pd. (Adapted with permission from Ref. [112])

on alumina membranes obtained from sol-gel plasma-enhanced ALD methods that were used to create hydrogen selective membranes. EDX was used to characterize the prepare membranes' surfaces [113].

5.6 Electrical Measurement

Materials are classified based on their electrical properties such as conductors, insulators, semiconductors, and superconductors. Different materials have varying electrical qualities; some, like metals, are good electrical conductors, while others, like plastics, are good insulators. All types of materials cannot be measured for their electrical characteristics using a single method or tool. A crucial physical characteristic of all materials is their electrical characteristics [114].

5.6.1 Electrical Conductivity and Resistivity

The ability of a solid substance to resist the flow of electricity is one of its most crucial electrical properties. (Ωm) and (S/m), respectively, are the SI units for conductivity and resistivity. A material has low resistivity if electrons can move through it without much resistance. A substance has a high resistance if electricity flows through it with substantial difficulty [114]. The opposite of conductivity is resistivity. Accordingly, a high resistance is equivalent to a poor conductivity. The electrical characteristics of materials can be demonstrated using both electrical conductivity and electrical resistivity. It should be noted that conductivity can be computed based on the sample's dimensions [114, 115].

The electrical conductivity range of solid materials is astounding, spanning over a magnitude of 27. Moreover, one approach to categorize dense materials is to do so in accordance with the manner in which the material conducts electricity within itself. These substances can be categorized as conductors, semiconductors, or insulators. Excellent conductors are metals with order of 10^{-7} (Ωm) resistivities. Aluminum and copper are two excellent examples of conductors. On the other end of the spectrum are electrical insulators, which have extraordinarily high resistivities between 10^{10} (Ωm) and 10^{20} (Ωm). Examples of an insulator are materials like glass, wood, and plastic that are nonconductors of current. Semiconductors are materials having intermediate resistivities of between 10^{-4} (Ωm) and 10^{6} (Ωm). They divided the materials into conductors and insulators. Examples of semiconductors are germanium and silicon [116]. The quantity of electrons accessible to take part in the conduction process has a significant impact on the electrical conductivity of all conductors, semiconductors, and many insulating materials. The conduction process can only be carried out by electrons that are in the conduction band. But not every atom's electrons will speed up when an electric field is present. The number of electrons readily accessible for electrical conduction in a given material depends on the configuration of electron levels or states of energy and the means in which these statuses are filled by electrons [114–116].

The resistance as defined by Ohm's law is the ratio of voltage divided by the applied current. Electrical resistivity measurement requires the measurement of voltage and current. The definitions of resistance and resistivity are distinct. The type of the material determines the resistivity, whereas the amount of the material determines the resistance. Resistance, which is measured in Ohms (Ω), is, in general, the ability of a substance to resist the flow of an electrical current. Resistance measurements are used to calculate resistivity. The ohmmeter (Ωm) is the unit of resistivity. The evaluation of the surface resistance of a substance that comes in contact with electrodes is known as sheet resistance or surface resistivity. Bulk resistivity is determined by measuring the resistance across the two electrodes put into the sample's two ends. Bulk resistivity, measured in Ωm, is also known as volume resistivity [114–117].

5.6.2 Methods for Electrical Resistivity Measurements

It has been recommended to test the electrical resistance using a variety of models and techniques. Contact resistance and the sample's shape, regardless of whether it takes the shape of a thin film, single crystal, tiny crystallite, or powder pellet, are factors influencing the applicability of various procedures and the precision that may be achieved [118]. In electrical measuring, 2-probes (ammeter or voltmeter measurements) could be utilized for greater resistive samples, while 4-probes (potential probe measurements) are able to be deployed for low resistive and single crystal samples, while the methods of Montgomery and van der Pauw can be employed for larger and pellets samples [118, 119].

5.6.2.1 Two-Probe Electrical Resistivity Measurements Method

The straightforward and well-known two-probe method is used to assess resistivity. The usage of a digital multimeter to measure the resistance can serve as an illustration of this technique. The resistivity of a material can be determined with 2-point probe, by estimating the resistance and taking the sample configuration into account. With the proper tool, such as a caliper, the physical size of the sample being tested can be calculated. The wires will be attached to the sample's two sides as demonstrated in Fig. 5.15 [114].

An ammeter and a voltage source are required for this procedure. A current ***I*** flows across the sample as a consequence of a voltage source applying a voltage ***V*** across it. An alternative would be to apply current to the sample, measuring the voltage caused across the sample bar with a voltmeter connected in parallel to the current source. The ammeter, which is connected in series with the voltage source and sample, measures the volume of current ***I*** that flows across the sample. There should not be much voltage drop through the ammeter. The two sites at which the probes are attached to the sample are the basis for the method's name, "two-probe."

The resistance (R) of the sample is given by Eq. 5.2:

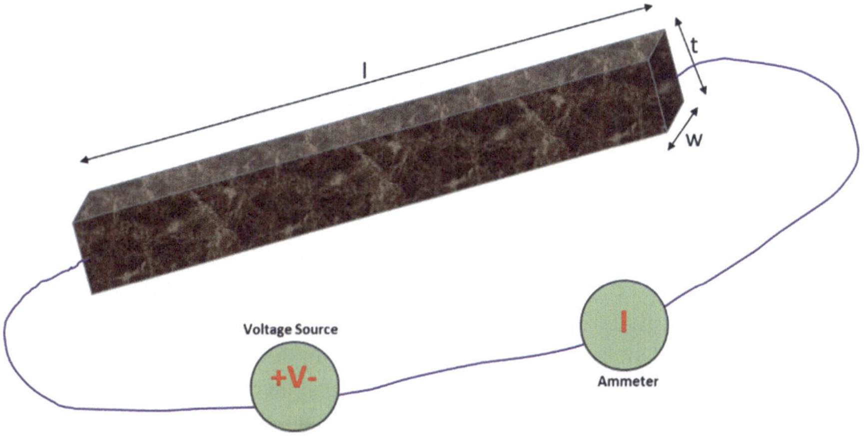

Fig. 5.15 Two-probe method for measuring electrical resistivity

$$R = V/A \tag{5.2}$$

where V stands for voltage (V), I for current (A), and R for resistance (Ω).

The two-point method for measuring resistivity yields the following results as shown in Eq. 5.3.

$$P = R\frac{wt}{I} \tag{5.3}$$

where R is the measured resistance (Ω), w, t, and I are the sample's physical dimensions, and is the ρ resistivity (Ωm).

This approach comes with several drawbacks. The two-probe approach is frequently unreliable in practice. Within the contact interface of the probes and the sample, there is typically some contact resistance, as well as lead resistance in the measuring apparatus. The additional resistances render the measurement of resistivity in low resistance measurements unreliable. Additionally, improper connections between electrodes and a sample result in other mistakes that increase resistance in addition to the resistance of the sample itself [114, 115].

5.6.2.2 Four-Probe Electrical Resistivity Measurements Method

The resistivity of both the bulk specimen and the single crystal can be assessed using the four-probe approach. Here, current flows through the outside contacts that are at the sample's margins. The potential difference between the inner contacts is estimated. This technique is best for moderate and accurate resistance measurements because it can remove the impact of contact resistance between the electrical connections and sample [115]. Four-probes are attached to the sample using this technique, as shown in Fig. 5.16. In the common kind, an ammeter is used to measure the current that travelled across the sample after a current source was generated to create a steady current through the sample's end. Estimating the voltage V created across the inner half of the sample simultaneously with a voltmeter. Alternatively, a voltage source might apply a voltage across the sample's outer portion while simultaneously measuring the sample's current flow with an ammeter connected in series with a voltmeter.

The evaluation of resistivity by 4-point method is as follows:

$$P = \frac{Vwt}{lI} \tag{5.4}$$

where ρ is the resistivity (Ωm), V is the measured voltage (V), W and t are the sample's width and height (m), I is the current flowing through the sample (A), and l is the distance between the voltmeter probe's two inner contacts (m). The four places at which the probes are attached to the sample are the basis for the method's name, 4-probe method. It should be noted that this technique only uses the distance

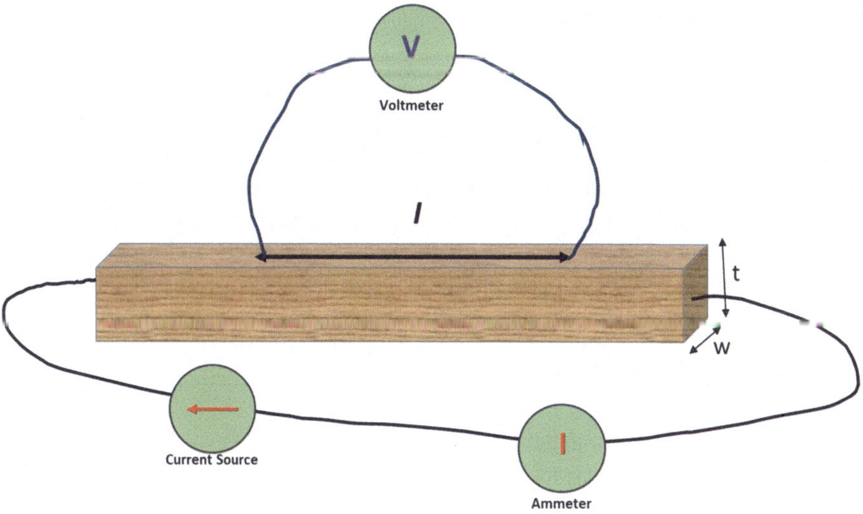

Fig. 5.16 Two-probe method for measuring electrical resistivity

l between two voltmeter probes to measure resistance rather than the entire length of the sample [114, 115].

In a recent study, the electrical properties of calcined silicon-titanium layered oxide (Si-Ti LO) thin films was reported by Kalderon et al. (2021). In this research, they used a combined Molecular layer deposition (MLD) and ALD approach to create layer-by-layer depositions of Si-Ti LO thin films. They studied the sheet resistance at diverse processing temperature in the scale of 450–1050 °C neath argon (Ar). They reported tunable electrical conductivities of the Si–Ti LO thin films. They concluded that the graphitic and Magnéli oxide phases were produced in conductive thin films by thermally treating Si-Ti LO thin films. Due to the unique stoichiometry and Si-Ti layer structure attained by the ALD process, as well as the breaking of Si-C bonds and Red-Ox reactions between the phases of TiO_2 and Si suboxide, which result in the conductive Magnéli phase. A lower onset temperature for Magnéli phase formation, and higher conductivity were attained [120].

In another study by Jiang, et al. (2008), an electrode/catalyst for solid oxide fuel cells was prepared by ALD deposition of Pt thin films. Electrical characterization (I–V measurements) was made to ascertain how the thickness of the platinum film at various operational temperatures affects the performance of the fuel cell. They concluded that platinum thin films can serve as effective electrode materials, and current collector grids that can greatly enhance the performance of fuel cells [121].

5.7 Optical Properties

Most laboratories that are concerned with the quantification and identification of inorganic and organic substances covering a wide variety of items and activities now use ultraviolet (UV) and visible (VIS) spectrophotometry as their preferred choice. They are equally important in a variety of manufacturing industries, geological research, animal husbandry and fisheries, food safety, and environmental monitoring, to name a few. They are used in manufacturing, research, and quality, with a persistent emphasis on life science and pharmaceutical contexts [122]. The optical properties of a substance are the modifications that light experiences when it interacts with it. The substance's macroscopic and microscopic characteristics, including the makeup of its outer surface and its electronic structure, have an impact on these optical qualities.

The optical properties of materials are frequently utilized to explore other features of the substance since it is typically simpler to discover how a material affects light than to directly examine its microscopic and macroscopic qualities. The reflection, transmission, and absorption characteristics of materials are frequently used to explain their optical properties. Materials are subjected to optical radiation in order to measure these critical parameters, which is then partially reflected, transmitted, or absorbed by the samples. A spectrophotometer is the most used tool for determining these optical characteristics [123].

5.7.1 Ultraviolet–Visible Spectroscopy (UV-vis)

UV-vis is a simple, low-cost, nondestructive, flexible analytical method appropriate for a significant class of organic/inorganic species. UV-vis measures the wavelength-dependent transmittance or absorbance of light passing across a substance [124]. The attenuation of a light beam after it has gone through a sample or after it has been reflected off of a sample surface is measured in UV-vis. The terms visible (VIS), ultraviolet (UV), and near-infrared (NIR) spectroscopy are used to refer to a range of measurements of transmittance, absorption, and reflectance in these spectral regions. Measurements of such can be performed at multiple wavelengths or throughout a wide spectral range [124]. UV-vis is one of the more popular characterization and analytical methods in science. The linear connection between absorber and absorbance concentration makes UV-vis specifically effective for conducting quantitative measurements. Visible and ultraviolet photons have sufficient energy to transport electrons into higher energy states in molecules and materials. Exploring the electrical properties of materials and their ancestors using UV-vis is useful for both fundamental research and the development of usable materials [125].

The applied range for UV-vis spectroscopy is between 200 and 800 nm; infrared is beyond 800 nm, while vacuum UV is less than 200 nm. The color of matter is determined by how well it can emit and absorb light, and the human eye is able to differentiate up to ten million different colors. The UV spectrophotometers operate by shining a light source through a sample and records the amount of light that is transferred to the detector on the other side. Most data graphs exhibit wavelength (in nanometers) on the x-axis and on the y-axis absorbance, with the baseline at the bottom and the peaks pointing upward. The greatest peak—λ_{max} of the transmittance curve, which shows the amount of light that is absorbed at each wavelength, is what we are particularly enthusiastic about [126]. There are distinctive absorbance bands for chromophores, and their different energy levels, and consequently the absorbance wavelength and intensity may change if their surroundings are altered by the addition of another substance or by a physical change, such as an increase in temperature [127]. Band shifts to shorter wavelengths (blue shift) and band shifts to extended wavelengths (red shift) are known as hypsochromic shifts and bathochromic shifts, respectively. Hyperchromism and hypochromism are changes in an absorption band's peak intensity, respectively [126–128].

The time needed to prepare for a UV-Vis spectroscopic measurement is minimal to none. The majority of the time required for the experiment is determined by the duration it requires for the light to warm up. The actual sampling process can take a few minutes or less, depending on the scanning speed and scope. Time spent on analysis is similarly quick; plotting the data in a spreadsheet application only takes a few minutes [129]. The basis for UV-visible spectroscopy is the shift of electrons from a lower energy orbital (lowest unoccupied molecular orbital—LUMO) to a higher energy unoccupied orbital (highest occupied molecular orbital—HOMO) in organic compounds that absorb light. The energy of the absorbed light wavelength should match the ΔE of the HOMO–LUMO energy gap's value [130]. There are various UV-Vis configurations existing, such as absorption, transmission, and diffuse reflectance. In a transmission setup, the user directs a collimated beam of light toward the sample of concern, hereafter described to as the working sample. Samples must be transparent to some extent. When light strikes the working sample, distinctive wavelengths that correspond to electronic transitions in the specimen are partially absorbed. The light that has been transmitted is collected by a spectrometer, which then measures the output against a reference value of 100% transmission (I_0). The reference measurement needs to account for any material support's absorbance, such as a glass slide or a cuvette holder. One or two beam setups can be used to perform transmission reference measurements [129].

The diffuse reflectance spectroscopy (DRS) measures diffusely reflected light from the specimen rather than the light that would have been transmitted in a diffuse reflectance setup. Films or powders are the samples that perform best in this configuration. A conventional setup for a DRS measurement culminates in using an integrated sphere for capturing all photons reflected (in all directions) from the specimen in question. The light source is connected to an input port, while the output port is connected to a signal meter that gathers diffusely reflected light, and an aperture that can be used to place the reference or working samples for measurement are all features of a conventional integrating sphere. There can be specular and diffuse reflections. When an incident radiation beam contacts a sample and reflects at an angle equal to the angle of incidence, this phenomenon is known as specular reflection.

Due to the lack of an absorption process, specularly reflected light provides very little to no real information about the electrical states of the samples. In the case of DRS, a portion of the photons from the incident beam are reemitted (reflected) at various non-incident angles after partial absorption on the sample surface. Specular reflectance will enhance noise, reduce measurement accuracy, and can cause erroneous peaks in the data during DRS measurements. A specular reflectance sink (or "plug") can be found inside an integrating sphere to reduce its contribution. In a non-absorbing matrix, dispersion of powders can further boost diffuse reflectance while lessening specular reflection. Examples of common non-absorbing matrix components are KBr, KCl, and Ba_2SO_4 [131–143]. Research and industrial laboratories frequently use UV-Vis spectroscopy as a tool to evaluate reaction kinetics for quantitative and qualitative analysis. Research on photocatalysis benefits greatly from the semiconductor band gap measurement by UV-vis diffuse reflectance spectroscopy. Furthermore, it allows for the characterization of the coordination state of oxides, which is crucial for the enhancement of oxide catalysts [126].

Singh et al. (2016) describe the development of photocatalysts with high surface-area synthesized by ALD deposition of TiO_2 on fibrous nano silica (KCC-1). The KCC-1/TiO_2 series' multiple Ti coordination sites were examined using DRS in relation to the ALD cycles before and after heat treatment. They noticed that when the number of ALD cycles increased, the intensity of the 250–370 nm band also increased, demonstrating the metamorphosis of Ti sites from tetrahedral to octahedral. They also performed UV-DRS analysis on the KCC-1-based materials to investigate the band gap alteration. Tauc plots used to measure the band gaps of these materials. They also reported the band gap value for the as-prepared samples reduced from 3.53 to 3.28 eV as the ALD deposition cycles increases from 1 to 60. The band gap changed as a consequence of the splitting of energy levels, indicating that molecule Ti sites in TiO_2 had grown bulkier [131].

Fabrication of TiO_2 on porous graphitic carbon nitride (P-g-C_3N_4) via ALD for enhanced solar-driven hydrogen evolution was expressed by Liu et al. (2018). In this investigation, UV-vis spectrometry was utilized to evaluate the absorption edges in diffused reflectance mode with a 750–350 nm scanning range. They reported an absorption edge of about 418 and 427 nm for P-g-C_3N_4 and g-C_3N_4, which correspond to 2.97 eV and 2.90 eV, band gaps respectively. The quantum effect might be to blame for the wider band gap of g-C_3N_4. Figure 5.17 demonstrates the two materials' optical absorption spectra [132].

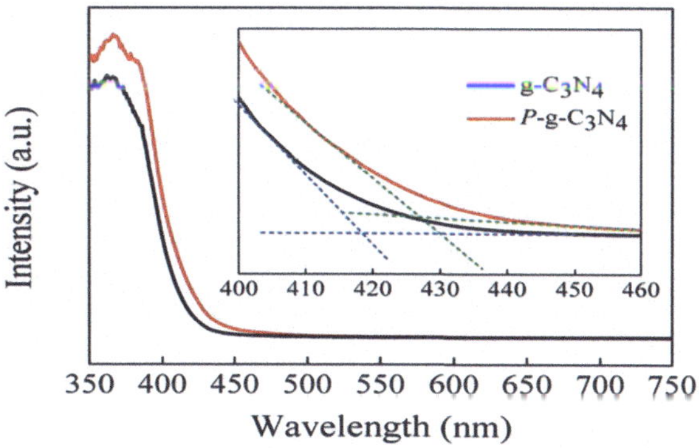

Fig. 5.17 Comparison between P-g-C_3N4 and g-C_3N_4. UV-vis diffuse reflectance spectra. (Reproduced with permission from Ref. [132])

References

1. H.K. Charles, J.A. Weiner, N.A. Blum, Materials characterization and analysis-applications to microelectronics. J. Hopkins APL Tech. Dig. **6**(3), 237–249 (1985)
2. P.J. Smith, Materials and chemical analysis of electronic devices, in *Microelectronics Manufacturing Diagnostics Handbook*, (Springer US, Boston, MA, 1993), pp. 396–424
3. D. Liu, Z. Shadike, R. Lin, K. Qian, H. Li, K. Li, . . ., B. Li, Review of recent development of in situ/operando characterization techniques for lithium battery research. Adv. Mater. **31**(28), 1806620 (2019)
4. K.W. Powers, S.C. Brown, V.B. Krishna, S.C. Wasdo, B.M. Moudgil, S.M. Roberts, Research strategies for safety evaluation of nanomaterials. Part VI. Characterization of nanoscale particles for toxicological evaluation. Toxicol. Sci. **90**(2), 296–303 (2006)
5. Z. Zhang, C.F. Cheung, C. Wang, J. Guo, Modelling of surface morphology and roughness in fluid jet polishing. Int. J. Mech. Sci. **242**, 107976 (2023)
6. E. Liu, G. Xu, N. Wang, Analysis of surface morphology of Ti-5553 based on wavelet transform. Integr. Ferroelectr. **217**(1), 129–140 (2021)
7. J.L. Hu, W. Ma, Y.Z. Pan, Z.H. Cheng, S.G. Yu, J. Gao, Z. Zhang, C.X. Wan, C.X. Qiu, Insights on the mechanism of Fe doped ZnO for tightly-bound extracellular polymeric substances tribo-catalytic degradation: The role of hydration layers at the interface. Chemosphere **276**, 130170 (2021)
8. A. Khan, S.M. Rangappa, S. Siengchin, A.M. Asiri, *Biofibers and Biopolymers for Biocomposites* (Springer, Cham, 2020)
9. H. Bruining, *Physics and Applications of Secondary Electron Emission: Pergamon Science Series: Electronics and Waves – A Series of Monographs* (Elsevier, 2016)
10. A.V. Phelps, Z.L. Petrovic, Cold-cathode discharges and breakdown in argon: Surface and gas phase production of secondary electrons. Plasma Sources Sci. Technol. **8**(3), R21 (1999)
11. G.D. Danilatos, Foundations of environmental scanning electron microscopy. Adv. Electron. Electron. Phys. **71**, 109–250 (1988)
12. P. Meredith, A.M. Donald, B. Thiel, Electron-gas interactions in the environmental SEM's gaseous detector. Scanning **18**, 467–473 (1996)
13. V. Neděla, E. Tihlaříková, J. Runštuk, J. Hudec, High-efficiency detector of secondary and backscattered electrons for low-dose imaging in the ESEM. Ultramicroscopy **184**, 1–11 (2018)
14. M. Abd Mutalib, M.A. Rahman, M.H.D. Othman, A.F. Ismail, J. Jaafar, Scanning electron microscopy (SEM) and energy-dispersive X-ray (EDX) spectroscopy, in *Membrane Characterization*, (Elsevier, 2017), pp. 161–179

15. C.H. Rohaida, C.T. Foo, O. Nor Azillah Fatimah, S. Nor Azwin, R. Mohd Saari, M.Y. Meor Sulaiman, A. Yusof, Field emission scanning electron microscope (Fe-Sem) facility in BTI. Mater. Charact., 1–4 (2016)
16. K. Arunachalam, M. Davoodbasha, Imaging bacteria and biofilm by field emission scanning electron microscopy, in *Analytical Methodologies for Biofilm Research*, (Springer Nature. 2021), pp. 205–222
17. T.R. Hellstern, D.W. Palm, J. Carter, A.D. DeAngelis, K. Horsley, L. Weinhardt, W. Yang, M. Blum, N. Gaillard, C. Heske, T.F. Jaramillo, Molybdenum disulfide catalytic coatings via atomic layer deposition for solar hydrogen production from copper gallium diselenide photocathodes. ACS Appl. Energy Mater. **2**(2), 1060–1066 (2019)
18. J.R. Scheffe, M.D. Allendorf, E.N. Coker, B.W. Jacobs, A.H. McDaniel, A.W. Weimer, Hydrogen production via chemical looping redox cycles using atomic layer deposition-synthesized iron oxide and cobalt ferrites. Chem. Mater. **23**(8), 2030–2038 (2011)
19. C. Wang, H. Tissot, M. Soldemo, J. Lu, J. Weissenrieder, Inverse single-site Fe1(OH)X/Pt (111) model catalyst for preferential oxidation of CO in H2. Nano Res. **15**(1), 709–715 (2022)
20. J. Lu, B. Liu, N.P. Guisinger, P.C. Stair, J.P. Greeley, J.W. Elam, First-principles predictions and in situ experimental validation of alumina atomic layer deposition on metal surfaces. Chem. Mater. **26**(23), 6752–6761 (2014)
21. S. Hettler, M. Dries, J. Zeelen, M. Oster, R.R. Schröder, D. Gerthsen, High-resolution transmission electron microscopy with an electrostatic Zach phase plate. New J. Phys. **18**(5), 053005 (2016)
22. S. Hettler, M. Dries, T. Schulze, M. Oster, C. Wacker, R.R. Schroder, D. Gerthsen, High-resolution transmission electron microscopy with Zach phase plate. Microsc. Microanal. **21**(S3), 1581–1582 (2015)
23. J.C. Spence, *High-Resolution Electron Microscopy* (OUP Oxford, Oxford, 2013)
24. O. Scherzer, The theoretical resolution limit of the electron microscope. J. Appl. Phys. **20**(1), 20–29 (1949)
25. J. Cazaux, Recent developments and new strategies in scanning electron microscopy. J. Microsc. **217**(1), 16–35 (2005)
26. X. Zhang, S. Han, B. Zhu, G. Zhang, X. Li, Y. Gao, . . ., W. Liu, Reversible loss of core–shell structure for Ni–Au bimetallic nanoparticles during CO_2 hydrogenation. Nat. Catal. **3**(4), 411–417 (2020)
27. S.M. Kim, A. Armutlulu, W.C. Liao, D. Hosseini, D. Stoian, Z. Chen, . . ., C. Müller, Structural insight into an atomic layer deposition (ALD) grown Al_2O_3 layer on Ni/SiO_2: Impact on catalytic activity and stability in dry reforming of methane. Cat. Sci. Technol. **11**(23), 7563–7577 (2021)
28. R. Ramesh, D.K. Nandi, T.H. Kim, T. Cheon, J. Oh, S.H. Kim, Atomic-layer-deposited MoN x thin films on three-dimensional Ni foam as efficient catalysts for the electrochemical hydrogen evolution reaction. ACS Appl. Mater. Interfaces **11**(19), 17321–17332 (2019)
29. G. Haugstad, *Atomic Force Microscopy: Understanding Basic Modes and Advanced Applications* (Wiley, Hoboken, 2012)
30. P. Eaton, P. West, *Atomic force microscopy* (Oxford University Press, Oxford, 2010)
31. G.M. McClelland, R. Erlandsson, S. Chiang, Atomic force microscopy: General principles and a new implementation, in *Review of Progress in Quantitative Nondestructive Evaluation. Review of Progress in Quantitative Nondestructive Evaluation*, ed. by D.O. Thompson, D.E. Chimenti, vol. 6A, (Springer, Boston, MA, 1987). https://doi.org/10.1007/978-1-4613-1893-4_148
32. H. Yu, N.A.A. Rahim (eds.), *Imaging in Cellular and Tissue Engineering* (CRC Press, Boca Raton, 2013)
33. M. Gaczynska, P.A. Osmulski, AFM of biological complexes: What can we learn? Curr. Opin. Colloid Interface Sci. **13**(5), 351–367 (2008). https://doi.org/10.1016/j.cocis.2008.01.004
34. S. Hussain, H. Erikson, N. Kongi, A. Tarre, P. Ritslaid, A. Kikas, V. Kisand, J. Kozlova, J. Aarik, A. Tamm, V. Sammelselg, Platinum sputtered on Nb-doped TiO_2 films prepared

by ALD: Highly active and durable carbon-free ORR electrocatalyst. J. Electrochem. Soc. **167**(16), 164505 (2020)
35. K.E. Elers, T. Blomberg, M. Peussa, B. Aitchison, S. Haukka, S. Marcus, Film uniformity in atomic layer deposition. Chem. Vap. Depos. **12**(1), 13–24 (2006)
36. T. Suntola, Surface chemistry of materials deposition at atomic layer level. Appl. Surf. Sci. **100**, 391–398 (1996)
37. H. Siimon, J. Aarik, Thickness profiles of thin films caused by secondary reactions in flow-type atomic layer deposition reactors. J. Phys. D. Appl. Phys. **30**(12), 1725 (1997)
38. A.T. Sivagurunathan, S. Adhikari, D.H. Kim, Strategies and implications of atomic layer deposition in photoelectrochemical water splitting: Recent advances and prospects. Nano Energy **83**, 105802 (2021)
39. S. Selvaraj, H. Moon, J.Y. Yun, D.H. Kim, Iron oxide grown by low-temperature atomic layer deposition. Korean J. Chem. Eng. **33**, 3516–3522 (2016)
40. B. Choi, H.U. Kim, N. Jeon, Uniformity of HfO_2 thin films prepared on trench structures via plasma-enhanced atomic layer deposition. Nano **13**(1), 161 (2022)
41. H. Van Bui, A.W. Groenland, A.A. Aarnink, R.A. Wolters, J. Schmitz, A.Y. Kovalgin, Growth kinetics and oxidation mechanism of ALD TiN thin films monitored by in situ spectroscopic ellipsometry. J. Electrochem. Soc. **158**(3), H214 (2011)
42. P. Bulkin, D. Daineka, D. Kouznetsov, B. Drevillon, Deposition of optical coatings with real time control by the spectroellipsometry. EPJ Appl. Phys. **28**(2), 235–242 (2004)
43. D.E. Aspnes, Spectroscopic ellipsometry – a perspective. J. Vac. Sci. Technol. A **31**(5), 058502 (2013)
44. J.N. Hilfiker, G.K. Pribil, R. Synowicki, A.C. Martin, J.S. Hale, Spectroscopic ellipsometry characterization of multilayer optical coatings. Surf. Coat. Technol. **357**, 114–121 (2019)
45. D. Gonçalves, E.A. Irene, Fundamentals and applications of spectroscopic ellipsometry. Quím. Nova **25**, 794–800 (2002)
46. J.A. Woollam, P.G. Snyder, Fundamentals and applications of variable angle spectroscopic ellipsometry. Mater. Sci. Eng. B **5**(2), 279–283 (1990)
47. H. Tompkins, E.A. Irene, *Handbook of Ellipsometry* (William Andrew, Heidelberg, 2005)
48. P.S. Hauge, Recent developments in instrumentation in ellipsometry. Surf. Sci. **96**(1–3), 108–140 (1980)
49. R. Kurrat, B. Wälivaara, A. Marti, M. Textor, P. Tengvall, J.J. Ramsden, N.D. Spencer, Plasma protein adsorption on titanium: Comparative in situ studies using optical waveguide lightmode spectroscopy and ellipsometry. Colloids Surf. B: Biointerfaces **11**(4), 187–201 (1998)
50. K. Vedam, Spectroscopic ellipsometry: a historical overview. Thin Solid Films **313**, 1–9 (1998)
51. B. Johs, C.M. Herzinger, J.H. Dinan, A. Cornfeld, J.D. Benson, Development of a parametric optical constant model for Hg1− xCdxTe for control of composition by spectroscopic ellipsometry during MBE growth. Thin Solid Films **313**, 137–142 (1998)
52. B.D. Johs, J. Hale, N.J. Ianno, C.M. Herzinger, T.E. Tiwald, J.A. Woollam, Recent developments in spectroscopic ellipsometry for in-situ applications, in *Optical Metrology Roadmap for the Semiconductor, Optical, and Data Storage Industries II*, vol. 4449, (2001), pp. 41–57
53. M.J. Müller, K. Komander, C. Höhn, R. van de Krol, A.C. Bronneberg, Growth of Bi_2O_3 films by thermal-and plasma-enhanced atomic layer deposition monitored with real-time spectroscopic ellipsometry for photocatalytic water splitting. ACS Appl. Nano Mater. **2**(10), 6277–6286 (2019)
54. U. Kilic, A. Mock, D. Sekora, S. Gilbert, S. Valloppilly, G. Melendez, N. Ianno, M. Langell, E. Schubert, M. Schubert, Precursor-surface interactions revealed during plasma-enhanced atomic layer deposition of metal oxide thin films by in-situ spectroscopic ellipsometry. Sci. Rep. **10**(1), 10392 (2020)

55. M. Labbe, K. Cadien, D.G. Ivey, Growth of multiple Island layers during iron oxide atomic layer deposition: An electron microscopy and spectroscopic ellipsometry investigation. J. Phys. Chem. C **126**(46), 19883–19894 (2022)
56. E.A. Marquis, M. Bachhav, Y. Chen, Y. Dong, L.M. Gordon, A. McFarland, On the current role of atom probe tomography in materials characterization and materials science. Curr. Opinion Solid State Mater. Sci. **17**(5), 217–223 (2013)
57. A.M. Cuevas, H.P. Gravie, Portable energy dispersive X-ray fluorescence and X-ray diffraction and radiography system for archaeometry. Nucl. Instrum. Methods Phys. Res., Sect. A **633**(1), 72–78 (2011)
58. H. Khan, A.S. Yerramilli, A. D'Oliveira, T.L. Alford, D.C. Boffito, G.S. Patience, Experimental methods in chemical engineering: X-ray diffraction spectroscopy – XRD. Can. J. Chem. Eng. **98**(6), 1255–1266 (2020)
59. R. Svoboda, Reaction/crystallization kinetics studied via in situ XRD: Experimental conditions versus methods of kinetic analysis. Philos. Mag. **99**(23), 2941–2956 (2019). https://doi.org/10.1080/14786435.2019.1648899
60. E.O. Ortega, H. Hosseinian, I.B.A. Meza, M.J.R. López, A.R. Vera, S. Hosseini, *Material Characterization Techniques and Applications* (Springer, Singapore, 2022)
61. M. Eckert, Max von Laue and the discovery of X-ray diffraction in 1912. Ann. Phys. **524**(5), A83–A85 (2012). https://doi.org/10.1002/andp.201200724
62. S.V. Borisov, N.V. Podberezskaya, X-ray diffraction analysis: A brief history and achievements of the first century. J. Struct. Chem. **53**(1), 1 (2012). https://doi.org/10.1134/S0022476612070013
63. G. Hildebrandt, The discovery of the diffraction of X-rays in crystals – a historical review. Cryst. Res. Technol. **28**(6), 747–766 (1993). https://doi.org/10.1002/crat.2170280602
64. H. Stanjek, W.J.H.I. Häusler, Basics of X-ray diffraction. Hyperfine Interact. **154**, 107–119 (2004). https://doi.org/10.1023/B:HYPE.0000032028.60546.38
65. C. Tsuboi, F. Kimura, T. Tanaka, T. Kimura, Single-crystal X-ray diffraction analysis of microcrystalline powders using magnetically oriented microcrystal suspensions. Cryst. Growth Des. **16**(5), 2810–2813 (2016)
66. K.D. Harris, E.Y. Cheung, How to determine structures when single crystals cannot be grown: Opportunities for structure determination of molecular materials using powder diffraction data. Chem. Soc. Rev. **33**(8), 526–538 (2004)
67. P., E. Imoisili, K.O. Ukoba, T.-C. Jen, Physical, mechanical and thermal properties of high frequency microwave treated plantain (Musa paradisiacal) fibre/MWCNT hybrid epoxy nanocomposites. J. Mater. Res. Technol. **9**(3), 4933–4939 (2020). https://doi.org/10.1016/j.jmrt.2020.03.012
68. S. Tariq, A.R. Raza, M. Khalid, S.L. Rubab, M.U. Khan, A. Ali, . . ., A.A.C. Braga, Synthesis and structural analysis of novel indole derivatives by XRD, spectroscopic and DFT studies. J. Mol. Struct. **1203**, 127438 (2020). https://doi.org/10.1016/j.molstruc.2019.127438
69. P.E. Imoisili, I.T. Dagogo, A.V. Popoola, A.E. Okoronkwo, Effect of high-frequency microwave radiation on the mechanical properties of plantain (Musa paradisiaca) fibre/epoxy biocomposite. J. Phys. Sci. **29**(3), 23–35 (2018). https://doi.org/10.21315/jps2018.29.3
70. P.E. Imoisili, T.-C. Jen, Mechanical and water absorption behaviour of potassium permanganate ($KMnO_4$) treated plantain (Musa Paradisiaca) fibre/epoxy bio-composites. J. Mater. Res. Technol. **9**(4), 8705–8713 (2020). https://doi.org/10.1016/j.jmrt.2020.05.121
71. P.E. Imoisili, T.-C. Jen, Modelling and optimization of the impact strength of plantain (Musa paradisiacal) fibre/MWCNT hybrid nanocomposite using response surface methodology. J. Mater. Res. Technol. **13**, 1946–1954 (2021). https://doi.org/10.1016/j.jmrt.2021.05.101
72. E. Charles Nwanna, P.E. Imoisili, S. Bitire, T.C. Jen, Green synthesis of $CuNb_2O_6$ thin film using *allium cepa* as catalyst. Energ. Source. Part A **44**(2), 3610–3627 (2022). https://doi.org/10.1080/15567036.2022.2067602

73. P.E. Imoisili, J. Ren, T.C. Jen, Single-atom catalysts for lithium sulfur batteries via atomic layer deposition process. Electrochem. Commun. **135**, 107215 (2022). https://doi.org/10.1016/j.elecom.2022.107215
74. P.E. Imoisili, T.-C. Jen, Microwave-assisted sol-gel template-free synthesis and characterization of silica nanoparticles obtained from South Africa coal fly. Nanotechnol. Rev. **11**, 1–11 (2022). https://doi.org/10.1515/ntrev-2022-0476
75. E.C. Nwanna, P.E. Imoisili, T.-C. Jen, Green synthesis preparation and evaluation of CuO doped SnO_2. J. Mater. Res. Technol. **17**, 1901 (2022). https://doi.org/10.1016/j.jmrt.2022.01.081
76. R.P. Hansen, R. Phillips, E. Eisenbraun, Substoichiometric TiO_2 thin films as a support catalyst for hydrogen fuel cells. ECS Trans. **35**(34), 17 (2011)
77. J. Lee, S.H. Hong, J.Y. Jho, A hydrogen gas sensor employing vertically aligned TiO2 nanotube arrays prepared by template-assisted method. Sensors Actuators B Chem. **160**(1), 1494–1498 (2011)
78. P. Rostron, S. Gaber, D. Gaber, Raman spectroscopy, review. laser **21**, 24 (2016)
79. S. Stewart, R.J. Priore, M.P. Nelson, P.J. Treado, Raman imaging. Annu. Rev. Anal. Chem. **5**, 337–360 (2012). https://doi.org/10.1146/annurev-anchem-062011-143152
80. H. Mitsutake, R.J. Poppi, M.C. Breitkreitz, Raman imaging spectroscopy: History, fundamentals and current of the technique. J. Braz. Chem. Soc. **30**(11), 2243–2258 (2019)
81. N. Bost, F. Westall, F. Foucher, G. Guimbretière, Petrographical andmineralogical applications of Raman mapping, in *Raman Spectroscopy and Applications*, ed. by K. Maaz, (Intech, London, 2017)
82. J.R. Beattie, J.J. McGarvey, A.W. Stitt, Raman spectroscopy for the detection of AGEs/ALEs, in *Methods in Molecular Biology*, ed. by L. Galluzzi, I. Vitale, O. Kepp, G. Kroemer, vol. 965, (Springer, Belfast, 2013), pp. 297–312
83. K. Eberhardt, C. Stiebing, C. Matthäus, M. Schmitt, J. Popp, Advantages and limitations of Raman spectroscopy for molecular diagnostics: An update. Expert. Rev. Mol. Diagn. **15**(6), 773–787 (2015). https://doi.org/10.1586/14737159.2015.1036744
84. D. Chenery, H. Bowring, Infrared and Raman spectroscopic imaging in biosciences. Spectrosc. Eur. **15**(4), 8–14 (2003) [Online]. https://www.spectroscopyasia.com/system/files/pdf/IR_15_4.pdf
85. E.V. Formo, Z. Wu, S.M. Mahurin, S. Dai, In situ high temperature surface enhanced Raman spectroscopy for the study of interface phenomena: probing a solid acid on alumina. J. Phys. Chem. C **115**(18), 9068–9073 (2011)
86. E. Chason, T.M. Mayer, Thin film and surface characterization by specular X-ray reflectivity. Crit. Rev. Solid State Mater. Sci. **22**(1), 1–67 (1997)
87. D. Rebiscoul, A. van Der Lee, P. Frugier, A. Ayral, S. Gin, X-ray reflectometry characterization of SON 68 glass alteration films. J. Non-Cryst. Solids **325**(1–3), 113–123 (2003)
88. P. Bergese, E. Bontempi, L.E. Depero, A simple solution to systematic errors in density determination by X-ray reflectivity: The XRR-density evaluation (XRR-DE) method. Appl. Surf. Sci. **253**(1), 28–32 (2006)
89. A.C. Ferrari, B.K. Andrea Libassi, V. Tanner, J. Stolojan, L.M. Yuan, S.E. Brown, B.K. Rodil, J. Robertson, Density, sp 3 fraction, and cross-sectional structure of amorphous carbon films determined by X-ray reflectivity and electron energy-loss spectroscopy. Phys. Rev. B **62**(16), 11089 (2000)
90. A. Gibaud, M.S. Chebil, T. Beuvier, X-ray reflectivity. Surf. Sci. Tech. **51**, 191–216 (2013)
91. M. Yasaka, X-ray thin-film measurement techniques. Rigaku J. **26**(2), 1–9 (2010)
92. J. Hämäläinen, T. Hatanpää, E. Puukilainen, T. Sajavaara, M. Ritala, M. Leskelä, Iridium metal and iridium oxide thin films grown by atomic layer deposition at low temperatures. J. Mater. Chem. **21**(41), 16488–16493 (2011)
93. J.Y. Kim, S.M. George, Tin monosulfide thin films grown by atomic layer deposition using tin 2, 4-pentanedionate and hydrogen sulfide. J. Phys. Chem. C **114**(41), 17597–17603 (2010)

94. Y. Waseda, E. Matsubara, K. Shinoda, Y. Waseda, E. Matsubara, K. Shinoda, Fundamental properties of X-rays, in *X-Ray Diffraction Crystallography: Introduction, Examples and Solved Problems*, (2011), pp. 1–20
95. M.S. Shackley, An introduction to X-ray fluorescence (XRF) analysis in archaeology, in *X-Ray Fluorescence Spectrometry (XRF) in geoarchaeology*, ed. by M. Shackley, (Springer, New York, 2011). https://doi.org/10.1007/978-1-4419-6886-9_2
96. L. Löwemark, H.F. Chen, T.N. Yang, M. Kylander, E.F. Yu, Y.W. Hsu, T.Q. Lee, S.R. Song, S. Jarvis, Normalizing XRF-scanner data: A cautionary note on the interpretation of high-resolution records from organic-rich lakes. J. Asian Earth Sci. **40**(6), 1250–1256 (2011)
97. G.J. Weltje, R. Tjallingii, Calibration of XRF core scanners for quantitative geochemical logging of sediment cores: theory and application. Earth Planet. Sci. Lett. **274**(3–4), 423–438 (2008)
98. A. Kumar, Henry Moseley's law and it's applications. Int. J. Res. Anal. Sci. Eng. **3**(3), 8–8 (2023)
99. M.S. Shackley, C. Dillian, Thermal and environmental effects on obsidian geochemistry: Experimental and archaeological evidence, in *The Effects of Fire and Heat on Obsidian*, (2002), pp. 117–134
100. E.T. Hall, X-ray fluorescent analysis applied to archaeology. Archaeometry **3**(1), 29–35 (1960)
101. A. Rindby, Software for energy-dispersive X-ray fluorescence. X-Ray Spectrom. **18**(3), 113–118 (1989)
102. M.S. Shackley, Is there reliability and validity in portable X-ray fluorescence spectrometry (PXRF). SAA Archaeol. Rec. **10**(5), 17–20 (2010)
103. P.D. Bouey, Recognizing the limits of archaeological applications of non-destructive energy-dispersive X-ray fluorescence analysis of obsidians. Mater. Res. Soc. Proc. **185**, 309–320 (1990)
104. M. Scimeca, S. Bischetti, H.K. Lamsira, R. Bonfiglio, E. Bonanno, Energy dispersive X-ray (EDX) microanalysis: A powerful tool in biomedical research and diagnosis. Eur. J. Histochem. **62**(1), 2841 (2018)
105. C. Angeles-Chavez, J.A. Toledo-Antonio, M.A. Cortes-Jacome, Chemical quantification of Mo-S, W-Si and Ti-V by energy dispersive X-ray spectroscopy, in *X-ray Spectroscopy*, ed. by S.K. Sharma, (IntechOpen, London, 2012), pp. 119–136
106. M. Rathi, P. Ahrenkiel, J. Carapella, M. Wanlass, A standards-based method to determine elemental compositions using energy dispersive X-ray spectrometry for compound semiconductors. Microsc. Microanal. **18**(S2), 1046–1047 (2012)
107. M. Deutsch, O. Gang, K. Hämäläinen, C.C. Kao, Onset and near threshold evolution of the Cu K α X-ray satellites. Phys. Rev. Lett. **76**(14), 2424 (1996)
108. F. De Groot, High-resolution X-ray emission and X-ray absorption spectroscopy. Chem. Rev. **101**(6), 1779–1808 (2001)
109. E.A. Kenik, Evaluating the performance of a commercial silicon drift detector for X-ray microanalysis. Microscopy Today **19**(3), 40–46 (2011)
110. P. Schlossmacher, D.O. Klenov, B. Freitag, H.S. Von Harrach, Enhanced detection sensitivity with a new windowless XEDS system for AEM based on silicon drift detector technology. Microscopy today **18**(4), 14–20 (2010)
111. P. Echlin, C.E. Fiori, J. Goldstein, D.C. Joy, D.E. Newbury, *Advanced Scanning Electron Microscopy and X-Ray Microanalysis* (Springer Science+Business Media New York, 1986)
112. H. Feng, J.W. Elam, J.A. Libera, W. Setthapun, P.C. Stair, Palladium catalysts synthesized by atomic layer deposition for methanol decomposition. Chem. Mater. **22**(10), 3133–3142 (2010)
113. T.H.Y. Tran, W.G. Haije, V. Longo, W.M.M. Kessels, J. Schoonman, Plasma-enhanced atomic layer deposition of titania on alumina for its potential use as a hydrogen-selective membrane. J. Membr. Sci. **378**(1–2), 438–443 (2011)

114. M.M. Ghorbani, R. Taherian, 12 – Methods of measuring electrical properties of Material, in *Electrical Conductivity in Polymer-Based Composites: Experiments, Modelling, and Applications*, (Elsevier, 2018), p. 365
115. Heaney, M. B. Electrical conductivity and resistivity. Electrical Measurement, Signal Processing, and Displays, 7–1 (CRC Press, 2003)
116. W.D. Callister Jr., D.G. Rethwisch, *Fundamentals of Materials Science and Engineering: An Integrated Approach* (Wiley, Hoboken, NJ: John Wiley & Sons 2020)
117. M.D. Janezic, N. Paulter, J. Blendell, *Dielectric and Conductor-Loss Characterization and Measurements on Electronic Packaging Materials* (NIST Technical note, 1520, 2001)
118. Y. Singh, Electrical resistivity measurements: A review, in *International Journal of Modern Physics: Conference Series*, vol. 22, (World Scientific Publishing Company, Hong Kong 2013), pp. 745–756
119. N. Keller, T. Bein, Optoelectronic processes in covalent organic frameworks. Chem. Soc. Rev. **50**(3), 1813–1845 (2021)
120. B. Kalderon, D. Sarkar, K. Killi, T. Danzig, D. Azulay, O. Millo, G. Cohen-Taguri, R. Yerushalmi, Layered Si–Ti oxide thin films with tailored electrical and optical properties by catalytic tandem MLD-ALD. RSC Adv. **11**(56), 35099–35109 (2021)
121. X. Jiang, H. Huang, F.B. Prinz, S.F. Bent, Application of atomic layer deposition of platinum to solid oxide fuel cells. Chem. Mater. **20**(12), 3897–3905 (2008)
122. L. Evans, *UV-VIS Spectrophotometry: A Brief Background to Spectrophotometry* (Biochrom Ltd a division of harvad biosciencie, Inc, 2004), pp. 1–15
123. J. Mistrik, S. Kasap, H.E. Ruda, C. Koughia, J. Singh, Optical properties of electronic materials: Fundamentals and characterization, in *Springer Handbook of Electronic and Photonic Materials*, (2017), pp. 1–1
124. B.M. Tissue, Ultraviolet and visible absorption spectroscopy, in *Characterization of Materials*, (2002). https://doi.org/10.1002/0471266965.com059
125. R.X. Yang, M.K. Horton, J. Munro, K.A. Persson, High-throughput optical absorption spectra for inorganic semiconductors. arXiv preprint arXiv, 2209.02918 (2022)
126. F.S. Rocha, A.J. Gomes, C.N. Lunardi, S. Kaliaguine, G.S. Patience, Experimental methods in chemical engineering: Ultraviolet visible spectroscopy – UV-vis. Can. J. Chem. Eng. **96**(12), 2512–2517 (2018)
127. J.M. Antosiewicz, D. Shugar, UV–vis spectroscopy of tyrosine side-groups in studies of protein structure. Part 1: Basic principles and properties of tyrosine chromophore. Biophys. Rev. **8**(2), 151–161 (2016)
128. D.T. Burns, M.J. Walker, Origins of the method of standard additions and of the use of an internal standard in quantitative instrumental chemical analyses. Anal. Bioanal. Chem. **411**, 2749–2753 (2019)
129. Z. Chen, H.N. Dinh, E. Miller, *Photoelectrochemical Water Splitting*, vol 344 (Springer, New York, 2013), pp. 6–15
130. M. Sauer, J. Hofkens, J. Enderlein, Basic principles of fluorescence spectroscopy, in *Handbook of Fluorescence Spectroscopy and Imaging*, vol. 1, (2011), p. 30
131. R. Singh, R. Bapat, L. Qin, H. Feng, V. Polshettiwar, Atomic layer deposited (ALD) TiO_2 on fibrous nano-silica (KCC-1) for photocatalysis: nanoparticle formation and size quantization effect. ACS Catal. **6**(5), 2770–2784 (2016)
132. W.S. Liu, L.C. Wang, T.K. Chin, Y.C. Yen, T.P. Perng, Fabrication of TiO_2 on porous g-C_3N_4 by ALD for improved solar-driven hydrogen evolution. RSC Adv. **8**(54), 30642–30651 (2018)

Chapter 6
Future Prospects

6.1 Hydrogen, Countries, and Innovations Hubs

ALD is highly applicable to the design and production of novel energy and other devices, where nanoengineering surfaces and interfaces at the nanoscale are crucial for enhancing their performance [1]. Technologies for producing hydrogen with low emissions are developing rapidly; however, additional attention is required on demand and supply side technology. This has to do with market policies, regulations, and interventions. Many countries like the United States and Europe have clear policies on actions, and China is also doing well in deployment.

6.1.1 United States of America

Investing in hydrogen entails investing in the future, specifically in climate protection, creating skilled employment opportunities, and ensuring energy security. Hydrogen has been designated as the first of the U.S. Department of Energy's (DOE) "Energy Earthshots," an effort that attempts to speed up the development of more reasonably priced and dependable clean energy solutions. The DOE announced it seeks to reduce the cost of clean hydrogen by 80% to **$1** per **1 kg** in **1 decade ("1 1 1")**. How do we achieve this 80% to $1 per 1 kg in 1 decade ("1 1 1")?

P. O. Oviroh et al., *Emerging Atomic Layer Deposition for Hydrogen Energy*,
Springer Series in Materials Science 348,
https://doi.org/10.1007/978-3-031-67774-8_6

6.1.1.1 Inflation Reduction Act of 2022 Hydrogen and Fuel Cell Incentives

The hydrogen tax credit included in the Inflation Reduction Act (IRA) will assist in developing a clean hydrogen industry, which will be essential in reducing emissions from sectors that are harder-to-decarbonize, such as heavy industry and heavy transportation. As part of the IRA, the Clean Hydrogen Production Credit establishes a new incentive for producing clean hydrogen over 10 years, making it possible to get up to \$3.00/kg or 30% investment tax credit (ITC). The amount of carbon intensity determines the level of credit offered.

The Section 45V hydrogen production tax credit included in the IRA is undoubtedly a policy that addresses current challenges. It provides \$3/kg of hydrogen generated at its most profitable tier, which is available for the lowest-carbon output. That is about half of what it now costs, depending on equipment cost, production volume, and energy price, to create hydrogen using electrolyzers at a rate of \$5 to \$6/kg. With the tax incentive, clean hydrogen may become as cheap as \$1 to \$1.50/kg compared to gray hydrogen [2].

It is suggested that hydrogen producers be required to have a matching production and consumption of clean power down to an hourly basis. This will mitigate against the use of carbon intensive power sources. If the Inflation Reduction Act's tax credit and incentive regimes are not implemented effectively, clean hydrogen production could face challenges in its use, as it could be used inefficiently, undermining the decarbonization process.

The credit provides a varying incentive depending on the carbon intensity of the hydrogen production pathway. The credit measures emissions up to the point of production using the Argonne National Laboratory Greenhouse Gases, Regulated Emissions, and Energy Use in Technologies (GREET) Model. The incentive breakdown is detailed in the following Table 6.1.

The hydrogen hubs (H2Hubs), which are funded by the Bipartisan Infrastructure Law (BIL), will speed up the commercial-scale deployment of clean hydrogen, assisting in the production of clean, dispatchable power, developing new energy storage technology, and decarbonizing the transportation and heavy industries [3].

The DOE evaluated key areas for selecting H2Hubs, including technical merit and impact, financial and market viability, work plan, management team and project partners, and community benefit plan. These factors included the hub's ability to produce clean hydrogen, reduce emissions, and its financial and market viability [3] (Fig. 6.1).

Table 6.1 is the incentives breakdown for the clean hydrogen

Kg of CO_2 per kg of H_2	Credit value (\$)
4–2.5 kg CO_2	\$0.60/kg of H_2
2.5–1.5 kg CO_2	\$0.75/kg of H_2
1.5–0.45 kg CO_2	\$1.00/kg of H_2
0.45–0 kg CO_2	\$3.00/kg of H_2

Fig. 6.1 shows the H2huns in different locations within the United States [4]

Appalachian Hydrogen Hub: The primary objective of the Appalachian Hydrogen Hub is to generate affordable, environmentally friendly hydrogen while simultaneously sequestering carbon emissions inside the region's economical natural gas reserves. The strategic positioning of the hub and the presence of hydrogen pipes will effectively decrease the costs associated with transporting and storing hydrogen. The center's objective is to achieve a yearly reduction of 9 million metric tons of CO_2 emissions, comparable to the emissions produced by more than 2 million automobiles driven by petrol. Additionally, it will provide high-quality employment prospects for coal workers and forge collaborations with institutions and workforce organizations. The hub will furthermore create a Community Benefits Advisory Board to supervise the execution of the Community Benefits Plan and a Community Commitment Fund.

California Hydrogen Hub: The California Hydrogen Hub seeks to generate hydrogen from renewable energy and biomass, therefore lowering emissions from public transportation, heavy-duty haulage, and port activities. It will bring clean hydrogen into heavy-duty transportation via cargo handling equipment and drayage, preparing ports for possible export. The center will also include hydrogen for power generation, improving energy security and resilience. To encourage community involvement, the hub will have robust governance structures, strict environmental justice criteria, independent monitoring of Clean Building Practices (CBP), and monetary sanctions for noncompliance.

The Gulf Coast Hydrogen Hub: Using natural gas with carbon capture and electrolysis driven by renewable energy sources, the Gulf Coast Hydrogen Hub aims to establish a clean hydrogen economy. Through the development of salt cavern storage, a sizable open-access pipeline, and several hydrogen refueling stations, the hub hopes to lower the cost of hydrogen by utilizing plentiful renewable energy sources and natural gas. Ammonia, petrochemicals, refineries, naval fuel, fuel cell electric trucks, and industrial operations will all require hydrogen. To promote inclusive workforce development and community investment, the hub will also create a Community Advisory Board, utilizing partners in the community and education.

The Heartland Hydrogen Hub: Seeks to lower the price of clean hydrogen while decarbonizing fertilizer production in the agriculture industry. In addition to providing local farmers and farmer cooperatives with equity ownership, the hub intends to generate electricity using clean hydrogen. The hub will build a hydrogen network available to both present and future customers by utilizing pipeline infrastructure and open-access storage, to perhaps duplicate it throughout the Midwest.

The Mid-Atlantic Hydrogen Hub: aims to establish renewable hydrogen production facilities by harnessing the power of renewable energy and nuclear electricity. This initiative seeks to lower expenses and promote the widespread use of hydrogen technology. The hub tends to broaden the use of hydrogen in the heavy transportation, manufacturing, and combined heat and power industries. The hub would expand the hydrogen distribution infrastructure, enhance bus mechanic depots, and establish fueling stations to enhance the distribution of hydrogen to a larger number of end users, reducing storage and distribution costs and reducing carbon emissions.

The Midwest Hydrogen Hub, situated in a crucial U.S. industrial and transportation corridor, aims to promote decarbonization through strategic hydrogen uses like steel and glass production, power generation, refining, heavy-duty transportation, and sustainable aviation fuel. The hub will produce hydrogen using renewable, natural gas, and low-cost nuclear energy.

The Pacific Northwest Hydrogen Hub intends to produce clean hydrogen solely through electrolysis using renewable resources. This will lower the cost of the electrolyzer, increasing technological accessibility and lowering the cost of producing hydrogen. The Hub seeks to reduce about 1.7 million metric tons of CO_2 emissions a year, equivalent to 346,000 gasoline-powered automobiles off the road. Hard-to-decarbonize sectors, including heavy-duty transportation, agriculture, manufacturing, and seaports, will all undergo this shift. Additional hydrogen applications include the creation of fertilizer in agriculture, the manufacturing of generators, peak power, data centers, refineries, and drayage and cargo handling at seaports, where it will help reduce emissions from some of the region's highest-producing industries (Table 6.2).

Some important definitions:

Clean hydrogen refers to hydrogen produced by the process of electrolysis, which involves hydrogen extraction from liquid water using renewable or low-carbon emission energy sources such as wind, solar, or nuclear power. Clean hydrogen is

Table 6.2 is the summary of the hubs in the United States

Hub	Region	DOE ward ($) billions	Direct jobs
Heartland Hydrogen Hub (HH2H)	Midwest	Up to 0.925	3880
Pacific Northwest Hydrogen Hub (PNW H2)	West coast	Up to 1	10,000
Midwest Alliance for Clean Hydrogen (MachH2)	Midwest	Up to 1	13,600
Mid-Atlantic Clean Hydrogen Hub (MACH2)	Mid-Atlantic	Up to 0.750	20,000
Appalachian Hydrogen Hub (ARCH2)	Appalachian	Up to 0.925	21,000+
Gulf Coast Hydrogen Hub (HyVelocity Hydrogen Hub)	U.S. Gulf Coast	Up to 1.2	45,000
Alliance for Renewable Clean Hydrogen Energy Systems (ARCHES)	West coast	Up to 1.2	220,000

generated using thermal conversion methods while utilizing carbon capture and permanent storage (CCS) technology to reduce greenhouse gas emissions.

H2Hubs will enable the use of an innovative, eco-friendly fuel, and energy storage system, therefore assisting in the mitigation of carbon emissions in the economy.

Each geographical area in the country possesses distinct advantages in terms of the fuels; it can be utilized for hydrogen production. The Bipartisan Infrastructure Law acknowledges this fact by stipulating that, in addition to other criteria, a minimum of two Hubs must be situated in regions abundant in natural gas, at least one Hub must employ fossil fuels with carbon capture, one Hub must rely on nuclear power, and one Hub must utilize renewable energy as a feedstock.

Reduction of air pollution (e.g., particulate matter) through measures such as switching from diesel to fuel cell vehicles; cleaning up polluted sites that were previously used for production or end use; investing in workforce development and education; creating local jobs that are both permanent and of high quality; establishing new businesses; forming equity partnerships; and ensuring energy resilience.

In addressing risks associated with hubs, be it commitment to safe deployment and the mitigation of social, economic, technical, and environmental risks related to the H2Hubs. Projects must submit thorough risk assessments and management plans that include potential hazards, their impacts, and the actions taken to mitigate or eliminate them. A Hydrogen Safety Plan including the methodology for safely generating, storing, and conveying hydrogen will also be supplied.

6.1.2 Germany

Germany aims to substantially boost its hydrogen market, planning to double its domestic electrolyzer capacity from 5 GW to at least 10 GW by 2030. A separate

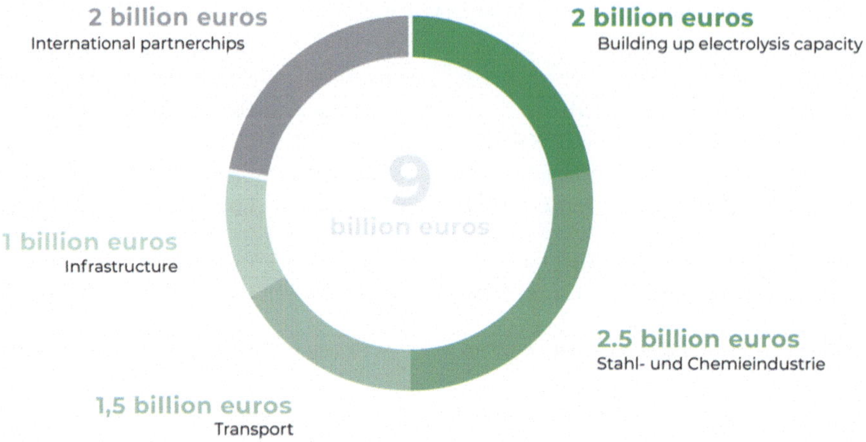

Fig. 6.2 Distribution of National Hydrogen Strategy (Nationale Wasserstoffstrategie) NWS federal funds by use (in €) [7]

import strategy will be developed to cover the remaining demand. An efficient hydrogen infrastructure will be developed, with a hydrogen start-up grid in Germany with more than 1800 km of repurposed and newly built hydrogen pipelines and 4500 km in Europe added by 2027/2028. By 2030, hydrogen and its derivatives will be extensively utilized in many sectors such as manufacturing, aviation, shipping, and electricity, hence enhancing energy supply security through the utilization of gas power [5].

Their plan is to develop a varied and reliable hydrogen supply, minimize reliance on new sources, implement sustainability criteria, and forge technological partnerships with nations that import hydrogen [6].

Hydrogen applications are expected in industries like chemical and steel, transport, and electricity for fuel cells or as a renewable fuel. However, by 2030, no broad hydrogen application is expected in the heating sector, although converting gas distribution networks to hydrogen and using decentralized hydrogen boilers should be legal and technically possible.

The 2020 stimulus package allocated €9 billion for the NWS, with €7 billion for the national hydrogen economy ramp-up and €2 billion for international hydrogen partnerships, including import infrastructure development and hydrogen production in third countries, to finance the first stage of market ramp-up by 2023 [7].

More details in the factsheets [6] (Fig. 6.2).

6.1.3 A Country in Middle East: Morocco

Morocco has been identified as one nation with great potential for exporting green hydrogen [8]. The wind is consistently available at a high level, ranging from 50% to

60%, with an average wind speed reaching 12 m/s. Additionally, Morocco enjoys enough sunlight, with 2500–3000 hours of sunshine annually [8].

From the presentation on their strategies by the green hydrogen organization [9], Morocco aims to establish an economic and industrial sector based on green molecules like hydrogen, ammonia, and methanol to reduce greenhouse gas emissions and support decarbonization in partner countries. The country's renewable energy potential will drive green hydrogen development, promote economic growth, strengthen energy security, and potentially export green energy.

The Moroccan Ministry of Energy, Mines and Environment has set a roadmap for green hydrogen, aiming for a demand of 30 TWh by 2030 and 307 TWh by 2050. The strategy is based on market and demand, cost reduction technology, and investment. Morocco's established renewable energy model and geographical positioning make it a promising candidate. The strategy includes three phases: short term (2020–2030), medium term (2030–2040), and long term (2040–2050). It also supports local ammonia production, which Morocco has previously relied on imported. The government aims to export green energy and establish a national industry based on green hydrogen.

The German Moroccan Energy Partnership (PAREMA) is developing a roadmap for green hydrogen production, with an investment of €300 million. Total energies aim to establish a 10 GW plant in Morocco.

As highlighted in a study on the "Technical-economic and socio-political assessment of hydrogen production from solar energy," a case study was based on Morocco. With decline in the costs of PV and wind, the results of the levelized cost of electricity $/kWh (LCOE) and levelized cost of hydrogen $/kg (LCOH) can be further reduced if the state provides subsidies, which could help greatly in reducing the CAPEX costs of the project [10], although Morocco's renewable energy resources and European proximity will enable green hydrogen expansion, reducing transportation costs. However, challenges that need more closer look include competitiveness, production capacity, and skill development for a mature industry. The country will need to invest in electrolyzer and fuel cells to meet the growing demand and expectations.

6.1.4 South Africa

Interest in hydrogen as a clean energy source is developing across the world, with South Africa having the ability to transform its economy while also supplying green hydrogen. South Africa's tremendous renewable energy potential might enable it to become a low-cost exporter of green hydrogen. Hydrogen (H_2) energy has the potential to be a crucial component of a sustainable energy future when produced from renewable resources and utilized in concert with other clean energy technologies [11]. This would allow for the development of new sectors, employment creation, and the earning of foreign exchange.

The abundant sunlight in South Africa (SA), which are required to produce green hydrogen, gives SA a comparative economic advantage. Due to its abundant solar radiation, SA is an ideal location for solar energy production [12]. The majority of South Africa receives more than 2500 hours of annual sunshine. The daily average solar radiation is between 4.5 and 6.5 kWh/m^2 [13]. In South Africa, the average annual 24-hour solar radiation is 220 W/m^2, which is greater than in many regions of the United States (about 150 W/m^2) and the European Union (about 100 W/m^2) [14]. In SA, Sasolburg Operations has installed a 3 MW solar farm and plan supplementing it with 69 MW renewable energy from a wind farm in 2024 to increase production to 3500 kg per day of green hydrogen [15].

Electrolyzer units up to 2 MW are available commercially. Electrolysis at low temperatures requires only electricity and water as inputs. Electricity consumption is approximately 45–50 kWh/kg H_2, while water consumption is 10–12 l/kg H_2 produced at an 80% conversion efficiency [12]. To achieve cost competitiveness, it is imperative to secure an economically efficient renewable power supply. The efficiency of the process usually ranges from 70% to 85%, depending on the kind of electrolyzer and its application. The alkaline water electrolyzer (AWE) is a century-old technology with a system efficiency of 50–78 kWh/kg. In contrast, the proton exchange membrane water electrolyzer (PEMWE) and anion exchange membrane water electrolyzer (AEMWE) are newer technologies with system efficiencies of 50–83 kWh/kg and 57–69 kWh/kg, respectively [16]. There is also need for innovations in SA context. In SA, SASOL has repurposed a 60 MW electrolyzer in Sasolburg to create green hydrogen for use in green mobility initiatives. The green hydrogen produced will underpin early local demand in heavy-duty mobility for commercial transportation and mining, city buses, green steel revitalization, and backup power generation. The focus is on developing major freight lines [17].

By the year 2030, the projected cost of producing green hydrogen using the LCOH method will be around $4/kg of hydrogen across various distribution hubs. Three green hydrogen hubs identified are Johannesburg hub, Durban hub, and Mogalakwena and Limpopo [18]. However, it is important to note that green hydrogen remains more costly than gray hydrogen, with a premium of $2–$2.5 associated with the creation of green hydrogen. These values are quite high to be competitive globally. Based on South Africa hydrogen valley 2021 reports [18], it is projected that by 2030, the cost of producing green hydrogen using the LCOH method will be around $4 $/kg H_2 at various locations in SA. South Africa has the advantage of the availability of critical minerals such as iridium and platinum which are critical minerals; in addition, they have higher solar insolation and good wind potential which is higher than many locations in the world.

Generating hydrogen using electricity is quite expensive. The major driver of the expensive price of green hydrogen generation is energy, at more than 70% of the cost on average, which will only get greater as a part of overall cost in the future. South Africa currently has energy deficit, to be able to use electricity from the grid; dedicated renewable sources are pertinent. Current national grid might not be able to sustain the energy requirement for electrolysis, for South Africa to provide green hydrogen at sustainable and economical costs.

Apart from the grid systems, there is need for infrastructure development. Transporting H_2 is difficult due to its volatile and extremely combustible nature. Hydrogen embrittlement is a concern as well. It should ideally be produced in close proximity to the areas where it will be consumed. Aside from pipelines, there are two current alternatives for the secure and cost-effective transportation of H_2: compressing and/or liquefying it to improve its density, and converting it into a more convenient carrier with a reconversion process before ultimate usage for long distances [16]. These pipelines must meet stringent safety requirements due to hydrogen's flammability. Building a network of hydrogen pipelines can help transport green hydrogen from production centers, such as renewable energy farms, to end users across the country. This would require significant investment in infrastructure but could facilitate widespread adoption of green hydrogen. SA has existing hydrogen infrastructure which it has been using for some industrial supply, however for a green hydrogen economy, that need to be scaled up to appreciable level to carter for all customers in the different demand sector.

There is need to grow and expand the market for hydrogen-powered mobility solutions, which is a crucial element of South Africa's green hydrogen economy plan. This initiative will reduce carbon emissions, attract investments, generate employment opportunities, and increase the demand for essential metals and raw materials, such as PGMs. A proof-of-concept demonstration which included a hydrogen mobility ecosystem, a sustainable option that has the potential to decarbonize the hard-to-abate and hard-to-electrify heavy-duty transportation sector, such as long-haul trucks, buses, and mining fleets, has been done [15, 17].

Government policies and regulations are required to meet the new challenges arising with having green hydrogen as a mainstream energy source; in addition, there is need for sustainable government investment to catalyze the sector, although in SA context the private sector need to take greater actions, considering the competitive nature on the available resources.

6.2 Material Innovation

Materials created using the atomic layer deposition (ALD) technique can be employed to address a wide range of environmental challenges [19] such as utilized in specialized applications to enhance the performance of photovoltaics [19–21], membranes [22–24], batteries [25, 26], or fuel cell technologies [27–30]. All these find application in green hydrogen energy.

A material being developed is a hydrogen-based fuel cell approach for eco-friendly car fuel, aiming to reduce carbon emissions and contribute to a sustainable future by using renewable electricity to extract hydrogen from steam, a "green" alternative to traditional fossil fuel methods.

The superprotonic hydrogen transport solution involves using ammonia, a mixture of hydrogen and nitrogen, to separate hydrogen from nitrogen for use in fuel cells. Ammonia offers advantages like pipelines and high transmission capacity.

Haile creates fuel cell-like devices using cesium dihydrogen phosphate as an electrolyte, allowing protons to move through, and removing hydrogen from ammonia. The material's other advantages include its affordability, non-toxicity, and earth abundance, making it a desirable sustainable energy technology [31].

6.3 Process Innovation

6.3.1 Methanol

Growing worldwide interest in renewable methanol is driven by the pressing need to reduce emissions of carbon dioxide (CO_2) from various energy sources in order to combat climate change. These alternatives, which may be produced from biomass or from green (renewable) hydrogen and carbon dioxide, have the potential to increase the use of methanol as a chemical feedstock and contribute to the goal of carbon neutrality in transportation and industry.

Methanol has several potential applications [32–34], including internal combustion engines, hybrids, fuel cells, and vessels. Its storage, transportation, and distribution are all made easy because it is a liquid at room temperature and pressure. It is compatible with regular fuels and can work with the current distribution system [34].

Renewable methanol is a very sustainable chemical derived from biomass or created using carbon dioxide and hydrogen generated from renewable power. It is sometimes referred to as bio-methanol and has an extremely low-carbon footprint [34].

Wind and solar photovoltaic (PV) are growing at very fast pace, offering clean and affordable electricity worldwide [35]. However, because of the intermittent nature, the power generated need to be dispatched immediately. When this happens, operators must disconnect the renewable source from the grid, wasting energy, and costing governments and operators. This energy source can be used to produce methanol which finds applications. Low-emission methanol has the potential to have a greater impact in reducing carbon emissions in sectors that now have few alternatives. It may be used as a raw material in the chemical industry or as a fuel in road and maritime transportation [36].

The utilization of waste streams from various industrial processes, such as black liquor from paper mills and municipal solid waste (MSW), presents a promising prospect for the production of bio-methanol. This approach offers advantages in terms of streamlining the transportation and handling of raw materials, thereby enhancing the economic viability of the plant. The co-generation of heat, electricity, or other chemicals has the potential to enhance the economic viability of bio-methanol production [34].

More efficient green hydrogen production, together with additional price drops for renewable power generation and electrolyzers, will determine the industry's future green hydrogen production costs.

Renewable methanol feedstocks are abundant and widespread, improving energy security and fiscal balances, and their economies through oil reliance mitigation [36].

6.3.2 Ammonia

Material research focuses on physisorbed or chemisorbed hydrogen storage, with metal-organic frameworks and complex metal hydrides being promising classes. A single material system meets stringent requirements, but functionalization through ALD offers unique characteristics for both [37].

ALD is a promising technique used to precisely design and synthesize catalytic materials at the atomic level [38]. ALD has been employed to create catalysts that exhibit enhanced activity, selectivity, and stability over a range of conditions, including high temperatures, gas and liquid phases, and corrosive environments [38]. Ammonia production in industries involves the use of catalysts, which are substances that accelerate chemical processes. Ammonia is a high-quality energy carrier that contains a significant amount of hydrogen. Nevertheless, the storage and use of energy through ammonia still face multiple challenges [39]. Ammonia factories will be required to modify or completely overhaul their production systems [40].

Furthermore, the engines will require reconfiguration in order to operate on the newly liquid fuel. Throughout the process, producers and consumers will encounter obstacles that they must overcome: Ammonia is a poisonous substance, and its combustion has the potential to generate greenhouse gas that is even more potent than carbon dioxide (CO_2) [40]. Despite its toxicity, there is already a comprehensive global infrastructure in place for the production, storage, and transportation of ammonia. Ammonia has several benefits, such as the use of pre-existing pipelines and a high transmission capacity.

Ammonia seems to be a very viable choice for scaling up and an essential fuel for transoceanic lengthy journeys that demand fuel with a high-energy density.

According to a 2021 International Energy Agency assessment, in order to achieve zero emissions by 2050, hydrogen-based fuels, which include ammonia, need make up around 30% of transportation fuels by that time, as opposed to essentially nothing at this point.

By 2030, hydrogen and ammonia are expected to be utilized as fuel sources for electricity generation. They will be predominantly used in conjunction with natural gas in gas turbines and with coal in coal-fired power plants. This utilization will prolong the lifespan of current assets, enhance the reliability of the electricity system, and decrease the overall expenses associated with transforming the electricity sectors in numerous countries [41]. Transforming hydrogen into other energy carriers, such as ammonia or synthetic hydrocarbon fuels, incurs much greater expenses. However, this process yields fuels that can be conveniently transported and stored and are frequently compatible with current infrastructure and end-use technologies [41].

To accelerate and expand the production of green ammonia, government policies will be necessary to provide subsidies for green hydrogen and promote the development of larger-scale operations.

6.3.3 Process Optimization

Clean hydrogen is gaining momentum, focusing on developing technologies to reduce costs, reduce resource usage, and expand its use in challenging sectors, addressing industry bottlenecks. The industry stands to gain a great deal from this innovation, which will eliminate long-standing bottlenecks that have prevented clean hydrogen from reaching its full potential. Spatial atomic layer deposition (SALD) nanotechnology has been used by to provide solutions that increase process and product efficiency in the electrolysis, fuel cell, battery, and solar cell sectors.

As commented in the interview with Bouman [42], SALD can help optimize material usage. For instance, the proton exchange membrane (PEM) electrolysis market is expected to grow at a compound annual growth rate (CAGR) of over 24% between 2023 and 2032 due to its high efficiency, compact design, and low-temperature operation, but its reliance on rare and expensive iridium catalysts. SALD technology can reduce iridium loading by 40 or more by applying a catalytic layer of iridium oxide or platinum to the electrolyzer's porous transport or gas diffusion layer, thereby reducing materials usage and costs. Perspectively put, PEM electrolysis currently requires about 500 g of PGM per 1 MW capacity, but with SALD, this can be reduced to 10 g per 1 MW. Spatial ALD reactors speed up process time by spatially separating gases, allowing a dynamic process. Scaling is more feasible than traditional batch reactors, which create longer, less economic processes. SALD offers benefits but can scale up to 100 times higher throughput. Spatial ALD enhances electrolyzer manufacturing efficiency by reducing waste and recycling unused iridium atoms, and reducing waste and promoting recycling for future cycles. With PEM as an important component of the green hydrogen market, SALD offers opportunities to speed up the process.

The versatility of ALD allows access to a wide range of chemicals, but this variety poses issues in establishing processing conditions for consistent and uniform film development while minimizing precursor usage [43]. Efficient optimization methods are required to autonomously determine processing settings for optimum ALD film growth. Both computational and experimental findings show that precursor concentration is the crucial factor influencing the cycle time and the precursor mass [44]. The average gas velocity of fluid flow determines ALD cycle time, and quick diffusion at low pressure optimizes the ALD process. SALD and atmospheric-pressure atomic layer deposition (TALD) are gaining more ground [45]; for ALD to find dominant application, there would be need for the development of both SALD and TALD. The reactor design should be tailored to each application to ensure its specific purpose is met.

6.3.4 Seawater for Green Hydrogen Production

There are several regions in the globe that have extensive coastlines and abundant wind resources, making them very suitable for generating enormous amounts of renewable energy. This makes it very ideal for the advancement of the green hydrogen energy sector. Currently, the established low-temperature water electrolysis technology, which includes both alkaline electrolyzers and proton exchange membrane (PEM)-based electrolyzers, uses high-purity water for the electrolysis process [46].

Both desalination, which removes salt from the water before it flows to conventional electrolyzers, and the use of saltwater directly for the electrolysis process are two methods that may be utilized for the creation of green hydrogen. According to Gao et al. [47], there are two distinct methods for producing hydrogen gas (H_2) using saltwater and renewable energy. Firstly, the saltwater undergoes purification through the reverse osmosis membrane module (two-step sea water electrolysis). Subsequently, the water splitting process takes place. The second is the one-step seawater electrolysis, in which seawater is electrolyzed after undergoing a basic pretreatment.

Gao et al. [47] stated that an effective and viable seawater electrolysis program must address the following issues:

- Metal cations (especially heavy metals) are deposited on the surface of the cathode to cause electrode deactivation.
- The Cl^- ions in the seawater will compete with the oxygen evolution reaction (OER) reaction; moreover, the generated Cl_2 will seriously corrode the seawater electrolyzer.
- Because there are quite limited kinds of buffer ions in seawater, the pH of seawater cannot be maintained constantly under high current densities.

Some of the challenges in using seawater for hydrogen generation are the low efficiency and the poor stability of the electrolysis systems. In addition, there are issues of cations present in seawater such as Mg^{2+} and Ca^{2+} and the high concentration of chlorine ion which are harmful [46].

From the study by Guo et al. [46] through direct electrolysis of actual seawater that had not been acidified or alkalized, this approach achieved performance comparable to a standard PEM electrolyzer running in high-purity water, as well as long-term stability exceeding 100 hours at 500 mA cm^{-2}. A seawater electrolyzer, designed with flow-type configuration and Lewis acid-modified electrodes (Cr_2O_3–CoOx), achieves the desired current density of 1.0 A cm^{-2} at an operating voltage of 1.87 V and a temperature of 60 °C, meeting industrial standards.

The major drawback is that due to electrode side reactions and corrosion from the difficulties of utilizing saltwater, this procedure is not as effective as pure water electrolysis and is not feasible in areas where seawater is limited (Fig. 6.3).

When it comes to green hydrogen production from sea water, some of the issues that need to be addressed are mechanism of corrosion resistance and protection of the oxygen evolution electrode, electrode deterioration brought on by chloride ions, the production of precipitates on the cathode, and the slow pace of the oxygen evolution reaction (OER) and the competing chlorine evolution reaction (CER) processes are

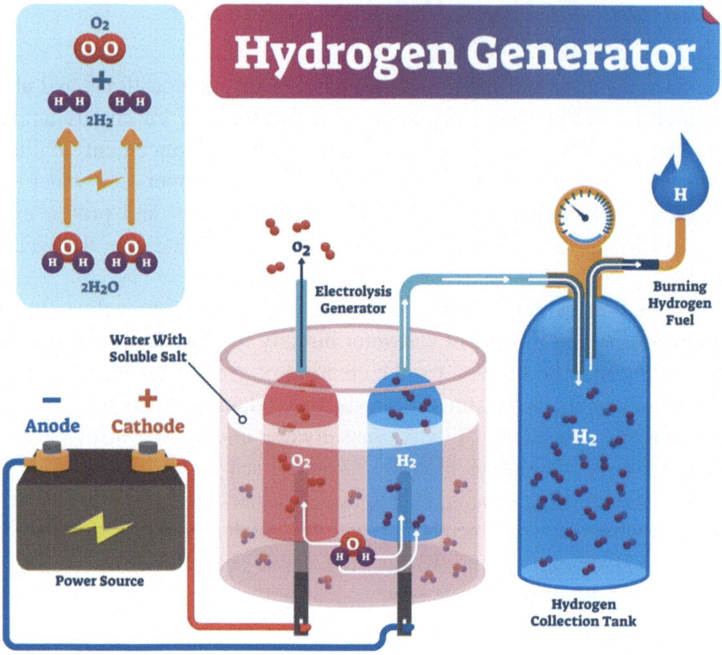

Fig. 6.3 Hydrogen generation from sea water

all factors that contribute to the degradation of the electrode. In addition, to make the required progress in the direct use of seawater for hydrogen production, there is the need for the development of a high-activity hydrogen evolution catalyst, anti-deposition hydrogen evolution electrode, and a corrosion-resistant oxygen evolution catalyst, design and integrate a high-performance electrolyzer, optimize mass and energy electrolytic seawater hydrogen production system and development and optimization of the integration of renewable energy sources. ALD can be applied to optimize the catalysts and electrodes.

Collaboration projects have the potential to enhance research exchanges, accelerate technological advancements in electrolytic seawater hydrogen production, foster future cooperation in hydrogen energy, facilitate long-term development of green energy technology, and sustain a positive cooperative relationship between different organizations. The development of this technology will create new employment opportunities across different skill levels, resulting in a good societal impact. The provision of electricity to the coastal region might also be advantageous for both urban and rural communities.

6.4 Industry 4IR

Industry 4.0 extends beyond the confines of factory settings. The system seamlessly integrates cyber-physical systems (CPS), enabling real-time communication and collaboration among machines, sensors, and humans. Cyber technologies such as the Internet of Things (IoT), cloud computing, artificial intelligence (AI), and big data analytics permeate every stage of the production process, blurring the lines between the physical and digital worlds [48–50].

The foundation of Industry 4.0 rests on respective key pillars, which includes;

(i) Interconnectivity: Machines, sensors, and people connect through networks, sharing data and enabling real-time collaboration and decision-making.
(ii) Automation: Advanced robots, cobots, and autonomous systems perform tasks with increased precision, efficiency, and flexibility.
(iii) Data-driven Insights: Vast amounts of data generated from sensors and processes are analyzed using AI and big data technologies, providing actionable insights for optimization and innovation.
(iv) Decentralization: Decision-making shifts from centralized control to individual machines and systems, leveraging real-time data for autonomous operations.

Industry 4.0 has revolutionized production by incorporating digital technology into industrial operations. Among the various technological advancements shaping Industry 4.0, hydrogen stands out as a significant trend. Hydrogen's capacity to improve energy generation, storage, and utilization has attracted significant interest as a crucial driver of sustainability and industrial change [51–53].

6.4.1 *Hydrogen Integration in Industry 4.0*

Hydrogen, the most abundant element in the universe, holds immense promise as a clean and versatile energy carrier. In the context of Industry 4.0, hydrogen is gaining traction primarily due to its role in decarbonizing various sectors, including manufacturing, transportation, and energy production. Unlike fossil fuels, hydrogen emits only water vapor when used in fuel cells or combustion processes, making it a viable solution for reducing greenhouse gas emissions and mitigating climate change [49, 54, 55].

The integration of hydrogen technologies within the framework of Industry 4.0 enables synergies between digitalization, automation, and decarbonization. Through smart grid integration, advanced sensors, and predictive analytics, Industry 4.0 facilitates the efficient production, distribution, and utilization of hydrogen across industrial sectors. Smart manufacturing processes can optimize energy consumption, minimize waste, and maximize resource efficiency, while hydrogen-powered fuel cells and electrolyzers provide clean energy solutions for industrial operations [54–57].

In addition, digital technology has the potential to expedite the green hydrogen transition in four key areas: digital twin, monitoring and control, advanced analytics, and certificate of origin. Artificial Intelligence of Things (AIoT) solutions, with increased investment, government support, and skilled workforce, can accelerate the transition to green hydrogen, reducing CAPEX and OPEX by 15–25%, potentially enabling 100% green hydrogen power in key industries [58].

6.4.2 Applications of Hydrogen in Industry 4.0

When it comes to the generation of renewable energy, the Internet of Things (I4.0) technologies are being utilized to enhance the efficiency, dependability, and scalability of renewable energy systems [59]. Wind, photovoltaic, and concentrated solar power can generate large amounts of electric power, but they are not developed enough or expensive to handle the challenge. In times of excess electricity from wind farms, excess electricity can be used to produce hydrogen through electrolysis. Green hydrogen, produced from electrolysis of water powered by solar or wind, is essential for climate neutrality and can store surplus renewable power, decarbonize hard-to-electrify sectors, and replace fossil fuels as a zero-carbon feedstock in chemicals and fuel production. IoT sensors and data analytics can optimize solar and wind energy systems, while artificial intelligence algorithms can predict and prevent equipment faults [59, 60]. Power grids may be made more efficient, reliable, and environmentally friendly by the implementation of smart grids, which make use of modern sensors, control systems, and communication technologies.

Using sensors connected to the Internet of Things (IoT) and data analytics to monitor and regulate energy systems in real time will result in the development of smart grid technologies, which will enhance the distribution of electricity. The construction of a smart grid model that is both scalable and adaptive, which can be expanded to optimize energy use and grid stability while simultaneously minimizing energy losses, is another aspect that will have an influence upon the green hydrogen industry. It is possible to optimize the integration of solar, battery, and hydrogen systems into the energy grid through the use of data analytics and machine learning. Estimating energy usage and optimizing the deployment of renewable resources may be accomplished through the utilization of machine learning algorithms by utilizing historical data. With this invention, the usage of renewable energy will be maximized, which will make the shift to renewable energy, thus green hydrogen economy more sustainable and cheaper. Localized energy production and distribution, which is made possible by intelligent energy management systems, will enhance access to energy in other areas. Internet of Things (IoT)-enabled smart meters and data analytics will allow consumers to monitor their energy consumption. The integration of renewable energy such as solar, wind, and green hydrogen with energy storage technologies will have the effect of increasing grid dependability, hence reducing the number of power outages and guaranteeing that even remote locations have a reliable supply of electricity. The scientific impact will be visible in the enhanced quality of life that will be experienced by communities who are able to obtain access to energy sources that are both dependable and sustainable.

6.4.3 *Benefits of Hydrogen Energy in Industry 4.0*

(i) Environmental Sustainability: A method to reach carbon neutrality in industrial processes may be established through the utilization of hydrogen, which in turn will helps to reduce emissions of greenhouse gases and mitigate the effects of climate change. The move toward energy systems that are based on hydrogen is in line with global efforts to transition to an economy with low-carbon emissions and to meet sustainability targets that have been stated in initiatives such as the Paris Agreement [49, 53, 54].

(ii) Energy Security: Hydrogen, which is a flexible energy carrier, can contribute to increased energy security for countries by increasing the energy mix and decreasing reliance on fossil fuels, which are limited in supply. Countries are able to improve their energy resilience and lower their susceptibility to supply disruptions through the use of renewable energy sources for the generation of hydrogen [55, 61].

(iii) Technological Innovation: Innovation is encouraged across a wide range of industries as a result of the incorporation of hydrogen technologies into Industry 4.0. This drives research and development efforts in areas such as electrolysis, fuel cells, hydrogen storage, and distribution infrastructure. In addition to enhancing the effectiveness and functionality of hydrogen systems, these technical improvements also contribute to the expansion of the economy and the creation of new employment opportunities [57, 61, 62].

(iv) Resilience and Reliability: In particular, hydrogen is useful for decentralized and off-grid applications because it improves reliability of the system and resilience. Hydrogen also offers backup power, grid stability, and energy resilience [61, 63, 64].

(v) Economic Growth: The hydrogen economy provides new prospects for innovation, the creation of jobs, and economic growth, which in turn will drives investments in research, infrastructure, and the expansion of industrial production. Through the utilization of digitization and automation, Industry 4.0 improves the economic feasibility and scalability of hydrogen technologies, which in turn speeds up their acceptance and market penetration [57, 61–63].

6.4.4 *Challenges and Considerations*

Despite its potential, the integration of hydrogen in Industry 4.0 faces several challenges and considerations:

(i) Cost and Competitiveness: The economic feasibility and competitiveness of hydrogen technologies are hindered by the substantial expenses associated with electrolyzers, hydrogen generation, and infrastructure implementation. The cost of green hydrogen generation presently exceeds that of conventional fuels. Ongoing research and development have a pivotal role in the lowering of costs [57, 61–65].

(ii) Scalability and Infrastructure: The expansion of hydrogen production and the establishment of a resilient hydrogen infrastructure, involving production facilities, distribution networks, and refueling stations, need substantial financial investments, legislative backing, and effective collaboration among the appropriate stakeholders [57, 66].

(iii) Technological Maturity: The technologies of electrolysis, fuel cells, and hydrogen storage are still in the process of developing, and there is a need for ongoing research to increase their performance, reliability, and cost-effectiveness even further [57, 61–66].

(iv) Standardization: The establishment of standards for the production, transportation, and utilization of hydrogen is very necessary in order to guarantee safety, interoperability, and the growth and expansion of the market [57, 61–66].

(v) Policy and Regulation: Policies, regulatory frameworks, and market incentives that are inconsistent with one another may be a barrier to the deployment and acceptance of hydrogen technology. This necessitates the need for harmonization and alignment on both the national and international levels [57, 61–66].

6.4.5 Outlook and Prospects

The future of hydrogen integration in Industry 4.0 is optimistic due to the growing recognition of decarbonization, energy transition, and sustainable development. Technological advancements, cost reductions, and supportive policies will drive innovation and market adoption [61–66]. Industry 4.0 represents a shift toward digitalization, automation, and decarbonization, and offers a transformative opportunity for sustainable industrial transformation. By integrating hydrogen technologies across sectors, it can lead to cleaner, more efficient, and resilient industrial ecosystems [49, 52–54].

6.5 Commercialization

The commercialization of hydrogen energy represents a pivotal opportunity to transition toward a sustainable and low-carbon energy future. Hydrogen, as an abundant element and versatile energy carrier, holds the potential to decarbonize sectors such as transportation, industry, and power generation [63–68].

Hydrogen energy has diverse applications across multiple sectors. In the transportation sector, hydrogen fuel cell vehicles offer zero-emission mobility, with fast refueling times and long driving ranges. In industry, hydrogen serves as a feedstock for chemical processes, including ammonia production and refining operations. Additionally, hydrogen can be utilized in power generation through fuel cells, providing reliable and decentralized energy solutions. Furthermore, hydrogen has the potential to integrate with renewable energy systems, storing excess electricity and contributing to grid stability and flexibility [65–67].

When it comes to green hydrogen production, electrolyzers are very important components that need considerations. An electrolyzer unit consists of an electrolytic cell with two electrodes and a membrane, along with other components like pumps, vents, storage tanks, and a power supply. Electrolysis occurs when an electric current is applied across electrolytes. A very important component in the electrolyzers is the stack. Cell stacks facilitate the conversion of chemical energy into electricity (and vice versa) by use of an electrochemical process that involves an anode and a cathode. Electrons pass through an external connection between the anode and the cathode in a PEM cell stack, whereas protons are carried through the electrolyte membrane in a PEM cell stack.

There has been a push to improve active surface area and current density while maintaining efficiency in an effort to save costs. An increase in the current density leads to a progressive rise in the cell voltage, stack voltage, and the accompanying power, resulting in an increase in hydrogen generation [69]. An increase in current density raises the rate at which hydrogen is produced, which lowers the cost of producing hydrogen [70–72]. Several factors can influence the current density in an electrolyzer unit, including the applied voltage, the electrolyte type, the temperature, and the surface area of the electrodes. The current density in an electrolyzers can be optimized through voltage adjustment, use of suitable electrolyte, and good operating conditions such as parameter temperature and electrode surface area [73].

In order to improve the performance of fuel cells, it is crucial to design electrodes with superior electrocatalytic properties. This is because a considerable amount of energy is lost during the charge transfer process, which takes place either on the surface of the electrodes or at the interfaces between the electrodes and the electrolytes [74]. When it comes to the design and engineering of high-performance fuel cell systems, ALD is considered to be a key fabrication/manufacturing method. ALD has found applications in improving both the surface areas of the electrodes and also membrane. It has the capability to enhance the structural and surface characteristics of specifically designed electrodes by employing different materials deposited by ALD [75].

ALD has been used to improve materials as demonstrated by Laube et al. [76] The accumulation of ALD loadings results in an increase in the current density. In their work on "Overpotential analysis of alkaline and acidic alcohol electrolyzers and optimized membrane-electrode assemblies," Sapountzi et al. [77] found out that using ALD as a consequence of optimizing the morphological electrode properties, it was possible to significantly increase the efficiency of catalyst utilization by almost ~30 times when compared to the commercial Pt on carbon cloth anodes. When compared to commercial electrodes, ALD leads to a more efficient utilization of the catalyst.

The process of splitting molecules involves power, which drives ion flow and pushes hydrogen and oxygen out of an electrolyzer cell. If power stops, hydrogen and oxygen can form bubbles, which can block current flow and degrade catalysts. Electrolyzers have safety systems to prevent explosions, but production shutdowns are never ideal [78].

ALD can impact current density in electrolyzers by modifying catalyst properties and electrode structure [76]. ALD allows precise control over the deposition of thin films on surfaces, enhancing catalytic activity and stability [22, 23, 38, 79, 80]. It can also deposit catalytic materials like platinum, iridium, or nickel-based catalysts onto electrode surfaces, lowering overpotentials and improving efficiency [38, 76, 81, 82]. ALD-deposited thin films also serve as protective coatings, preventing corrosion or degradation of electrode materials [83]. Precise control over electrode surface morphology optimizes surface area for electrochemical reactions [84, 85], influencing current density. ALD-deposited materials also reduce resistive losses within the electrolyzer cell, allowing higher current densities without excessive voltage drops.

Membranes are physical barriers that enable mass species movement selectively. These have found applications in separation systems for gas or liquid purification and extraction, water treatment and desalination, catalytic reactors, and numerous environmental/recycling applications processes. Additionally, they have a crucial function in many different systems which includes desalination, batteries, sensors, fuel cells, electrolyzers, and barrier layers [22, 23, 86–91].

Atomic layer deposition is used to provide ultralow loading and high performance of a platinum catalyst for a polymer electrolyte membrane fuel cell anode [92].

From the study of Song et al., [92] the hydrogen oxidation reaction rate in polymer electrolyte membrane fuel cells (PEMFCs) can be reduced by reducing Pt loading below 0.025 mg cm^{-2}. An atomic layer deposition (ALD) technique is used to create membrane electrode assemblies (MEAs) with low Pt loading in the anode layer. The ALDPt NPs are distributed uniformly on the carbon surface, enhancing the performance of ALDPt-MEAs. The 50ALDPt-MEA shows excellent H2–air PEMFC activity and durability, while the 20ALDPt-MEA has a significantly higher surface area and better stability.

Improving ion conductivity, hydroxyl ion selectivity, and strong durability at higher operating temperatures (>60 °C) is crucial for practical implementation of improved AEM membranes and ionomers [93].

The cost and performance of electrolyzer technologies vary, with alkaline and polymer electrolyte membrane (PEM) electrolyzers being the most advanced and commercial [94]. Alkaline electrolyzers have the lowest installed cost [95–97], while PEM electrolyzers have a smaller footprint and higher current density [95, 97]. Solid oxide electrolyzers have the highest electrical efficiency. A reduced stack footprint of around 60% for PEM results in a 20–24% reduction in facility footprint. According to estimates, the amount of investment required each GW of manufacturing capacity is between EUR 45 and 69 million [98]. Gaps in cost and performance are expected to narrow over time as electrolysis technologies continue to evolve

[99]. Innovation is crucial for cost reduction and electrolyzer performance improvement. Goals include standardizing manufacturing, improving efficiency, and increasing durability to scale up production, reduce electricity consumption, and extend equipment lifespan.

6.5.1 Framework for Hydrogen Energy Commercialization

(i) Policy and Regulatory Frameworks:
Policy support plays a pivotal role in driving the commercialization of hydrogen energy. Governments worldwide are implementing various initiatives to promote hydrogen deployment, including financial incentives, research and development funding, carbon pricing mechanisms, and regulatory frameworks. Moreover, international collaboration and harmonization of standards are essential to facilitate trade, investment, and technology transfer in the global hydrogen market. Establishing a supportive policy environment can unlock investment opportunities, stimulate innovation, and accelerate the transition toward a hydrogen economy [67, 68, 100–102]

(ii) Market Dynamics:
The market for hydrogen energy is evolving rapidly, driven by technological innovation, climate goals, and market dynamics. As the costs of renewable energy and electrolysis technologies continue to decline, the competitiveness of hydrogen energy is expected to improve, driving demand across sectors. Market players, including energy companies, automotive manufacturers, and industrial conglomerates, are increasingly investing in hydrogen projects and forming strategic partnerships to capitalize on emerging opportunities. Moreover, the emergence of hydrogen hubs and clusters facilitates collaboration, infrastructure development, and value chain integration, driving market growth and innovation [68, 100–102].

6.5.2 Some Challenges with Green Hydrogen

Green hydrogen's success relies on extensive and coordinated support throughout the entire value chain. Currently, the cost of creating green hydrogen is three times higher compared to the production of other forms of hydrogen. Additionally, the feasibility of producing this gas relies on the availability of renewable energy sources. The integration of cost reductions for electricity and electrolyzers, along with enhanced efficiency and extended operating lifetime, can result in a significant 80% decrease in the cost of hydrogen [99]. As highlighted in the report by Council of Engineers for the Energy Transition (CEET) [16], there are barriers that need to be overcome for the of green hydrogen. They include

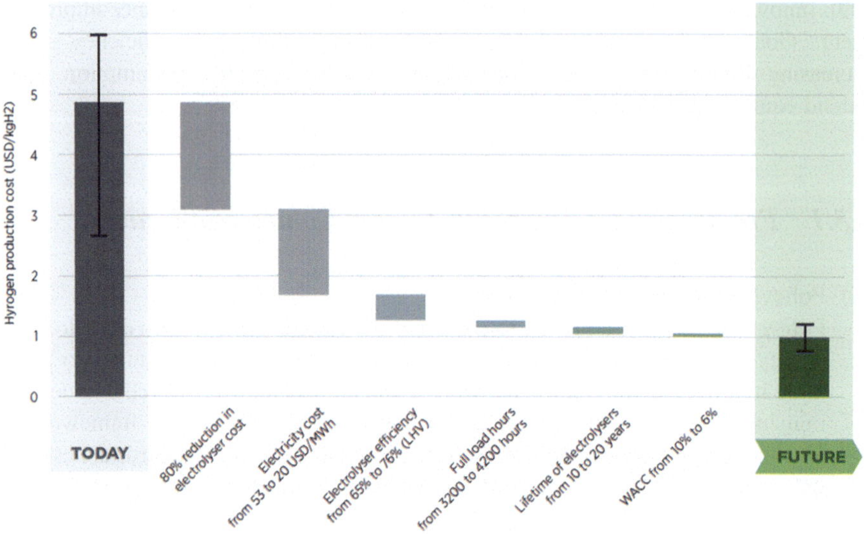

Fig. 6.4 Step changes for achieving green hydrogen competitiveness

- Technological barriers to RH2 production
- Manufacturing capacity and supply chain
- Infrastructure barriers
- Commercialization barriers
- Land availability issue and water access
- Skill workforce access
- Regulatory barriers

Commercialization Barriers: The cost of electricity can account for 40–80% of the overall cost of producing renewable hydrogen (RH2) in certain parts of the world. Moreover, the production costs of RH2 are primarily influenced by the significant initial investment required for the electrolyzer, which has a high capital expenditure (ranging from US$1–2 million per MW, depending on the electrolyzer technology). RH2 production is currently no-cost competitive unless public funding is provided. Its competitiveness depends on cost reduction drivers, including electrolyzer CAPEX reduction and electricity costs. The optimal combination of RES and electrolyzers is discussed in IRENA 2022 report on green hydrogen. H2 logistics, demand characteristics, and supply stability determine competitiveness. Heavy investment is needed to change industrial processes, with conversion costs varying depending on equipment and size (Fig. 6.4).

These are scenario based on IRENA analysis. More details of this are the green hydrogen cost reduction: scaling up electrolyzers to meet the 1.5 °C climate goal [99] (Fig. 6.5).

The cost of producing green hydrogen varies based on the deployment of electrolyzers and is influenced by the electricity price. Assuming a constant

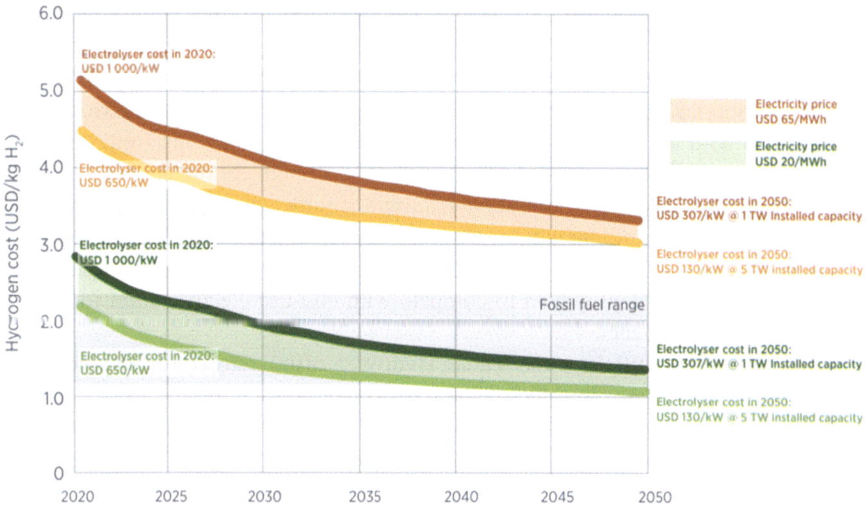

Fig. 6.5 Cost of producing green hydrogen as a function of electrolyzer deployment, using a low electricity price of USD 20/MWh and an average price of USD 65/MWh, constant over the years 2020–2050 [99]

electricity price of USD 65/MWh and USD 20/MWh from 2020 to 2050, the cost of green hydrogen production may be determined.

As highlighted in the World Economic Forum (WEF) [58], some of the current challenges of green hydrogen are as follows:

1. Limited Understanding of Optimal Design and Return on Investment, Limiting Bankability:

   Without an adequate supply of water and renewable energy to utilize as feedstock, a means of storing and transporting the product, and an offtaker to purchase and consume the product, an electrolyzer used to produce green hydrogen will be useless.
2. Scarcity of Skilled Workforce and High Operation Costs: Finance, infrastructure, and policy are commonly cited as significant obstacles in energy transition. However, the local energy industry has an additional challenge in the form of a deficiency in the workforce skills. It is imperative for governments and organizations to allocate resources toward training and education initiatives aimed at cultivating a proficient workforce capable of driving innovation, promoting energy efficiency, and fostering economic growth. The absence of a competent labor force demands the recruitment of foreign workers from different parts of the world, resulting in inevitable rise of costs for executing projects. In addition to project implementation, there is a significant need for individuals who possess expertise in the formulation of financially viable projects. It is crucial to emphasize the significance of comprehending the complex nature of the energy transition. Capacity building should include not only the establishment and upkeep of renewable energy systems including solar, wind, bioenergy, geothermal, and

green hydrogen, but also enhancing the readiness of the environment for investing in renewable energy. An incorporation of renewable energy and sustainability concepts should be implemented through experiential learning within the educational system. The focus should be on cultivating a skilled workforce equipped with the necessary expertise to bridge the gap resulting from outdated educational programs and the fast advancement of engineering technology.

3. High-Energy Losses: Green hydrogen suffers significant energy losses throughout its supply chain, including 30–35% during electrolysis, 13–25% during liquefying, and 40–50% in fuel cells. These inefficiencies require significant renewable energy deployment to compete with end-use electrification for green hydrogen electrolyzers.
4. Green Hydrogen Offtakers and Value: Hydrogen offtake structures may be more challenging to finance in sectors where hydrogen is not currently used at scale. Heavy transport, electricity generation, industrial processes, mining, and heating are some of the emerging use cases for hydrogen. Fuel cells offer advantages over batteries for hydrogen refueling, making public transport projects the most compelling model for financing. Long-haul trucking requires a more dispersed refueling network, and financing may become viable only after infrastructure plans are established with government support. Hydrogen can also be used for long-duration storage in renewable electricity plants, and for industrial processes, where creditworthy manufacturers can be counterparties in bankable offtake arrangements [103].

Others include

5. Cost Competitiveness: Currently, green hydrogen is more expensive than traditional fossil fuels due to the high upfront cost of electrolyzers and renewable energy infrastructure. Continued advancements in technology and economies of scale are crucial for cost reduction [68, 100–102].
6. Infrastructure Gap: A robust infrastructure for hydrogen production, transportation, storage, and refueling stations is essential. Significant investments are needed to build this infrastructure and create a seamless value chain [68, 100–102].
7. Standardization and Regulations: Clear and harmonized standards are needed for hydrogen production, transportation, and use to ensure safety, interoperability, and market confidence. Regulatory frameworks need to adapt and incentivize green hydrogen adoption [68, 100–102].
8. Public Perception and Awareness: Lack of awareness and potential concerns about safety can hinder wider public acceptance of hydrogen technology. Education and transparency are key to building trust and fostering adoption [68, 100–102].

Electrolyzers, which are the bane of green hydrogen, are getting more attention. For instance, Toyota in partnership with Chiyoda in Japan plans mass production of 5 MW PEM. Toyota and Chiyoda are partnering to develop a large-scale electrolysis system for hydrogen production, combining their expertise in fuel cell technology

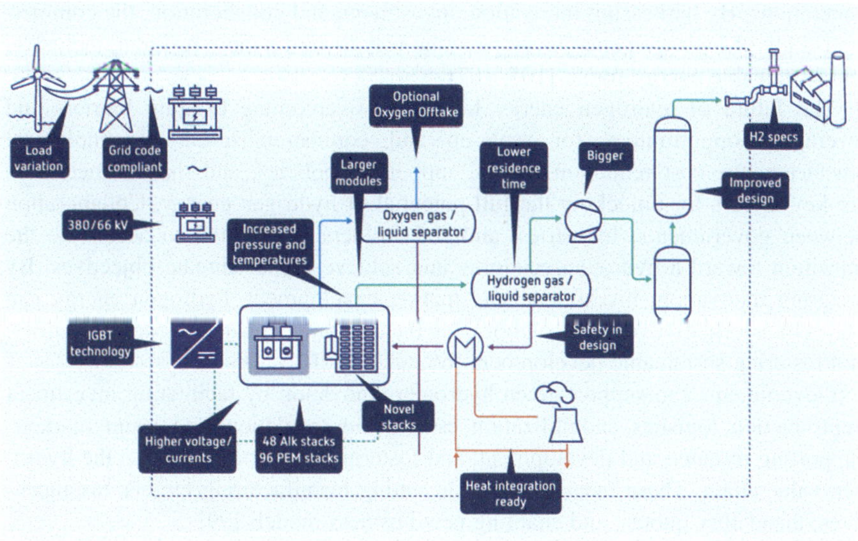

Fig. 6.6 Example of the optimizations for the advanced design of a green hydrogen plant with a capacity of one gigawatt (1-GW) based on alkaline water electrolysis (AWE) technology

and plant design. The 5 MW electrolyzers, each capable of producing 100 kg of hydrogen per hour, will be combined to create a standard package for large-scale electrolysis systems. This will allow for lower costs, increased production efficiency, and more stable quality for green hydrogen production. The equipment, occupies only half the floor area of conventional ones, offers ease of maintenance, faster shipping, shorter construction times, and lower construction costs. The system will be introduced at Toyota's Honsha plant in 2025 and expanded to the 10 MW class for verification and development [104] (Fig. 6.6).

Key optimizations for the advanced design of a 1-GW green hydrogen plant utilizing alkaline water electrolysis (AWE) technology [94].

6.5.3 Future Perspectives for Hydrogen Energy

In the race for a sustainable future, green hydrogen emerges as a frontrunner, offering a clean and versatile fuel with the potential to decarbonize various industries. The commercialization of hydrogen energy holds immense promise as a clean, versatile, and sustainable energy solution. Despite facing challenges related to cost, infrastructure, and policy support, hydrogen energy presents unprecedented opportunities for decarbonizing various sectors and addressing climate change. With concerted efforts from governments, industries, and stakeholders, hydrogen energy can emerge as a cornerstone of the global energy transition, driving economic growth, enhancing energy security, and preserving the environment for future generations. By harnessing innovation, investment, and collaboration, the commercialization of hydrogen energy can pave the way toward a more resilient, inclusive, and sustainable energy future [105–107].

The future of hydrogen energy hinges on overcoming existing barriers and leveraging opportunities for scale-up and commercialization. Technological advancements, cost reduction efforts, supportive policies, and market incentives are key drivers for unlocking the full potential of hydrogen energy. Collaboration between governments, industries, and stakeholders is essential to accelerate the transition toward a hydrogen economy and achieve global climate objectives. By fostering innovation, investment, and market development, hydrogen energy can play a transformative role in decarbonizing the economy, enhancing energy security, and fostering sustainable development [66, 68, 100–102, 105–107].

Governments can support green hydrogen innovation by facilitating investment in production, logistics, and utilization, establishing regulations and design markets, supporting research and development, and fostering coordination across the hydrogen value chain. These measures include setting manufacturing targets, tax incentives, mandatory quotas, and enabling new business models [99].

References

1. M. Weber et al., Assessing the environmental Impact of atomic layer deposition (ALD) processes and pathways to lower it. ACS Mater. Au **3**, 274–298 (2023)
2. J. St. John. The great 'green hydrogen' battle. *Canary Media* (2023), https://www.canarymedia.com/articles/hydrogen/the-great-green-hydrogen-battle
3. Department of Energy, Regional clean hydrogen hubs selections for award negotiations, https://www.energy.gov/oced/regional-clean-hydrogen-hubs-selections-award-negotiations
4. P. Bukirwa, C. Gamage, M. McClellan, H.A. Sheerazi, G. Westler, *Delivering Equitable and Meaningful Community Benefits Via Clean Hydrogen Hubs* (Rocky Mountain Institute. https://rmi.org/delivering-equitable-and-meaningful-community-benefits-via-clean-hydrogen-hubs/?utm_medium=email&utm_source=spark&utm_content=spark&utm_campaign=2024_02_01&utm_term=title-1, 2024)
5. National Hydrogen Strategy Update, (2023), https://www.energypartnership.jp/fileadmin/user_upload/japan/media_elements/national-hydrogen-strategy-update.pdf
6. J. Eckardt, J. Hoehne, B. Stenzel, *Factsheet: Germany's Updated National Hydrogen Strategy* (July 2023), https://adelphi.de/system/files/document/Factsheet%20Germany%27s%20National%20Hydrogen%20Strategy%20Update%202023_0.pdf
7. Hydrogen Action Plan Germany 2021–2025, (2021), https://www.wasserstoffrat.de/fileadmin/wasserstoffrat/media/Dokumente/EN/2021-07-02_NWR-Hydrogen_Action_Plan.pdf
8. B.E. Lebrouhi, B. Lamrani, Y. Zeraouli, T. Kousksou, Key challenges to ensure Morocco's sustainable transition to a green hydrogen economy. Int. J. Hydrog. Energy **49**, 488–508 (2024)
9. Green Hydrogen Organisation, Morocco, Green Hydrogen Organisation, https://gh2.org/countries/morocco
10. A. Berrada, M.A. Laasmi, Technical-economic and socio-political assessment of hydrogen production from solar energy. J. Energy Storage **44**, 103448 (2021)

11. X. Zhu, L. Chen, Y. Liu, Z. Tang, Atomically precise Au nanoclusters for electrochemical hydrogen evolution catalysis: progress and perspectives. Polyoxometalates **2**, 9140031 (2023). https://doi.org/10.26599/POM.2023.9140031
12. J.E. Hoffmann, On the outlook for solar thermal hydrogen production in South Africa. Int. J. Hydrog. Energy **44**, 629 (2019)
13. SOLSRGIS, Solar resource maps and GIS data for 180+ countries | Solargis. *Solargis* (2019), Preprint at https://solargis.com/resources/free-maps-and-gis-data
14. Mineral and Resources Department, S. A. Renewable Energy, (2023), https://www.energy.gov.za/files/esources/renewables/r_solar.html#:~:text=The%20southern%20African%20region%2C%20and,Europe%20and%20the%20United%20Kingdom
15. SASOL, Hydrogen ecosystem proof-of-concept demonstration – A highlight at smarter mobility Africa conference, SASOL (2023), https://www.sasol.com/media-centre/media-releases/hydrogen-ecosystem-proof-of-concept-demonstration-a-highlight-at-smarter-mobility-africa-conference
16. B.G. Pollet, C. La Marca, M. Ali Zerouali, *Issue Brief Hydrogen Technology* (2023), https://irp.cdn-website.com/6f2c9f57/files/uploaded/202312%20CEET%20Brief%20Hydrogen.pdf
17. NRF, *Sasols' Energy Transition – Collaboration on the Hydrogen Economy* (National Research Foundation, 2022), https://www.nrf.ac.za/sasols-energy-transition-collaboration-on-the-hydrogen-economy/
18. Impact, E., *South Africa Hydrogen Valley* (Final report, 2021), Preprint at https://www.dsti.gov.za/images/2021/Hydrogen_Valley_Feasibility_Study_Report_Final_Version.pdf
19. C. Marichy, M. Bechelany, N. Pinna, Atomic layer deposition of nanostructured materials for energy and environmental applications. Adv. Mater. **24**, 1017–1032 (2012)
20. J.A. van Delft, D. Garcia-Alonso, W.M.M. Kessels, Atomic layer deposition for photovoltaics: applications and prospects for solar cell manufacturing. Semicond. Sci. Technol. **27**, 074002 (2012)
21. D. Muñoz-Rojas, J. MacManus-Driscoll, Spatial atmospheric atomic layer deposition: a new laboratory and industrial tool for low-cost photovoltaics. Mater. Horiz. **1**, 314–320 (2014)
22. P.O. Oviroh, T.-C. Jen, J. Ren, A. van Duin, Towards the realisation of high permi-selective MoS_2 membrane for water desalination. NPJ Clean Water **6**, 14 (2023)
23. P.O. Oviroh, R. Akbarzadeh, D. Pan, R.A.M. Coetzee, T.-C. Jen, New development of atomic layer deposition: processes, methods and applications. Sci. Technol. Adv. Mater. **20**, 465 (2019)
24. P.O. Oviroh et al., Nanoporous MoS_2 membrane for water desalination: a molecular dynamics study. Langmuir **37**, 7127–7137 (2021)
25. S. Karimzadeh, B. Safaei, C. Yuan, T.-C. Jen, Emerging atomic layer deposition for the development of high-performance lithium-ion batteries. Electrochem. Energy Rev. **6**, 24 (2023)
26. X. Meng, X. Yang, X. Sun, Emerging applications of atomic layer deposition for lithium-ion battery studies. Adv. Mater. **24**, 3589–3615 (2012)
27. M. Cassir, A. Ringuedé, L. Niinistö, Input of atomic layer deposition for solid oxide fuel cell applications. J. Mater. Chem. **20**, 8987 (2010)
28. C. Liu, C. Wang, C. Kei, Y. Hsueh, T. Perng, Atomic layer deposition of platinum nanoparticles on carbon nanotubes for application in proton-exchange membrane fuel cells. Small **5**, 1535–1538 (2009)
29. X. Jiang, H. Huang, F.B. Prinz, S.F. Bent, Application of atomic layer deposition of platinum to solid oxide fuel cells. Chem. Mater. **20**, 3897–3905 (2008)
30. J.H. Shim, C.-C. Chao, H. Huang, F.B. Prinz, Atomic layer deposition of Yttria-stabilized zirconia for solid oxide fuel cells. Chem. Mater. **19**, 3850–3854 (2007)
31. R. Kendall, A green hydrogen innovation for clean energy, *MIT News*, (2023), https://news.mit.edu/2023/green-hydrogen-innovation-clean-energy-sossina-haile-1127

32. A. Al-Qahtani et al., Electricity grid decarbonisation or green methanol fuel? A life-cycle modelling and analysis of today's transportation-power nexus. Appl. Energy **265**, 114718 (2020)
33. S. Sollai, A. Porcu, V. Tola, F. Ferrara, A. Pettinau, Renewable methanol production from green hydrogen and captured CO_2: a techno-economic assessment. J. CO_2 Util. **68**, 102345 (2023)
34. Irena and Methanol Institute, *Innovation Outlook Renewable Methanol*, (2021), https://www.irena.org/-/media/Files/IRENA/Agency/Publication/2021/Jan/IRENA_Innovation_Renewable_Methanol_2021.pdf?rev=ca7ec52e824041e8b20407ab2e6c7341
35. International energy Agency. *Renewables 2023 – Analysis*. (2023)
36. C. Hobson, C. Márquez, *Renewable Methanol Report*, vol 103 (Methanol Institute, Singapore, 2018)
37. N. Leick et al., *Atomic Layer Deposition for Materials-Based H_2 Storage: Opportunities and Limitations*, vol MA2021-02 (2021)
38. B.J. O'Neill et al., Catalyst design with atomic layer deposition. ACS Catal. **5**, 1804–1825 (2015)
39. C. Chen, Y. Zhou, H. Fang, X. Peng, L. Jiang, Progress and challenges in energy storage and utilization via ammonia. Surf. Sci. Technol **1**, 13 (2023)
40. N. Jones, From fertilizer to fuel: can 'Green' ammonia be a climate fix? in *YaleEnvironment360* (2022), https://e360.yale.edu/features/from-fertilizer-to-fuel-can-green-ammonia-be-a-climate-fix
41. S. Bouckaert, A.F. Pales, C. McGlade, U. Remme, B. Wanne, Laszlo Varro. Davide D'Ambrosio. and Thomas Spencer, in *Net Zero by 2050 A Roadmap for the Global Energy Sector* (2021). https://iea.blob.core.windows.net/assets/deebef5d-0c34-4539-9d0c-10b13d840027/NetZeroby2050-ARoadmapfortheGlobalEnergySector_CORR.pdf
42. A. Bouman, SparkNano: nanotechnology with the potential for global impact. SparkNano. https://26070417.fs1.hubspotusercontent-eu1.net/hubfs/26070417/Nanotechnology_with_the_Potential_for_Global_Impact.pdf
43. N.H. Paulson, A. Yanguas-Gil, O.Y. Abuomar, J.W. Elam, Intelligent agents for the optimization of atomic layer deposition. ACS Appl. Mater. Interfaces **13**, 17022 (2021)
44. Z. Deng, W. He, C. Duan, B. Shan, R. Chen, Atomic layer deposition process optimization by computational fluid dynamics. Vacuum **123**, 103 (2016)
45. M. Chen et al., Atmospheric-pressure atomic layer deposition: recent applications and new emerging applications in high-porosity/3D materials. Dalton Trans. **52**, 10254–10277 (2023)
46. J. Guo et al., Direct seawater electrolysis by adjusting the local reaction environment of a catalyst. Nat. Energy **8**, 264–272 (2023)
47. F.Y. Gao, P.C. Yu, M.R. Gao, Seawater electrolysis technologies for green hydrogen production: challenges and opportunities. Curr. Opin. Chem. Eng. **36** (2022), Preprint at https://doi.org/10.1016/j.coche.2022.100827
48. M.M. Kabir et al., Machine learning-based prediction and optimization of green hydrogen production technologies from water industries for a circular economy. Desalination **567**, 116992 (2023)
49. A.V. Sheveleva, V.I. Solomos, *Smart Green Innovations in Industry 4.0* (Springer Nature Switzerland, Cham, 2023). https://doi.org/10.1007/978-3-031-45830-9
50. G. Cheng et al., Analysis and prediction of green hydrogen production potential by photovoltaic-powered water electrolysis using machine learning in China. Energy **284**, 129302 (2023)
51. L.N. de Menezes, M.C. de Alencar Lira, L.S. Neiva, IoT and knowledge economy: two strong pillars of industry 4.0. Sci. Cum Ind. **9**, 10–15 (2021)
52. A.A. Wagire, A.P.S. Rathore, R. Jain, Exploration of pillars of industry 4.0 using latent semantic analysis technique, in *Intelligent Manufacturing and Energy Sustainability: Proceedings of ICIMES 2019*, (Springer, 2020), pp. 711–719

53. Z. Suleiman, S. Shaikholla, D. Dikhanbayeva, E. Shehab, A. Turkyilmaz, Industry 4.0: clustering of concepts and characteristics. Cogent Eng **9**(1), 2034264 (2022)
54. P. Fragiacomo, F. Piraino, M. Genovese, Insights for industry 4.0 applications into a hydrogen advanced mobility. Procedia Manuf **42**, 239–245 (2020)
55. J.O. Abe, A.P.I. Popoola, E. Ajenifuja, O.M. Popoola, Hydrogen energy, economy and storage: review and recommendation. Int. J. Hydrog. Energy **44**, 15072–15086 (2019)
56. B. Nastasi, S. Mazzoni, D. Groppi, A. Romagnoli, D. Astiaso Garcia, Optimized integration of hydrogen technologies in Island energy systems. Renew. Energy **174**, 850–864 (2021)
57. N. Sazali, Emerging technologies by hydrogen: a review. Int. J. Hydrog. Energy **45**, 18753–18771 (2020)
58. L.A. Sylvie Ouziel, 4 technologies that are accelerating the green hydrogen revolution. World Econ. Forum. (2021), https://www.weforum.org/agenda/2021/06/4-technologies-accelerating-green-hydrogen-revolution/
59. P. Onu, A. Pradhan, C. Mbohwa, The potential of industry 4.0 for renewable energy and materials development – The case of multinational energy companies. Heliyon **9**, e20547 (2023)
60. S.G. Pease et al., An intelligent real-time cyber-physical toolset for energy and process prediction and optimisation in the future industrial internet of things. Futur. Gener. Comput. Syst. **79**, 815 (2018)
61. F. Dawood, M. Anda, G.M. Shafiullah, Hydrogen production for energy: an overview. Int. J. Hydrog. Energy **45**, 3847–3869 (2020)
62. M. Genovese et al., Power-to-hydrogen and hydrogen-to-X energy systems for the industry of the future in Europe. Int. J. Hydrog. Energy **48**, 16545–16568 (2023)
63. A. Roy, S. Pramanik, A review of the hydrogen fuel path to emission reduction in the surface transport industry. Int. J. Hydrog. Energy **49**, 792–821 (2024)
64. P.P. Edwards, V.L. Kuznetsov, W.I.F. David, Hydrogen energy. Philos. Trans. R. Soc. A Math. Phys. Eng. Sci. **365**, 1043–1056 (2007)
65. M. Yue et al., Hydrogen energy systems: a critical review of technologies, applications, trends and challenges. Renew. Sust. Energ. Rev. **146**, 111180 (2021)
66. M. Ball, M. Wietschel, The future of hydrogen – Opportunities and challenges. Int. J. Hydrog. Energy **34**, 615–627 (2009)
67. W. Liu et al., Trends and future challenges in hydrogen production and storage research. Environ. Sci. Pollut. Res. **27**, 31092–31104 (2020)
68. Q. Hassan, S. Algburi, A.Z. Sameen, H.M. Salman, M. Jaszczur, Green hydrogen: a pathway to a sustainable energy future. Int. J. Hydrog. Energy **50**, 310–333 (2024)
69. M. Shahabuddin, M.A. Rhamdhani, G.A. Brooks, Technoeconomic analysis for green hydrogen in terms of production, compression, transportation and storage considering the Australian perspective. Processes **11**(7), 2196 (2023)
70. T.A.H. Ratlamwala, I. Dincer, Electrochemical, energy, exergy, and exergoeconomic analyses of hybrid photocatalytic hydrogen production reactor for Cu–Cl cycle. Green Energy Tech. (2018). https://doi.org/10.1007/978-3-319-62572-0_44
71. L. Zhang, Z. Shi, Y. Lin, F. Chong, Y. Qi, Design strategies for large current density hydrogen evolution reaction. Front. Chem. **10**, 866415 (2022)
72. P. Trinke, B. Bensmann, R. Hanke-Rauschenbach, Current density effect on hydrogen permeation in PEM water electrolyzers. Int. J. Hydrog. Energy **42**, 14355–14366 (2017)
73. S. Niroula, C. Chaudhary, A. Subedi, B.S. Thapa, Parametric modelling and optimization of alkaline electrolyzer for the production of green hydrogen. IOP Conf. Ser. Mater. Sci. Eng **1279**, 012005 (2023)
74. J.H. Shim et al., Atomic layer deposition for surface engineering of solid oxide fuel cell electrodes. Int. J. Precis. Eng. Manuf.-Green Technol. **6**, 629–646 (2019)
75. A.T. Sivagurunathan, S. Adhikari, D.-H. Kim, Strategies and implications of atomic layer deposition in photoelectrochemical water splitting: recent advances and prospects. Nano Energy **83**, 105802 (2021)

76. A. Laube et al., PEM water electrolysis cells with catalyst coating by atomic layer deposition. Int. J. Hydrog. Energy **46**, 38972–38982 (2021)
77. F.M. Sapountzi et al., Overpotential analysis of alkaline and acidic alcohol electrolysers and optimized membrane-electrode assemblies. Int. J. Hydrog. Energy **44**, 10163–10173 (2019)
78. PXiSE Energy Solutions, Green hydrogen: understanding the innovation and policy landscape. *PXiSE Energy Solutions* (2023), https://www.utilitydive.com/spons/green-hydrogen-understanding-the-innovation-and-policy-landscape/700382/
79. H. Zhang, C.L. Marshall, Atomic layer deposition: catalytic preparation and modification technique for the next generation. Chin. J. Catal. **40**, 1311 (2019), Preprint at https://doi.org/10.1016/S1872-2067(19)63321-8
80. J. Camacho-Bunquin et al., Catalyst synthesis and evaluation using an integrated atomic layer deposition synthesis-catalysis testing tool. Rev. Sci. Instrum. **86**, 084103 (2015). https://doi.org/10.1063/1.4928614
81. F. Wu, L. Zhou, D. Guo, B. Xi, Efficient modulation of electrocatalyst interfaces by atomic layer deposition: fundamentals to application. Adv. Energy Sustain. Res **3**(7), 2200026 (2022)
82. S. Haghverdi Khamene, C. van Helvoirt, M.N. Tsampas, M. Creatore, Electrochemical activation of atomic-layer-deposited nickel oxide for water oxidation. J. Phys. Chem. C **127**, 22570–22582 (2023)
83. L. Santinacci, Atomic layer deposition: an efficient tool for corrosion protection. Curr. Opin. Colloid Interface Sci. **63**, 101674 (2023)
84. Y. Hu, J. Lu, H. Feng, Surface modification and functionalization of powder materials by atomic layer deposition: a review. RSC Adv. **11**, 11918–11942 (2021)
85. W. Guo et al., Surface and Interface modification of electrode materials for lithium-ion batteries with organic liquid electrolyte. Front. Energy Res **8**, 170 (2020)
86. M. Weber, A. Julbe, A. Ayral, P. Miele, M. Bechelany, Atomic layer deposition for membranes: basics, challenges, and opportunities. Chem. Mater. **30**, 7368–7390 (2018)
87. G. Amy et al., Membrane-based seawater desalination: present and future prospects. Desalination **401**, 16 (2017)
88. B.E. Logan, M. Elimelech, Membrane-based processes for sustainable power generation using water. Nature **488**. Preprint at 10.1038/nature11477 (2012). https://www.nature.com/articles/nature11477
89. J.P.K. Abal, Water desalination by MoS_2 nanoporous membrane: a molecular dynamics analysis (2020), http://hdl.handle.net/10183/211475
90. Y.X. Shen, P.O. Saboe, I.T. Sines, M. Erbakan, M. Kumar, Biomimetic membranes: a review. J. Membr. Sci. **454**, 359–381 (2014)
91. D. Cohen-Tanugi, J.C. Grossman, Nanoporous graphene as a reverse osmosis membrane: recent insights from theory and simulation. Desalination **366**, 59 (2015). https://doi.org/10.1016/j.desal.2014.12.046
92. Z. Song et al., Ultralow loading and high-performing Pt catalyst for a polymer electrolyte membrane fuel cell anode achieved by atomic layer deposition. ACS Catal. **9**, 5365 (2019). https://doi.org/10.1021/acscatal.8b04504
93. C. Li, J.-B. Baek, The promise of hydrogen production from alkaline anion exchange membrane electrolyzers. Nano Energy **87**, 106162 (2021)
94. van 't Noordende, H, P. Ripson, *A One-GigaWatt Green-Hydrogen Plan Costs: Advanced Design and Total Installed-Capital Cost* (Institute for Sustainable Process Technology (ISPT), The Netherlands 2022)
95. M. Carmo, D.L. Fritz, W. Maier, D. Stolten, (Invited) Alkaline Water Electrolysis Vs. PEM Water Electrolysis – Exploring Their Full Performance. ECS meeting abstracts MA2015-01, (2015)
96. S. Krishnan et al., Present and future cost of alkaline and PEM electrolyser stacks. Int. J. Hydrog. Energy **48**, 32313 (2023)

97. A.H. Reksten, M.S. Thomassen, S. Møller-Holst, K. Sundseth, Projecting the future cost of PEM and alkaline water electrolysers; a CAPEX model including electrolyser plant size and technology development. Int. J. Hydrog. Energy **47**, 38106 (2022)
98. J. Cihlar et al., *Hydrogen Generation in Europe: Overview of Key Costs and Benefits* (Publications Office of the European Union, Luxembourg, 2020)
99. IRENA, Green hydrogen cost reduction: scaling up electrolysers to meet the 1.5°C climate goal (2020), https://www.irena.org/-/media/Files/IRENA/Agency/Publication/2020/Dec/IRENA_Green_hydrogen_cost_2020.pdf
100. M.G. Rasul, M.A. Hazrat, M.A. Sattar, M.I. Jahirul, M.J. Shearer, The future of hydrogen: challenges on production, storage and applications. Energy Convers. Manag. **272**, 116326 (2022)
101. M.S. Iqbal, Z. Abdul Rahim, Understanding the potential of hydrogen as a game-changing energy source. Chem. Eng. Technol. **46**, 2577–2582 (2023)
102. S. Bhagavathy, J. Thakur, *Green Hydrogen: Challenges for Commercialization* (IEEE Smart Grid, 2021)
103. R. Crouch, *Financing Hydrogen Projects Brings Unique Challenges*, vol https://www.nortonrosefulbright.com/en/knowledge/publications/cd725de6/financing-hydrogen-projects-brings-unique-challenges (Norton Rose Fulbright, 2021)
104. L. Collins, *Toyota to Mass-Produce Hydrogen Electrolysers in Partnership with Chiyoda*, vol https://www.hydrogeninsight.com/electrolysers/toyota-to-mass-produce-hydrogen-electrolysers-in-partnership-with-chiyoda/2-1-1593648 (HydrogenInsight, 2024)
105. Q. Hassan, A.Z. Sameen, H.M. Salman, M. Jaszczur, A.K. Al-Jiboory, Hydrogen energy future: advancements in storage technologies and implications for sustainability. J Energy Storage **72**, 108404 (2023)
106. E.S. Hanley, J.P. Deane, B.P.Ó. Gallachóir, The role of hydrogen in low carbon energy futures – a review of existing perspectives. Renew. Sust. Energ. Rev. **82**, 3027–3045 (2018)
107. P.-A. Le et al., The current status of hydrogen energy: an overview. RSC Adv. **13**, 28262–28287 (2023)

The manufacturer's authorised representative in the EU is Springer Nature Customer Service Centre GmbH, Europaplatz 3, 69115 Heidelberg, Germany. If you have any concerns regarding our products, please contact ProductSafety@springernature.com

Printed and bound by CPI Group (UK) Ltd, Croydon, CR0 4YY
01/10/2026
02225961-0012